The Papers of Benjamin Henry Latrobe

EDWARD C. CARTER II, EDITOR IN CHIEF

SERIES II

THE ARCHITECTURAL AND ENGINEERING DRAWINGS

The Engineering Drawings of Benjamin Henry Latrobe

The Engineering Drawings of Benjamin Henry Latrobe

Edited with an Introductory Essay by Darwin H. Stapleton

Published for The Maryland Historical Society

by Yale University Press   New Haven and London, 1980

EDITORIAL BOARD

Whitfield J. Bell, Jr., American Philosophical Society
Lyman H. Butterfield, Emeritus, *The Adams Papers*
Rhoda M. Dorsey, Goucher College
Alan Gowans, University of Louisville
Jack P. Greene, The Johns Hopkins University
Walter Muir Whitehill, Emeritus, Boston Athenaeum (deceased)
Samuel Wilson, Jr., New Orleans, Louisiana

Copyright © 1980 by Yale University. All rights reserved. This book may not be reproduced, in whole or in part, in any form (beyond that copying permitted by Sections 107 and 108 of the U.S. Copyright Law and except by reviewers for the public press), without written permission from the publishers.

Designed by John O. C. McCrillis and set in Monotype Baskerville type. Printed in the United States of America by The Murray Printing Co., Westford, Massachusetts.

Published in Great Britain, Europe, Africa, and Asia (except Japan) by Yale University Press, Ltd., London. Distributed in Australia and New Zealand by Book & Film Services, Artarmon, N.S.W., Australia; and in Japan by Harper & Row, Publishers, Tokyo Office.

The research connected with this volume of *The Papers of Benjamin Henry Latrobe* has been generously supported by the National Science Foundation. This volume is published with the financial assistance of the National Historical Publications and Records Commission.

Library of Congress Cataloging in Publication Data

Latrobe, Benjamin Henry, 1764–1820.
   The engineering drawings of Benjamin Henry Latrobe.
   (The papers of Benjamin Henry Latrobe: Series II, The architectural and engineering drawings; v. 1)
Sponsored by the Maryland Historical Society.
   Bibliography: p.
   Includes index.
   1. Engineering drawings. 2. Latrobe, Benjamin Henry, 1764–1820. 3. Civil engineering—United States—History—Sources. I. Stapleton, Darwin H. II. Maryland Historical Society. III. Series.
TA175.L27 1979 624'.092'4 78-15405 ISBN 0-300-02227-1

CONTENTS

Foreword	vii
Edward C. Carter II	
Editor's Note	ix
Short Titles and Abbreviations	xiii
Chronology	xv
Illustrations	xvii
The Engineering Practice of Benjamin Henry Latrobe	
Introduction	3
Transportation	8
Waterworks	28
River Control	41
Industrial Works	48
Conclusion	61
The Susquehanna River Survey Map	73
Stephen F. Lintner, with Darwin H. Stapleton	
Map Data	89
The Engineering Drawings	111
Chelmer and Blackwater Navigation	113
Naval Drydock and Potomac Canal Extension	116
Chesapeake and Delaware Canal	125
Washington Canal	130
Dismal Swamp Canal	137
James River Canal and Railroad	139
National Road (Cumberland Road)	141
Philadelphia Waterworks	144
New Orleans Waterworks	199
Mud Island Bar	205
Fascine Works	208
Washington Navy Yard Steam Engine	214
Machinery	237
Appendix: Additional Drawings by Benjamin Henry Latrobe	243
Bibliography	248
Index	253

FOREWORD

FROM THE MOMENT of his arrival in Norfolk, Virginia, from England in March 1796 until his death from yellow fever in New Orleans a quarter of a century later, Benjamin Henry Latrobe (1764–1820) was continually involved in the study, development, and promotion of American technology. Although today Latrobe is primarily remembered as America's first great professional architect, he himself was just as eager to be known as an engineer. Thus, he often signed his plans "Engineer," a practice he extended even to the title page of "An Essay on Landscape," a treatise on landscape drawing that Latrobe wrote and illustrated in Richmond and Philadelphia (1798–99).[1]

Although Latrobe's engineering skills and achievements were recognized by many of his contemporaries, modern historians have generally been unaware of his various contributions to early American technology. The reasons for their seeming lack of appreciation are not difficult to discover. Latrobe's architectural achievement has undergone a steady process of rediscovery and re-evaluation during this century; also a substantial, if much-reduced, remnant of his architectural *oeuvre* survives, notably his great interiors at the United States Capitol and his masterpiece, the Roman Catholic Cathedral of Baltimore. In addition, the memory of Latrobe's engineering largely was obscured by the advent of the great age of canal building; today the only physical evidence of his projects consists of a few archaeological remains. The manuscript documents that survive from Latrobe's engineering career are numerous, varied, and scattered, and not until they were assembled in a single archive, such as *The Papers of Benjamin Henry Latrobe*, was a serious, full-scale study of them likely to have been undertaken. An additional factor was the late development of the discipline of the history of American technology. It emerged as a distinct field only after the Second World War and quite naturally has only recently produced the requisite body of scholars, historiography, methodologies, and conceptual tools for investigating such central problems as the origin of American engineering and the transfer of technology from Europe to the United States.

It is our hope that *The Engineering Drawings of Benjamin Henry Latrobe*, edited by Darwin H. Stapleton, will contribute to the growing comprehension of the complex nature and scope of early American technology. The basic scholarly purpose of this volume is that of the entire edition of the Latrobe Papers—to publish accurately historical documents, clearly and precisely annotated, in order to facilitate research, publication, and teaching. As this volume's modest title suggests, and a reading of its constituent parts will reveal, the editor has conformed admirably to this philosophy. His introductory essay on "The Engineering Practice of Benjamin Henry Latrobe"—which is designed to place the engineering drawings in historical perspective—is characterized by descriptive clarity and speculative restraint. Throughout the remainder of the volume, the primary focus is on the documents themselves.

An initial survey of the Maryland Historical Society's Latrobe collections in 1970 and a review of writings by and about Latrobe informed the editor in chief of the central role played by both technology and science in the architect-engineer's career and personal life. By the conclusion of the project's collection phase in August 1973, it was evident that we required the full-time services of a historian of technology. The editor in chief sought assistance from an institution whose particular research interests and strengths were nineteenth-century technological and economic history, the Eleutherian Mills-Hagley Foundation of Greenville, Wilmington, Delaware. Walter J. Heacock, the foundation's general director, and Richmond D. Williams, director of the Eleutherian Mills Historical Library, responded to our request promptly and generously. We were offered facilities similar to those for resident scholars at the Library and convenient access to its unique reference and research collections. Eugene S. Ferguson, curator of technology at that institution and professor of the history of technology at the University of Delaware, cooperated in the selection of an assistant editor for technology.

Darwin H. Stapleton was appointed in September 1974 and worked at the Eleutherian Mills Historical Library until he joined the Case Western Reserve University faculty in August 1976. Initially, his work was supported by grants from the Rockefeller Foundation and the United States Capitol Historical Society; since October 1975, Dr. Stapleton's salary and other expenses connected with technological research and the preparation of this volume have been funded by National Science Foundation grants SOC75-17947 and SOC75-17947 A01.

Charles E. Brownell, assistant editor for architectural history, collaborated in the early planning of this volume. At first it was thought that the architectural and engineering drawings could be published as a single book. This plan was abandoned when the inherent differences in the two sets of drawings, revealed in the process of annotation, began to emerge. This entire process, however, was most valuable for establishing an analogous structure for both

1. BHL, *Journals*, 2:467.

Foreword

The Engineering Drawings and *The Architectural Drawings of Benjamin Henry Latrobe.*

Of all Latrobe's engineering drawings, none offered greater problems of analysis, annotation, and graphic presentation than his beautifully rendered Susquehanna River Survey Map. Fortunately, M. Gordon Wolman, chairman of the department of geography and environmental engineering of The Johns Hopkins University and a noted specialist in river system behavior, offered his assistance in finding a suitable consultant. Ultimately, Stephen F. Lintner was selected to annotate the map's historical, scientific, and technical aspects.[2] His fieldwork was augmented by a short-term residence fellowship at the Hermon Dunlap Smith Center for the History of Cartography at The Newberry Library, Chicago.

To the editor in chief falls the agreeable duty of acknowledging the valuable scholarly, technical, and financial assistance we have received from other institutions and members of their faculties or staffs in preparing *The Engineering Drawings of Benjamin Henry Latrobe* for publication.[3] In addition to those individuals and organizations noted above, we would like to thank the officers and staffs of the four cosponsors of this edition: the Maryland Historical Society, National Endowment for the Humanities, National Historical Publications and Records Commission, and the National Science Foundation. At NSF, we are greatly indebted to Ronald J. Overmann, associate program director for history and philosophy of science, for the continued, informed interest he has shown in our work. We particularly desire to draw attention to the fact that this volume is published with the financial assistance of the NHPRC.

We also wish to thank those librarians, manuscript and graphics curators, officers of private and public repositories, and individual citizens who allowed us to include literary or illustrative materials from their collections. (The ownership of each item so used is appropriately noted in the volume.) At the Hermon Dunlap Smith Center for the History of Cartography at The Newberry Library, David Woodward, director, and Robert W. Karrow, Jr., curator of maps, proved to be especially helpful in establishing the cartographic tradition and uniqueness of the Susquehanna River Survey Map. This volume has also benefited from the critical interest and expertise of Carroll W. Pursell, Jr., of the department of history, University of California, Santa Barbara, who read the first draft of the manuscript in its entirety.

Baltimore, April 1978

EDWARD C. CARTER II
Editor in Chief

---

2. The map provides the data base for his dissertation, which determines the sediment transport characteristics and the way in which the channel of this section of the Susquehanna River has behaved over the last 175 years. Stephen F. Lintner, "The Historical Physical Behavior of the Lower Susquehanna River, 1801 to 1976" (Ph.D. diss. in progress, Johns Hopkins University).

3. For a detailed expression of our gratitude to all the individual and institutional supporters of this edition of Benjamin Henry Latrobe's work over the years, see, BHL, *Journals*, 1:liv–lix.

EDITOR'S NOTE

FROM THE EARLIEST PLANNING of *The Engineering Drawings of Benjamin Henry Latrobe* it was intended that the drawings themselves would be treated as important historical evidence. The primary goal was to publish a catalogue of drawings in which each drawing was completely described and explained. There is little precedent for this treatment of historic technical drawings, even though some recent publications in the history of technology have included many technical drawings as illustrations, and there have been new editions of Renaissance machine books.[1] For this volume, art exhibit catalogues have provided the most useful model.[2]

It is possible that a significant new approach to the publication of technical drawings is beginning, particularly because the activities of groups such as the Historic American Engineering Record (Department of the Interior) and the Society for Industrial Archeology have accelerated the growth of collections of historic technical drawings.[3] (For this discussion, "technical drawings" include the paintings, sketches, and measured drawings made by technicians such as engineers.)

Eugene S. Ferguson has also provided a philosophical basis for the serious consideration of technical drawings as objects of historical study. In his publications and teaching he has been concerned with the means by which technically skilled persons express themselves and communicate with one another and has concluded that nonverbal thinking and expression are significant elements of technical skill and creativity. Ferguson contends that historians of technology must recognize the importance of nonverbal thought if they are rightly to perceive their discipline, and he has focused on the technical drawing as a unique documentation of the technician's nonverbal thought processes.[4] Yet few historical works explain, and fewer still interpret, the technical renderings that they contain. Even the annotated edition of Leonardo da Vinci's *Madrid Codices* has been criticized for its inadequate consideration of "the visual and spatial world" of Leonardo's drawings.[5]

The Engineering Drawings of Benjamin Henry Latrobe explores the historical record embedded in technical drawings. Separating Latrobe's engineering drawings on the basis of Ferguson's distinction between drawings that suggest new ideas and those that show established technique, we find that most of Latrobe's fall in the latter category.[6] The record reveals Latrobe's skill more than his imagination. Seldom do the drawings allow us to trace the conceptual development of his nonverbal design process. Sketches, notebooks, and field books are generally lacking, with an important exception. The two field books of the Susquehanna survey do survive and provide a fascinating insight into the manner in which Latrobe accumulated visual information to complement his surveying data. One may also judge from Latrobe's architectural practice, in which a much larger number of extant drawings permit a better tracing of the evolution of his ideas, that Latrobe used drawings as an aid in working out his engineering designs. Thus the preponderance of Latrobe's known engineering drawings are finished renderings (often meant to be seen by, or presented to, his employers), and they are superbly expressive of his mature designs. They reveal the practical knowledge, or "engineering vocabulary," which one of the most skillful and active American engineers possessed during an important era of the nation's technical and economic growth. This knowledge passed on to Latrobe's many pupils and associates, partly through these very drawings.

THE ENGINEERING PRACTICE OF BENJAMIN HENRY LATROBE

The introductory essay attempts to provide sufficient historical background for the engineering drawings, in part because the treatment of Latrobe's engineering practice by the standard biography, Talbot Hamlin's *Benjamin Henry Latrobe*, is generally sketchy, lacking in technical detail, and sometimes in error. The essay's arrangement under the headings of projects closely parallels the order of the engineering drawings in order to make the essay more useful to those studying the drawings. It is not a complete engineering biog-

---

1. E.g., Thomas Parke Hughes, *Elmer Sperry, Inventor and Engineer* (Baltimore: Johns Hopkins Press, 1971); Reese V. Jenkins, *Images & Enterprise: Technology and the American Photographic Industry* (Baltimore: Johns Hopkins Press, 1975); Augustino Ramelli, *The Diverse and Artificial Machines of Augustino Ramelli*, ed. Martha Teach Gnudi and Eugene S. Ferguson (Baltimore: Johns Hopkins Press, 1976).
2. E.g., Frederick Cummings, Allen Staley, and Robert Rosenblum, *Romantic Art in Britain: Paintings and Drawings, 1760–1860* (Philadelphia: Philadelphia Museum of Art, 1968).
3. HAER produces modern drawings of historic engineering works. See Donald E. Sackheim, comp., *Historic American Engineering Catalogue* (Washington: National Park Service, 1976); and Theodore Anton Sande, *Industrial Archeology: A New Look at the American Heritage* (Brattleboro, Vt.: The Stephen Greene Press, 1976).
4. Eugene S. Ferguson, "The Mind's Eye: Nonverbal Thought in Technology," *Science* 197 (26 August 1977): 827–36.
5. Bert S. Hall, review of *The Madrid Codices*, by Leonardo da Vinci and *The Unknown Leonardo*, ed. Ladislao Reti, in *Isis* 67 (September 1976): 466.
6. Ferguson, "The Mind's Eye," pp. 828–29.

Editor's Note

raphy, although all major areas of Latrobe's engineering practice are considered, including some not well represented in the drawings.

Readers who wish to know more about Latrobe's personal life, intellectual pursuits, and architectural career are urged to consult the other publications of the Latrobe Papers, especially the indispensable *Microfiche Edition*, and Hamlin's *Latrobe*.

THE SUSQUEHANNA RIVER SURVEY MAP

In the collection of Latrobe's engineering drawings, the Susquehanna map is extraordinary. Having dimensions of seventeen feet by two feet, it contains a detailed engineering and natural history record of the lower Susquehanna River in 1801. The Susquehanna map's size creates certain difficulties for book presentation, but the scale and scope of its execution yield the most exhaustive record of Latrobe's cartographic ability.

The vast amount of data recorded on the map that relates to geography and natural history makes it an incomparable statement of human ecology. Moreover, the supporting evidence includes the only extant Latrobe survey books and a number of his beautiful watercolors. Thus it appears as the first of the engineering drawings in this volume, not because of chronology or its significance as an engineering project, but because of its unique qualities as a document.

To explain the map, we have chosen to provide a separate introductory essay, followed by ten map sections annotated with selected drawings, watercolors, diagrams, and quotations.

THE ENGINEERING DRAWINGS

More than half of the known Latrobe engineering drawings are published here, and all are catalogued. There are a greater number of Latrobe architectural drawings, which appear in the other volume of this series. Since Latrobe made the distinction between his employment as an engineer and as an architect, there is little difficulty in categorizing drawings. However, a few of them, including several of the Philadelphia Waterworks, incorporate both architectural and engineering elements.

A number of factors determined whether an engineering drawing merited annotation and publication. The major criterion was significance of content. Those drawings were favored that are complete in detail, or form a part of a series of drawings that provide a complete view of a project. Some drawings and series that are not complete were still judged of significant historical interest and are also included. In the case of duplicate or similar drawings, only one is published.

The second criterion was how well the drawings could be reproduced. Some that presented serious photographic problems, such as the dismembered and damaged Jones' Falls survey map, could not be published. Writing that could not easily be read also weighed against some drawings. Yet almost all of the drawings have some elements that can be understood only by seeing the original; for instance, notations in brown ink inscribed on a dark background that come out poorly when photographed.

Finally, this selection of drawings reflects a liberal interpretation of the volume's title, *The Engineering Drawings of Benjamin Henry Latrobe*. Most of the drawings are signed by Latrobe, or show some feature, such as handwriting, attributable to him. Such drawings clearly came out of his office. The degree to which any of Latrobe's drawings may have been rendered under his direction by his pupils and assistants is uncertain. Thus drawings from Latrobe's office have no catalogue entry for "delineator."

Some drawings published here are not from Latrobe's office, and others are of doubtful authorship. Most of these non-Latrobe drawings are included because they were made under his general direction or for projects for which he was engineer and were partly or wholly based on his concepts and instructions. A few other drawings are copies of missing originals (such as engravings) that are thought to be faithful facsimiles. It is hoped that this volume is richer and more illustrative of Latrobe's engineering work than a literal adherence to the volume's title might have allowed, but it is still dominated by the high quality of Latrobe's drafting skill.

Readers should note that the drawings record Latrobe's intentions and that a completed project always differed somewhat. I have used verbs in the present tense to describe the contents of the drawings, and I have changed to the past tense to signal a shift to other historical evidence concerning a project's execution.

References at the end of a section of the drawings catalogue list the sources I found most helpful in preparing that section.

CITATION OF LATROBE DOCUMENTS

Virtually all of the documents cited in this volume that were written by, signed by, drawn by, or sent to Latrobe (including journals, letters, sketchbooks, maps, plans, drawings, and field books) may be consulted in Thomas

Editor's Note

E. Jeffrey, ed., *The Papers of Benjamin Henry Latrobe: The Microfiche Edition* (Clifton, N.J.: James T. White & Company, 1976). Thus in citing this manuscript material I have given only the repository of the original document. The interested reader can easily locate a Latrobe document in the *Microfiche Edition* through the use of the *Guide and Index* accompanying the edition. Reference has also been made in footnotes to several Latrobe documents accessioned since the filming of the *Microfiche Edition*, which took place in October 1974. These documents are cited by manuscript repository followed by the parenthetical phrase (not in microfiche ed.). It is hoped that these documents, and all others accessioned since the filming of the *Microfiche Edition*, will be published in an addendum to the *Microfiche Edition* to appear in the early 1980s.

ACKNOWLEDGMENTS

This volume represents a dedicated effort by the entire staff of *The Papers of Benjamin Henry Latrobe*, and I have relied upon their support, unstinting cooperation, and encouragement. The facilities and collections of the Eleutherian Mills Historical Library of Greenville, Wilmington, Delaware, provided the environment for the research and writing of the majority of this volume, and for that I am very grateful.

In addition to those institutions and individuals acknowledged in the Foreword, I wish to acknowledge certain other assistance. Among those who read and commented on all or portions of the manuscript were Charles E. Brownell, Lee W. Formwalt, John C. Van Horne, J. Frederick Fausz, Eugene S. Ferguson, and Judy Metro. Special thanks for reading and copy editing the entire manuscript go to Lawrence Kenney. Geraldine S. Vickers typed the manuscript and provided other vital assistance. I received various courtesies and contributions from Angeline Polites, Thomas E. Jeffrey, Carol Hallman, Susan Danko, Jean E. Carr, Thomas C. Guider, Stephanie Morris, Karen Peterson, Charles E. Peterson, and Stephen F. Lintner. Professional encouragement was extended by Edward C. Carter II, Morrell Heald, and Robert E. Schofield. My wife, Donna, gave me constant support and encouragement.

Cleveland, April 1978   DARWIN H. STAPLETON

For D.L.H.S.

SHORT TITLES AND ABBREVIATIONS

APS
American Philosophical Society, Philadelphia, Pa.

ASP
American State Papers: Documents, Legislative and Executive, of the Congress of the United States . . ., 38 vols. (Washington, 1832–61).

BHL
Benjamin Henry Latrobe.

BHL, *Journals*
Edward C. Carter II et al., eds., *The Virginia Journals of Benjamin Henry Latrobe, 1795–1798*, 2 vols. (New Haven: Yale University Press, 1977).

BHL, *Opinion*
Benjamin Henry Latrobe, *Opinion on a Project for Removing the Obstructions to a Ship Navigation to Georgetown, Col.* (Washington, 1812).

C & D Engineer's Reports, Hist. Soc. Del.
Reports of the Engineer, Chesapeake and Delaware Canal Company Papers, Historical Society of Delaware, Wilmington, Del.

Cope Diaries, Haverford College
Thomas P. Cope Diaries, Haverford College Library, Haverford, Pa.

DAB
Dictionary of American Biography.

DNB
Dictionary of National Biography.

Eleutherian Mills
Eleutherian Mills Historical Library, Greenville, Wilmington, Del.

Essex Record Office
Essex Record Office, Chelmsford, England.

Franklin Inst.
Franklin Institute, Philadelphia, Pa.

Hamlin, *Latrobe*
Talbot Hamlin, *Benjamin Henry Latrobe* (New York: Oxford University Press, 1955).

Haverford College
Haverford College Library, Haverford, Pa.

Hist. Soc. Del.
Historical Society of Delaware, Wilmington, Del.

Hist. Soc. Pa.
Historical Society of Pennsylvania, Philadelphia, Pa.

Jefferson Papers, LC
Papers of Thomas Jefferson, Manuscript Division, Library of Congress, Washington, D.C.

Journal, Comm. of Works, Salem Co. Hist. Soc.
Journal of the Committee of Works, Chesapeake and Delaware Canal Company, Salem County Historical Society, Salem, N.J.

Latrobe Journals
Benjamin Henry Latrobe Journals, Papers of Benjamin Henry Latrobe, Maryland Historical Society, Baltimore, Md.

Latrobe Sketchbooks
Benjamin Henry Latrobe Sketchbooks, Papers of Benjamin Henry Latrobe, Maryland Historical Society, Baltimore, Md.

LC
Library of Congress, Washington, D.C.

Letterbooks
Benjamin Henry Latrobe Letterbooks, Papers of Benjamin Henry Latrobe, Maryland Historical Society, Baltimore, Md.

MdHS
Maryland Historical Society, Baltimore, Md.

Minutes, Comm. of Survey, Hist. Soc. Del.
Minutes of the Committee of Survey, Chesapeake and Delaware Canal Company Papers, Historical Society of Delaware, Wilmington, Del.

Minutes, Wash. Canal Co., NA
Minutes of the Washington Canal Company, RG 351, National Archives, Washington, D.C.

NA
National Archives, Washington, D.C.

New Orleans Pub. Lib.
New Orleans Public Library, New Orleans, La.

Short Titles and Abbreviations

N-Y Hist. Soc.
   New-York Historical Society, New York, N.Y.

Pa. Hist. Mus. Comm.
   Pennsylvania Historical and Museum Commission, Harrisburg, Pa.

PBHL
   Papers of Benjamin Henry Latrobe, Maryland Historical Society, Baltimore, Md.

Phil. Archives
   Archives of the City and County of Philadelphia, City Hall, Philadelphia, Pa.

Salem Co. Hist. Soc.
   Salem County Historical Society, Salem, N.J.

Tulane
   Tulane University Library, New Orleans, La.

CHRONOLOGY

1764 *May 1*	Born at Fulneck, England, to the Rev. Benjamin and Anna Margaretta (Antes) Latrobe
1767 *June 13*	Enters the Moravian school at Fulneck in Yorkshire
1776 *September*	Leaves England to attend Moravian schools at Niesky and Barby in Germany
1783 *August 28*	Returns to England and establishes residence in London
c. 1786	Begins engineering studies under John Smeaton
c. 1786–1787	Works on Rye Harbour Improvement under Smeaton's chief assistant, William Jessop
c. 1788–1789	Helps construct the Basingstoke Canal under Jessop
c. 1789–1792	Works in the office of architect Samuel Pepys Cockerell
1793 *Spring*	Obtains first of several commissions from a group from Maldon, Essex, opposed to the intended route for the Chelmer and Blackwater Navigation
1795 *February 13*	BHL's testimony regarding the plans for the Chelmer and Blackwater Navigation is reported to Parliament
1795 *November 25*	Sails for America on the ship *Eliza*
1796 *mid-March*	Arrives at Norfolk, Virginia, and begins a two-and-a-half-year residence in that state
1796 *June 7–17*	Makes informal survey of the Appomattox River for the Upper Appomattox Navigation Company
1797 *June 6–27*	Visits property of the Dismal Swamp Land Company as an engineering consultant
1798 *July*	Draws plans for fortifications at Norfolk, Virginia
1798 *December 1*	Leaves Richmond to establish residence in Philadelphia, having been selected as architect of the Bank of Pennsylvania
1798 *December 29*	Writes *View of the Practicability and Means of Supplying the City of Philadelphia with Wholesome Water* in response to a request of John Miller, chairman of a joint committee of the Philadelphia city councils
1799 *March 2*	BHL's plan for a city waterworks is adopted by the Philadelphia city councils
1801 *January 27*	In a public ceremony, BHL's Centre Square engine pumps the first water of the Philadelphia Waterworks
1801 *September 20*	On the death of Frederick Antes, BHL becomes contractor—in addition to his former duties as engineer—of the Susquehanna River improvement and survey
1802 *November 8*	Accepts Thomas Jefferson's commission to design a naval drydock at Washington; BHL reports to Congress on December 4
1803 *July 5*	Begins survey for the Chesapeake and Delaware Canal Company
1804 *c. January*	Appointed engineer to the Navy Department
1804 *January 25*	Appointed engineer of the Chesapeake and Delaware Canal Company
1804 *February 5*	Submits report and plans to the first Washington Canal Company
1804 *May 2*	Ground broken for the Chesapeake and Delaware Canal feeder
1805 *May 18*	Submits plans for the development of the Washington Navy Yard
1805 *December*	All workers dismissed from Chesapeake and Delaware Canal because of financial difficulties
1807 *May 25/June 6*	Submits reports on the Mud Island bar to the Philadelphia Chamber of Commerce

Chronology

1807	*June*	Leaves Philadelphia to establish residence in Washington, D.C.
1808	*March 16*	Completes essay that is submitted to Congress as part of Albert Gallatin's report on internal improvements
1810	*January 24*	Becomes engineer to the second Washington Canal Company
1810	*Spring*	Petitions the territorial legislature for a waterworks franchise at New Orleans (the petition fails)
1810	*May 2*	Ground broken for the Washington Canal
1810	*June 6*	Appointed commissioner of the Columbia Turnpikes in the District of Columbia
1810	*September*	Directs installation of steam engine at the Washington Navy Yard
1810	*Fall*	Sends son Henry to New Orleans to obtain waterworks franchise from the city council
1811	*June 10*	Accepts New Orleans Waterworks franchise; it is renewed in 1813, 1815, 1817, and 1819
1813	*July–August*	Meets with Robert Fulton and agrees to be agent for the Ohio Steamboat Company at Pittsburgh
1813	*October 31*	Arrives in Pittsburgh
1814	*March*	Becomes contractor for the steam engine of the Steubenville, Ohio, woolen mill
1814	*May 13*	Launches the steamboat *Buffalo* at Pittsburgh
1814	*September 30*	Puts writ of attachment on *Buffalo* for recovery of his investment in the Ohio Steamboat Company after Fulton had removed him from the agency
1815	*April 20*	Returns to Washington
1816	*June 16*	Writes to the mayor of New Orleans about using fascine works to repair breaks in the levee
1816	*Fall*	Directs construction of masonry lock for the Washington Canal
1818	*January*	Moves to Baltimore after the conclusion of bankruptcy proceedings
1818	*May 31/June 6*	Submits plans for improvement of Jones' Falls to the Baltimore City Council
1818	*December 17*	Sails for New Orleans to direct the construction of the waterworks in person (his son Henry had died in New Orleans on 3 September 1817)
1819	*September 19*	Makes return trip to Baltimore to prepare for moving family to New Orleans
1820	*April*	Arrives back in New Orleans with family
1820	*September 3*	Dies in New Orleans of yellow fever while working on the waterworks

ILLUSTRATIONS

COLORPLATES
following page 136

1. Washington Canal—"Longitudinal Section of the upper Gate; Plan of the Lock Floor," c. 1810.
2. Washington Canal—"Transverse Section of the Lock," c. 1810.
3. Washington Canal—Plan and sections of a wooden lock, c. 1810.
4. Washington Canal—"Details of the Gates of the Locks," 20 May 1810.

following page 198

5. Philadelphia Waterworks—"Section of the Basin; Plan of the Basin Wall & of the Coffre-Dam," c. 1800.
6. Philadelphia Waterworks—"Plan of the Platform of the Basin Wall; Plan of the Basin Wall at high Water Mark," c. 1800.
7. Philadelphia Waterworks—"West Elevation of the Wall of the Basin; Section of the Basin and of the Coffre Dam," c. 1800.
8. Philadelphia Waterworks—"An Engine house and Water office," March 1799.
9. Washington Navy Yard Steam Engine—Forge hammer and gearing, 1 July 1811.
10. Washington Navy Yard Steam Engine—Cogwheels and cam ring, 10 August 1810.

FIGURES

The Engineering Practice of Benjamin Henry Latrobe

1. Portrait of Benjamin Henry Latrobe by Rembrandt Peale, c. 1816. — 4
2. BHL's engineering experience in southeast England, c. 1786–95. — 6
3. Map of Naval Drydock and Potomac Canal Extension, 1802. — 10
4. Map of Chesapeake and Delaware Canal and feeder, 1804. — 13
5. Map of Chesapeake and Delaware Canal feeder, 1804–5. — 15
6. Proposed swing bridge, 1804. — 15
7. Tools for canal excavation, 1804. — 16
8. English tools for canal excavation, c. 1800. — 16
9. Chesapeake and Delaware Canal feeder, 1976.
   a. Road bridge — 17
   b. Culvert — 18
   c. Canal trough. — 18
10. Map of the Washington Canal, 1810–17. — 20
11. Construction of the Eastern Lock of the Washington Canal, 1811. — 21
12. Map of the Philadelphia Waterworks, 1801. — 31
13. Schuylkill Engine House, c. 1830. — 32
14. Brick conduit of the Philadelphia Waterworks, 1975. — 32
15. Centre Square Engine House, 1812. — 33
16. Proposed design of the New Orleans Waterworks, 1815. — 39
17. The Potomac River at Washington, 1811. — 43
18. Map of the Jones' Falls Improvement, 1818. — 47
19. Rollers in an English rolling mill, c. 1800. — 49
20. An English rolling mill, c. 1800. — 49
21. Map of the Washington Navy Yard, 1815. — 54
22. BHL's plan of the Steubenville engine, 1814. — 56
23. Steubenville Woolen Mill, 1815. — 57
24. Plan of BHL's machine shop at Pittsburgh, 1814. — 59

The Susquehanna River Survey Map: Essay

25. Susquehanna River keelboat, 1801. — 75
26. Ark on the Susquehanna River, 1809. — 76
27. BHL's observations from the Blue Rock, 1801. — 80
28. "View at the Bear Island at the head of Niels falls," 13 November 1801. — 81
29. "Log and Plaister; Black & white stripe," 30 October 1801. — 81
30. "View Among the Bear Islands, Head of Niels Falls," undated. — 82
31. Composite of three of BHL's sketches of trees. — 87

THE SUSQUEHANNA RIVER SURVEY MAP SECTIONS AND ANNOTATION

Map Data — 89

Index for map sections 1–10. — 89
Scale for map sections 1–10. — 89

Illustrations

Section One

fig. 1. Panorama, looking north from Columbia and Wright's Ferry, 2 November 1801.   90
fig. 2. "View of the Susquehannah above Columbia."   90
fig. 3. "View of Chickisalunga Rocks."   90

Section Two

fig. 1. Panorama, looking east and south.   92
fig. 2. "Anderson's Mill drawn in a Fog."   92
fig. 3. Looking north from Turkey Hill.   92

Section Three

fig. 1. Two panoramic views from Kendrick's Bottom.   94
fig. 2. "Geo. Stoners on Pequai [Pequea] Creek."   94
fig. 3. "View of the Susquehannah."   94

Sections Four and Five

fig. 1. "Fulton's ferry," 11 November 1801.   97
fig. 2. Field notes, 11 November 1801.   97
fig. 3. Field notes, 10 November 1801.   98
fig. 4. "Stoney point."   99
fig. 5. "Susquehannah at Stoney point."   99
fig. 6. View of Lancaster Neck Mountain.   99
fig. 7. "View of Culleys falls."   99

Section Six

fig. 1. Detail of "Quiggler's Bottom."   100
fig. 2. Details from Montesson, *La science de l'arpenteur*, 1766, and Hayne, *Deutliche und ausfürliche Anweisung*, 1794.   100
fig. 3. Pen and ink study of a log cabin.   100
fig. 4. Pencil sketch of a log cabin.   100

Section Seven

fig. 1. BHL's and Hauducoeur's surveys, 1798–1801.   102
fig. 2. Detail from *A Map of the Head of Chesapeake Bay and Susquehanna River, 1799*.   102

Section Eight

fig. 1. Geology of the Susquehanna River.   104
fig. 2. Detail from *A Map of the Head of Chesapeake Bay and Susquehanna River, 1799*.   105

Section Nine

fig. 1. Detail of "Sauers point."   106
fig. 2. Detail of "Rock run."   106
fig. 3. Detail from Hayne, *Deutliche und ausfürliche Anweisung*.   106
fig. 4. Detail from *A Map of the Head of Chesapeake Bay and Susquehanna River, 1799*.   106

Section Ten

fig. 1. "From the ferry at Havre de Grace."   108
fig. 2. C. P. Hauducoeur, *A Map of the Head of Chesapeake Bay and Susquehanna River, 1799*.   108
fig. 3. Detail from *A Map of the Head of Chesapeake Bay and Susquehanna River, 1799*.   108

ENGINEERING DRAWINGS

Chelmer and Blackwater Navigation

1. "A Survey of the Course of the River Blackwater," 11 November 1793.   114
2. "A Map or Plan of the intended Improvement of the Navigation of the Rivers Blackwater and Chelmer," 1794.   115

Naval Drydock and Potomac Canal Extension

3. "No. I. Section of Locks; Plan of Locks," 4 December 1802.   119
4. "No. II. [Plan, sections and elevations of the drydock]," 4 December 1802.   120
5. Detail from Potomac Canal Extension No. I, [1802].   121
6. Detail from Potomac Canal Extension No. II, 1802.   122
7. Detail from Potomac Canal Extension No. IIII, 1802.   123
8. Detail from Potomac Canal Extension No. IIII, 1802.   124

Chesapeake and Delaware Canal

9. Plan of the Elk Creek aqueduct, [1804]. — 126
10. "South, or downstream Elevation of the Aqueduct over Elk creek," [1804]. — 127
11. "Transverse Section of the Aqueduct over Elk Creek," [1804]. — 128
12. "Elevation of the Aqueduct over the head of Cow run," [1804]. — 129

Washington Canal

13. "Plan of the Groundwork; Plan of the piling," c. 1810. — 132
14. "Longitudinal Section of the upper Gate; Plan of the Lock Floor," c. 1810. (See also colorplate 1.) — 133
15. "Transverse Section of the Lock," c. 1810. (See also colorplate 2.) — 134
16. Plan and sections of a wooden lock, c. 1810. (See also colorplate 3.) — 135
17. "Details of the Gates of the Locks," 20 May 1810. (See also colorplate 4.) — 136

Dismal Swamp Canal

18. "Sketch of Canals *executed* or *proposed* near Norfolk," 19 December 1803. — 138

James River Canal and Railroad

19. Map of James River Canal and proposed railroad, 19 December 1803. — 140

National Road (Cumberland Road)

20. Plan and sections for the National Road, 1814. — 143

Philadelphia Waterworks

21. "No. I. Section of the Works from the Schuylkill to the lower, or Schuylkill Engine house," [1799]. — 145
22. Basin, canal, tunnel, and Schuylkill Engine House, c. 1800. — 148
23. "Section of the Basin; Plan of the Basin Wall & of the Coffre-Dam," c. 1800. (See also colorplate 5.) — 149
24. "Plan of the Platform of the Basin Wall; Plan of the Basin Wall at high Water Mark," c. 1800. (See also colorplate 6.) — 150
25. "West Elevation of the Wall of the Basin; Section of the Basin and of the Coffre Dam," c. 1800. (See also colorplate 7.) — 151
26. "Section of the Archimedean Screw," c. 1800. — 153
27. "Schuylkill Engine House No. I; Plan of the Foundations," [1799]. — 155
28. "Schuylkill Engine House No. II; Ground-plan," [1799]. — 156
29. Plan of the Schuylkill engine, c. 1799. — 160
30. Section of the Schuylkill engine, c. 1799. — 161
31. Plan and sections of a cast-iron boiler, 1809. — 162
32. "Section of the Boiler," c. 1803. — 163
33. Section of the Schuylkill engine, c. 1799. — 164
34. "Section of the Schuylkill Engine House," c. 1799. — 166
35. "Plan [of the] Schuylkill Mills," 24 June 1800. — 169
36. Section of the Schuylkill mills, 25 June 1800. — 170
37. "Section Showing the Rollers—Schuylkill Mills," 24 June 1800. — 171
38. "Section parrallel to the Sheer block [of the] Schuylkill Works," 1 July 1800. — 172
39. "An Engine house and Water office," March 1799. (See also colorplate 8.) — 174
40. Plan of the foundation of the Centre Square Engine House, c. 1799. — 176
41. "Second, or Center Engine House, No. II; Plan of the Ground Story," [1799]. — 178
42. "Second, or Center Enge. He. No. V; Section from West to East," [1799]. — 180
43. Plan of the Centre Square engine, c. 1799. — 184
44. Plans and sections of the wooden boiler, 1809. — 185
45. Section of the Centre Square engine, c. 1799. — 186
46. Section of the air pump and condenser, 1809. — 187
47. Section of the Centre Square engine, c. 1799. — 188
48. East-to-west section of the flywheel and pump, c. 1799. — 190
49. North-to-south section of the flywheel and pump, c. 1799. — 191
50. Section of the pump, 1809. — 192
51. Plan of three reservoirs, c. 1805–8. — 194
52. Plan of the decking; section of a reservoir, c. 1805–8. — 195
53. Combined fireplug and hydrant, fireplug, and stopcock, 1876. — 197
54. Hydrant detail from New Orleans Waterworks drawing, c. 1811. — 198

Illustrations

New Orleans Waterworks

55. Map of New Orleans Waterworks, c. 1811. 200
56. "Details of a New mode of arranging the Side pipes & working Gear of a Steam Engine," 1812. 202
57. "A pier to cover the Suction pipe of the Pump," 29 January 1819. 204

Mud Island Bar

58. Map of the Delaware River below Philadelphia, 25 May 1807. 206
59. "Scheme for blocking up the Passage between Hog & Maiden Islands," 6 June 1807. 207

Fascine Works

60. "I"—the Dagenham Gap, 1816. 210
61. "II"—tools and materials, 1816. 211
62. "III"—layout of boone, 1816. 212
63. "IV"—construction of boone, 1816. 213

Washington Navy Yard Steam Engine

64. Plan of a steam engine, 28 September 1808. 217
65. Elevation of a steam engine, 29 September 1808. 218
66. Elevation of a steam engine, 29 September 1808. 219
67. Elevation of a beam, cranks, and flywheels, 30 September 1808. 220
68. Section of the cylinder and plans of parts, 1 October 1808. 221
69. "Nozzles," 30 September 1808. 222
70. Plan of a steam engine, 19 March 1810. 225
71. Elevation of a steam engine, 17 March 1810. 226
72. Elevation of a steam engine, 22 March 1810. 227
73. Elevation of the beam, cranks, and flywheels, 17 March 1810. 228
74. "Plan of Timber for the Engine," 16 January 1810. 229
75. "Side Elevation of the Timber for the Engine," 17 January 1810. 230
76. "End Elevation of the Timber for the Engine," 17 January 1810. 231
77. Elevations of the working gears, 18 January 1810. 232
78. Details of the flywheel and cam connection, 22 March 1810. 233
79. Forge hammer and gearing, 1 July 1811. (See also colorplate 9.) 235
80. Cogwheels and cam ring, 10 August 1810. (See also colorplate 10.) 236

Machinery

81. Machinery for a block mill, 23 November 1796. 238
82. "Hints for a block Mill," 23 November 1796. 239
83. Hoisting machine, 1804. 241
84. "Sketch of the landing place of the Columns of the House of Repr.," 3 May 1817. 242

The Engineering Practice of Benjamin Henry Latrobe

1. INTRODUCTION

BENJAMIN HENRY LATROBE's work as an engineer was largely forgotten soon after he died. Nineteenth-century histories of American engineering celebrated the great canal and railroad achievements of later years and found little reason to mention his name.[1] It has remained for historians of the twentieth century to assess his accomplishments and failures.

Talbot Hamlin's fine biography *Benjamin Henry Latrobe*, published in 1955, was the first major work to rekindle interest in Latrobe. It traced Latrobe's life with such detailed attention to his character and career that only an exceptional biography based on new materials could replace it. Hamlin critically examined Latrobe's architectural career and outlined the major themes of Latrobe's engineering practice. Yet he relegated much of his treatment of Latrobe's engineering activities to a subsidiary chapter, and a sustained account of what engineering skills Latrobe had and how he employed them is lacking in Hamlin's book.[2]

Other recent works bearing on Latrobe's engineering career are not biographical but examine closely certain aspects of his engineering. Nelson Blake's *Water for the Cities* includes an excellent chapter on the establishment of the Philadelphia Waterworks, especially concentrating on public support before, during, and after construction of the project. Carroll Pursell considered not only Latrobe's Philadelphia experience, but also his involvement with the Washington Navy Yard and the New Orleans Waterworks in *Early Stationary Steam Engines in America*. Daniel Calhoun's *The American Civil Engineer* does not focus as much on Latrobe's career as on those of others, but it recognizes that Latrobe was a leading engineer of his era, as well as one of the most articulate in attempting to define the profession during its early years.[3]

While these publications have extended the understanding of the engineering work of Benjamin Henry Latrobe beyond Talbot Hamlin's, their deliberate limits precluded full presentations of Latrobe's engineering career and considerations of its meaning for American technological history. However, the amassing of Latrobe's extant papers and many auxiliary materials by *The Papers of Benjamin Henry Latrobe* now makes it possible to present a much more complete story of Latrobe's engineering practice. This essay considers Latrobe's engineering training, his career in England and America, his techniques (e.g., his choices of materials and power sources), and his relationship to the American technical community. Special attention is given to those projects represented by the engineering drawings in the latter part of this volume. In this way the essay serves not only as an extended treatment of Latrobe's engineering practice, but also as a background for the engineering drawings.

Benjamin Henry Latrobe was born on 1 May 1764 to Benjamin Latrobe and Anna Margaretta (Antes) Latrobe, the headmaster and headmistress of the Moravian school at Fulneck in Yorkshire, England. Although his parents moved to London in 1768 so that his father could fill a new position, Latrobe remained at the Fulneck school for his early education. It was common among Moravians of this era to send their children away to school; they believed that a sound and thorough education under church authority was the best assurance that a person would know, understand, and believe the word of God. In 1776, at the age of twelve, Latrobe left England to receive further education in Germany, the world center of Moravian culture. First he went to Niesky in Silesia, and later to a seminary at Barby, in Saxony.[4]

Latrobe's father was a church official, one brother received religious training and became a Moravian missionary, and Latrobe himself went to a seminary. It seems likely, therefore, that his family intended his education to prepare him for the ministry. Indeed, the curricula of Moravian schools emphasized study of the Bible and classical languages, as well as regular religious devotions and rituals for all students. Latrobe might easily have taken a vocation in the church.

---

1. See, for example, J. Elfreth Watkins, "The Beginnings of Engineering," *Transactions*, American Society of Civil Engineers, 24 (1891): 309–80; Charles B. Stuart, *Lives and Works of Civil and Military Engineers of America* (New York, 1871). Watkins omits Latrobe from his list of important early American engineers (p. 341), while Stuart mentions him briefly (p. 243) in the biography of his son, Benjamin H. Latrobe, Jr.
2. Talbot Hamlin, *Benjamin Henry Latrobe* (New York: Oxford University Press, 1955).
3. Nelson Manfred Blake, *Water for the Cities*. Maxwell School Series, III (Syracuse, N.Y.: Syracuse University Press, 1956); Carroll W. Pursell, Jr., *Early Stationary Steam Engines in America* (Washington: Smithsonian Institution Press, 1969); Daniel Hovey Calhoun, *The American Civil Engineer: Origins and Conflict* (Cambridge, Mass.: M.I.T. Press, 1960).
4. Hamlin, *Latrobe*, pp. 3–4, 8–13. Latrobe was born at Fulneck, Yorkshire, England. He died 3 September 1820 at New Orleans, Louisiana; he married Lydia Sellon (c. 1761–93), 27 February 1790, and Mary Elizabeth Hazlehurst (1771–1841), 1 May 1800; his children were Lydia (1791–1878), Henry (1792–1817), John (1803–91), Julia (1804–90), Benjamin (1806–78).

Fig. 1. Portrait of Benjamin Henry Latrobe by Rembrandt Peale, c. 1816. Courtesy of Mrs. Gamble Latrobe.

On the other hand, Moravian schools taught a wider range of subjects, and at Niesky and Barby Latrobe embarked upon studies which were of fundamental importance to his later engineering career. He had regular instruction in mathematics, geometry, trigonometry, physics, natural history, drawing, and painting, all of which contributed directly to his later mastery of surveying, technical drafting, and hydrology. At Niesky Latrobe was also encouraged to choose a profession and apprentice himself to it.[5]

While in Germany Latrobe made his first attempt to acquire training in civil engineering. He spent some time when he was "scarce 17 years old" (in 1781, if his memory was correct) with a hydraulic engineer, Riedel, who built and repaired dikes along the rivers of Saxony.[6] This experience suggests that Latrobe had at least tentatively chosen a career in engineering. At this time as well, Latrobe considered a military career, possibly because most German engineers were trained at military academies.[7] But his Moravian background, with its tradition of pacifism, and his return to England, where engineering had much more distant relations with the military, may have helped to turn him in a different direction.

In 1783 Latrobe left Germany and probably toured Europe for some months. The details of how this journey was devoted to professional enlightenment are uncertain, but he did observe many great public buildings, and it appears that he did not neglect forts, canals, harbors, waterworks, and the like.[8]

In August, after an absence of seven years, Benjamin Henry Latrobe returned to England and established his residence in London. As his parents worked and lived there, London was a natural place to go, but it was also an excellent

5. Gillian Lindt Gollin, *Moravians in Two Worlds* (New York: Columbia University Press, 1967), pp. 67–89, 106; Heinrich C. G. Graf zu Lynar, *Nachricht von dem Ursprung und Fortgange, und hauptsächlich von der gegenwärtigen Verfassung der Brüder-Unität.* 2d ed. (Halle, 1781), pp. 146, 147, 151; Hermann Plitt, *Das Theologische Seminarium de Evangelischen Brüder-Unität in seinem Anfang und Fortgang* (Leipzig, 1854), p. 56; E. R. Meyer, *Schleiermacher's und C. G. von Brinkmann's Gang durch die Brudergemeine* (Leipzig: Friedrich Tanza, 1905), pp. 94, 98.

The last cited source is particularly valuable since it details the education of Carl Gustav von Brinkmann, who was born in the same year as Latrobe (1764) and was his classmate at Niesky and probably at Barby. "Brinkmann, Karl Gustav," *Allgemeine Deutsche Biographie*, 56 vols. (Leipzig, 1875–1900; reprint ed., Berlin: Duncker & Humblot, 1967–71), 47:236–38; "Report of Unity Schools, 1776–1777," *Gemein-Nachrichten*, 1777, 52d week, 1st part, manuscript collections, Moravian Archives, Bethlehem, Pa.

6. BHL to Mayor and Corporation of the City of New Orleans, 10 June 1816, Louisiana Division, New Orleans Pub. Lib. See also BHL to Seaton, 3 September 1817, Letterbooks.

7. BHL to Jefferson, 4 July 1807, Letterbooks; BHL to Eustis, 17 September 1812, Letterbooks; Frederick B. Artz, *The Development of Technical Education in France, 1500–1850* (Cambridge, Mass.: Society for the History of Technology and M.I.T. Press, 1966), p. 75n.

8. Edward C. Carter II, "Introduction," in BHL, *Journals*, 1:xix; BHL, "Remarks on the best form of a room for hearing and speaking," c. 1803, APS; BHL to Jefferson, 4 July 1807, Letterbooks.

location for an aspiring engineer to develop his skills. At first, possibly through his father's influence, Latrobe secured employment in the government's Stamp Office, which administered the distribution of tax stamps. He probably retained a position there for about nine years, but his duties were slight enough to permit him to have major professional engagements at the same time.[9]

The engineering training which Latrobe acquired in these years is only sketchily known from Latrobe's later brief references to it. His statements are consistent with other evidence, however, and can be made into a plausible sequence of events.[10]

About 1786 he came to the attention of John Smeaton, the foremost English engineer.[11] Latrobe may have sought a position in Smeaton's office, since apprenticeship under a practicing engineer was the standard means of entry into the profession. Possibly impressed by Latrobe's German experience and recent knowledge of continental engineering, as well as his drafting ability, Smeaton took him into his office.[12] In addition to preparing drawings,[13] Latrobe was sent to the Cambridgeshire and Lincolnshire fens "to report to him [Smeaton] on the situation of the *scouring works* as they are called; on the inefficacy of which he was then consulted."[14] These devices were undoubtedly wingworks and sluices erected in the rivers so that their currents could keep the channels free of silt and debris.

Latrobe's next assignment was to the Rye Harbour improvement project on the English Channel, for which his examination of the fens may have been preparation.[15] In 1786, on Smeaton's recommendation, the Rye Harbour commissioners appointed William Jessop, Smeaton's chief assistant, to direct the construction of a canal and dams to divert the River Rother into the new Rye Harbour. The harbor had a constant problem with silting, and Jessop's was one of a series of attempts to provide sufficient scouring action from the river's flow to keep the harbor open. But late in 1787 it was clear that the effort had failed, and the project was suspended.[16]

Jessop must have learned to respect Latrobe's abilities during the work at Rye because within a year Latrobe joined Jessop's staff on the Basingstoke Canal southwest of London. Possibly Latrobe participated in surveying and laying out the canal before cutting actually began in October 1788. Then he was employed on "a district" of the canal, perhaps as a resident engineer or under a resident engineer. He helped supervise the work of John Pinkerton, who was the contractor at Rye and at many other engineering projects. With two tunnels, many bridges and locks, and several wharves and warehouses to build, Pinkerton had a variety of tasks which provided Latrobe with valuable opportunities to learn about canal construction.[17]

Certainly Latrobe could not have had better preparation for his future engineering career than by serving under two of the most eminent civil engineers in England. To learn engineering by reading and studying alone was impossible. Few English engineering projects were extensively recorded (Smeaton's book on the Eddystone lighthouse was a later exception), and almost no practical texts were available in English, although the French published some works which Englishmen consulted.[18]

Latrobe did acquire further knowledge of engineering than what he learned under Smeaton and Jessop. He was acquainted with John Rennie, a prominent canal engineer; he was a friend of Humphrey Repton, a landscape architect (and his son John Repton, an architect); and he knew a London banker who promoted the construction of turnpikes.[19] He saw steam engines in operation

---

9. BHL to Arthur Young, 22 May 1788, The British Library; (Ackermann's) *Repository of the Arts* 2d series, 7 (1821): 33; Charles William Janson, *The Stranger in America* (London, 1807), p. 204n.

10. What follows concerning Latrobe's English training is a reconstruction, often without strong supporting evidence. The statements in the text are frequently qualified to reflect their suggestive nature.

11. *DNB*, s.v. Smeaton, John.

12. BHL, *Opinion*, p. 17; BHL to Moore, 20 January 1811, Letterbooks; BHL to Seaton, 3 September 1817, Letterbooks. See also BHL to Madison, 8 April 1816, Letterbooks, where he mentions his "education . . . in England under a very eminent Engineer."

By 1786 Smeaton was spending less time in the field, and more time in his office. Joseph Banks et al., eds., *Reports of the late John Smeaton, F.R.S.*, 3 vols. (London, 1812), 1:xx.

13. BHL mentioned making a drawing (for himself) of Smeaton's design for a dredge at Cronstadt (Kronshtadt) in Russia. BHL to President and Directors of the Washington Canal Company, 5 April 1817, Letterbooks.

14. BHL to Moore, 20 January 1811, Letterbooks.

15. BHL, *Opinion*, p. 17.

16. D. Swann, "The Engineers of English Port Improvements, 1660–1830: Part II," *Transport History* 1 (1968): 260; Abraham Rees, ed., *The Cyclopaedia*, 1st American ed., 40 vols. (Philadelphia, 1805–24), s.v. Canal, River Rother; L. T. C. Rolt, *Great Engineers* (New York: St. Martin's Press, 1963), p. 47.

17. BHL to Gilpin, 19 August 1805, Letterbooks; Charles Hadfield, *The Canals of South and South East England* (New York: Augustus M. Kelley, 1969), pp. 152–53; Charles Hadfield, *British Canals*, 4th ed. (New York: Augustus M. Kelley, 1969), p. 41; P. A. L. Vine, *London's Lost Route to Basingstoke* (Newton Abbot, England: David & Charles, 1968), pp. 45, 46, 52; 10 June 1786, Minutes of the Rye Harbour Commissioners, East Sussex Record Office, Lewes, England.

18. D. S. L. Cardwell, *Turning Points in Western Technology* (New York: Science History Publications, 1972), pp. 123–24; John Smeaton, *A Narrative of the Building and a Description of the Construction of the Edystone [sic] Lighthouse with Stone* (London, 1791).

19. BHL to Smith, 30 July 1805, Letterbooks; BHL to Madison, 16 April 1816, Letterbooks; *DNB*, s.v. Repton, Humphrey; BHL, "To the Editor of the Emporium [on the construction of the National Road]," *Emporium of the Arts and Sciences*, new series, 3 (1814): 296; BHL to Burney, 10 March 1789, Berg Collection, New York Public Library, New York (not in microfiche ed.).

at waterworks and coal mines, observed Royal Navy yards, and took careful notes on Smeaton's published experiments concerning hydraulic cement.[20] He read British technical journals and French engineering works.[21]

Soon after leaving the Basingstoke Canal, perhaps early in 1789, Latrobe began two or three years' work in the office of Samuel Pepys Cockerell, a prominent English architect. While there he assisted Cockerell on the House for the First Lord of the Admiralty in Whitehall, London. Then, beginning in 1792, he had several private and public architectural commissions.[22]

As an architect Latrobe worked in a profession which was at the time recognized as being distinct from engineering. Architecture was often identified as a branch of the fine arts. Yet architecture and engineering required many similar skills. Making accurate drawings, organizing and directing the labor of skilled craftsmen, and knowing intimately the properties of materials were equally necessary. Thus Latrobe's contemporary, Thomas Telford, was trained as an architect but found his profession and fame as an engineer.[23] In Benjamin Henry Latrobe's case, it seems wise to regard his English years of technical training and development not as divided between architecture and engineering, but as devoted to two complementary and overlapping professional spheres.

Latrobe's final engineering work in England was the support of a Parliamentary petition concerning the route of the Chelmer and Blackwater Navigation in Essex. The Strutt family of Essex employed him to plan a route which would permit seaborne cargoes to be transferred to the navigation at the seaport of Maldon rather than at a new basin downriver from Maldon, as John Rennie's survey had proposed. Latrobe produced two maps, accompanied by reports, which showed that providing an equivalent terminal for the navigation at Maldon was technically feasible. Accordingly, he was called to testify at Parliamentary hearings on the subject in 1793 and 1795, but his

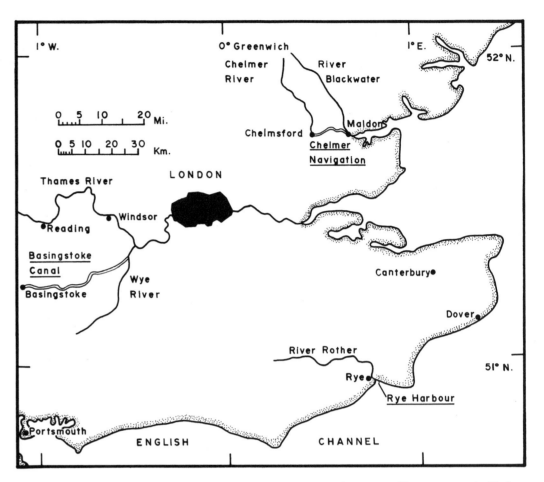

Fig. 2. Map of southeast England. BHL's early engineering experience, c. 1786–95, was on the Basingstoke Canal, at Maldon on the Chelmer and Blackwater Navigation, and at Rye Harbour.

Stephen F. Lintner

ideas apparently had no effect. Eventually the navigation included a secondary terminal at Maldon. Latrobe in the meantime had left for America and took no part in construction.[24]

Latrobe's employment as a canal consultant and his appearances in Parliament suggest that he had considerable credibility as an engineer. John Rennie,

---

20. BHL, notebook, c. 1793–94, Osmun Latrobe Papers, LC; BHL to Ogden, 3 March 1814, Letterbooks; BHL, *Answer to the Joint Committee of the Select and Common Council of Philadelphia* (Philadelphia, 1799), pp. 5–7; BHL to Smith, 29 August 1805, Letterbooks.

21. BHL to Chequiere, 12 November 1804, Letterbooks; BHL to Eustis, 17 September 1812, Letterbooks; BHL, *Opinion*, p. iv; BHL to Cooper, 8 February 1814, Letterbooks; BHL to Ingersoll, 17 January 1814, Letterbooks.

22. (Ackermann's) *Repository* 2d series, 7 (1821): 33; BHL to Antes, 8 April 1798, printed in BHL, *Journals*, 2:367–68; BHL to Cutting, 11 January 1814, Letterbooks; *DNB*, s.v. Cockerell, Samuel Pepys. For other attempts at a reconstruction of this phase of Latrobe's career, see Hamlin, *Latrobe*, pp. 42–47; and, based on a more extensive collection of Latrobe materials, Carter, "Introduction," in BHL, *Journals*, 1:xx–xxi.

23. L. T. C. Rolt, *Thomas Telford* (London: Longmans, Green and Co., 1958), pp. 9, 12, 15, 19, 33.

24. BHL, "Report upon the practicability and advantage of making the River Blackwater navigable," 30 December 1793, Essex Record Office; BHL, "Report upon the plan of the intended improvement of the rivers Blackwater and Chelmer," 28 October 1794, Essex Record Office; bill, William Birch to John Strutt, October 1793–May 1795, Terling Place Papers, Essex Record Office; *Journals of the House of Commons*, 50 (1795): 184; Rees, *Cyclopaedia*, s.v. Canal; John Booker, *Essex in the Industrial Revolution* (Chelmsford, England: Essex County Council, 1974), pp. 141–42; BHL to Strutt, 24 February 1794 and 4 October 1794, Terling Place Papers, Essex Record Office.

whose survey and report he was called upon to dispute, was established as one of the premier engineers in Britain, and it is probable that only someone with recognized experience and capability in the field could have even testified against Rennie's plan in Parliament. Moreover, the surviving documents which Latrobe prepared to support the case indicate a high level of skill. The maps are exquisite renderings which accomplish their purpose of explanation very well, and his reports were written with assurance and clarity.[25]

Thus, although our knowledge of Benjamin Henry Latrobe's English career is limited, it seems clear that by the time he left for the United States he had the training and experience expected of an English engineer. As England was undergoing a period of great public works superintended by engineers this education placed Latrobe in the forefront of European engineering development. The United States could have had no better source from which to acquire an immigrant engineer.

Latrobe's reasons for leaving England are uncertain, but obviously had something to do, on the one hand, with personal trials resulting from his wife's death in 1793, and, on the other, with his admiration for America and the American Revolution. Moreover, his mother was an American, and had willed some land to him in Pennsylvania. So, in the late fall of 1795 he took passage on a ship, the *Eliza*, for the United States, consigning his large professional library and some of his drafting and surveying instruments to another vessel.[26] One of the greatest hindrances to his American career was that his library was lost when the ship carrying it was captured and condemned by the French.[27] He personally carried a few instruments, however, such as a theodolite made by the celebrated English instrument maker Jesse Ramsden which may have been the first of its type in America.[28]

Latrobe intended to continue his profession as architect and engineer in the United States, but he may not have understood the rudimentary state of the engineering art there. In 1796, when he landed in Norfolk, Virginia, his new homeland was a prosperous nation dependent upon agriculture and oceanic trade. There was only a scant tradition of building public works. Harbors and transportation, major sources of employment for engineers in England in the eighteenth century, had required little or no human design in a country adequately supplied with deep water ports and many rivers navigable some distance into the interior. That was particularly so while the population was concentrated within the first hundred miles from the coast, as it was until after the Revolutionary War.

That state of affairs was changing slowly. By the mid-1790s, the time of Latrobe's arrival, settlement had pushed further west and there was considerable, though often misguided, activity in the field of internal improvements. In New England and the Middle States turnpike companies had improved or begun to improve roads. The Lancaster Turnpike from Philadelphia to Lancaster was the greatest of these efforts. Canals were projected or begun in almost every state. Most of them were shortcuts around rapids or falls of otherwise unobstructed rivers, but a few were cross-country canals intended to direct trade along a new route. Such were the Middlesex Canal of Massachusetts and the Santee Canal of South Carolina, both of which were completed within a few years of the start of construction, and the Schuylkill and Susquehanna Canal, which was never finished.[29] In promoting turnpikes and canals Americans were both reacting to the possibilities of a burgeoning economy and population, and imitating the developments which were occurring in England.

There was considerable confusion in America concerning who should direct engineering enterprises. If the British example was to be followed exactly, experienced engineers had to be employed in an advisory capacity and large salaries had to be offered to attract them. That is what the managers of the Schuylkill and Susquehanna Canal actually did, contracting with William Weston to come from England in 1792 and to assume direction of their work. Partly to divide the burden of his salary, and partly because the managers had interests in the other Pennsylvania works, Weston's time was also allotted to the Delaware and Schuylkill Canal, the Conewago Canal, and the Lancaster Turnpike.

In two or three years the canals' construction came to a halt because of financial difficulty, and Weston began to assist in other American projects, including the Middlesex Canal, Western Inland Lock Navigation Company, James River Canal, and the Schuylkill Permanent Bridge. Weston demonstrated considerable ability and greatly impressed the American managers of these works, who paid him high salaries and consultation fees. By the time he returned to England in 1799, he had given valuable training to a number

---

25. See maps and discussion in this volume, pp. 113–15 below.
26. Hamlin, *Latrobe*, pp. 49–53.
27. BHL to Chequiere, 12 November 1804, Letterbooks.
28. BHL to Gilpin, 10 October 1803, Letterbooks.

29. Curtis P. Nettels, *The Emergence of a National Economy, 1775–1815* (New York: Holt, Rinehart and Winston, 1962), pp. 251–53, 261–62; Richard Shelton Kirby and Philip Gustave Laurson, *The Early Years of Modern Civil Engineering* (New Haven: Yale University Press, 1932), pp. 40–44, 62–55.

of Americans who had worked with him.[30] Weston's American career might have set important precedents for Latrobe but it did not. Somehow Latrobe never won the acceptance that Weston did, and he found that people objected to his fees, which were similar to those willingly paid to Weston.[31]

Other than Weston, there were few men in America at the time of Latrobe's arrival who claimed to be engineers, and fewer perhaps who deserved the title. There were some French military engineers who served in the Revolution and remained afterward, but among them only Pierre L'Enfant engaged in any significant work besides fortifications. Some skilled Englishmen, such as Christopher Colles, Josiah Hornblower, and James Brindley (who claimed to be a nephew of the English engineer John Brindley), were on the fringes of engineering for varying periods, but none sustained confidence in his ability. Indeed, company directors or promoters themselves were frequently relied upon to direct construction. Dissatisfied with the independence or questionable ability of candidates for engineering positions, directors believed that their interests were best represented by one of their own—by a person who had a financial commitment to the work.[32] Such was the state of American engineering when Latrobe arrived at Norfolk in 1796.

He was hardly off the ship before his engineering knowledge was recognized and he did some engineering consultation related to the Dismal Swamp Canal.[33] From then until his death he was almost constantly engaged in American engineering projects.

2. TRANSPORTATION

Considering that most of Latrobe's engineering experience in England was with waterways, it comes as no surprise that much of his American engineering work concerned canals, navigations, and other modes of transportation. The widespread interest in transportation improvement in the United States during this period made Latrobe's experience valuable and widely solicited. Yet his transportation projects yielded little satisfaction because few of them attained complete success in his lifetime.

---

30. Richard Shelton Kirby, "William Weston and his Contribution to Early American Engineering," *Transactions of the Newcomen Society* 16 (1935-36): 111-27. For Weston's services to the James River Canal, see BHL to Wood, 14 February 1798, Executive Papers of Governor James Wood, Virginia State Library, Richmond, Va., printed in BHL, *Journals*, 2:362-65.

31. See comment by Daniel Calhoun, *The American Civil Engineer*, p. 16.

32. Ibid., pp. 9, 16.

33. BHL may have had letters of introduction which gave his professional claims credibility and secured him immediate employment. BHL, *Journals*, 1:79.

Appomattox Navigation

About three months after Latrobe's landing at Norfolk, Virginia, in 1796, he was commissioned by the directors of the Upper Appomattox Navigation to report on necessary improvements of the Appomattox River. Very little had been accomplished since the company had been chartered in 1795, and the directors needed an expert opinion on how to proceed. The purpose of the company was to make the rapid lower section of the river (just above Petersburg, Virginia) navigable for craft coming down from the gentler upper reaches. Accordingly Latrobe and a few directors floated down the final miles of the river for three days to examine it, but springtime high water made it difficult for Latrobe to make complete measurements.

Had he been able to make a more detailed report it still seems unlikely that Latrobe's suggestions would have had much effect on the navigation's development. Over fifteen years later it was reported that "the work has been in progress for some time, and will be completed in a few years."[1] In the meantime Latrobe had been asked to give advice to the Lower Appomattox Navigation's directors, and he responded with a lengthy letter giving general principles for river improvement in North America.[2]

The Appomattox navigation companies were typical of many companies chartered for river improvement in the late eighteenth and early nineteenth centuries which were sustained only by local interest. Capital was chronically short, and channel excavation and lock and dam construction proceeded erratically. Partial completion of the works meant that tolls were low or uncollectible, yielding little return on the stock, and making the enterprises unattractive to new investors.

Susquehanna River

Latrobe directed the navigational improvement of the lower Susquehanna River in 1801. The Susquehanna River drains the central area of the state of Pennsylvania, and the river and its tributaries were important routes of travel and commerce for early settlers. Potentially the Susquehanna was the most important commercial river on the Atlantic coast because agriculture and lumbering were developing along its shores and branches. However, the last segment of the river—from Columbia, Pennsylvania to the Conowingo Falls in Maryland—was full of rapids, islands, and obstructing rocks. Taking any sort of craft loaded with grain, whiskey, iron, or other products of the interior

---

1. Rees, *Cyclopaedia*, s.v. Canal.

2. BHL to Bate, 21 November 1809 and 3 February 1810, Letterbooks.

down that part of the river was highly risky. Only the expense of shipping goods by wagon and the possibility of getting a good price at the rising ports of Havre de Grace and Baltimore tempted men to take the risk.

The first attempts to improve the river involved the private construction of canals around the Conewago Falls in Pennsylvania and the Conowingo Falls in Maryland in the 1790s. Then in 1801 the Pennsylvania legislature appropriated $10,000 to clear a channel from Columbia to the Maryland border, and awarded a contract to direct the work to Colonel Frederick Antes, who was Latrobe's uncle.[3] Antes called on Latrobe to assist him in the work, but died in Lancaster shortly after the two began operations in September 1801.

Under Latrobe's direction various subcontractors cleared or blasted away projecting rocks to make a channel close to the shore. As he stated later, "My great principle was to cut the channel over a rapid, close to the shore, so as to make the simple rule of hugging the shore in all stages of the river, a sufficient guide to the Boatmen."[4] By the end of November all that could be done with the money appropriated was finished, although in his report to the governor Latrobe noted that it would take an annual appropriation of $20,000 for five years to clear the lower river for traffic. All that he had accomplished was to make the Susquehanna somewhat safer during the annual springtime rush to market.[5]

Perhaps a more significant result of Latrobe's efforts was the large map (seventeen feet long, two feet wide) which he prepared to delineate the channel which had been cleared in the Susquehanna. Although the original is lost, a duplicate which he prepared has survived, and is a great work of cartographic art. It combines the surveying data which he gathered with that of Christian Hauducoeur, a Maryland engineer, to show all of the river from Columbia to its mouth at Havre de Grace. Not only did Latrobe show islands, rapids, and rocky areas along with the channel, but he also included the vegetation, geology, and agricultural development of much of the shoreline.

The map is a superb record of the state of the river valley in 1801, and provides a base for the study of how it has changed to the present.[6]

Naval Drydock and Potomac Canal Extension

Latrobe was called to Washington late in 1802 by President Thomas Jefferson to design a drydock for United States Navy frigates. It was Latrobe's first commission in his long association with public works in Washington. Under an act of March 1801 which sought to reduce the navy from the size it had attained during the quasi war with France, and in accord with his party's principles of economy and peace, Jefferson was overseeing a radical reduction in the fleet.[7] Yet he did not wish to destroy the navy and developed a plan (congenial to his architectural avocation) for building a huge covered drydock at the Washington Navy Yard in which ships could be stored and made ready for service in much shorter time than new ships could be constructed. Jefferson prepared some sketches of his design, and in the summer of 1802 directed Thomas Tingey, commandant of the Washington Navy Yard, to examine Tiber Creek and Stoddart's Spring for their suitability as sources of the water required to float ships in and out of the drydock.[8]

Not until 2 November 1802, however, did Jefferson write to Latrobe asking him to come to Washington to "calculate the expense of the whole, that we may lead the legislature [i.e., Congress] to no expense in the execution of which they shall not be apprised in the beginning."[9] Within two weeks Latrobe was in Washington working on the design. For the drydock itself he had little to do but to interpret and put on paper the ideas which Jefferson had already formulated. But Latrobe was fascinated with the possibility of drawing water for the drydock from the Little Falls canal of the Potomac [River] Company, rather than from the source already examined, and devoted considerable time to surveying the route of a canal from Little Falls to the Navy Yard.[10] He reasoned that a canal connected to the Potomac Company's works would be of great economic importance to the Navy Yard by carrying stores from the interior directly to the yard; the usual method was transshipment in George-

---

3. *Pennsylvania Archives*, ed. Gertrude MacKinney, 9th series, 10 vols. (Harrisburg, Pa.: Department of Property and Supplies, 1931), 3: 1755–56.
4. BHL to Bate, 21 November 1809, Letterbooks.
5. BHL to McKean, 28 November 1801, Susquehanna River Improvement File, Pennsylvania Historical and Museum Commission, Harrisburg, Pa. (printed in Susquehanna Canal Company, *Report of the Governor and Directors to the Proprietors of the Susquehanna Canal* [Baltimore, 1802], pp. 13–20). For a general statement on the navigability of the river in this era, see James Weston Livingood, *The Philadelphia-Baltimore Trade Rivalry, 1780–1860* (Harrisburg, Pa.: Pennsylvania Historical and Museum Commission, 1947), pp. 37–38.

6. See the detailed examination of the map, pp. 73–109 below.
7. Samuel Eliot Morison, *The Oxford History of the American People* (New York: Oxford University Press, 1965), p. 362.
8. [Thomas Jefferson], *Message from the President of the United States, Transmitting Plans and Estimates of a Dry Dock, for the Preservation of our Ships of War* (Washington, 1802), pp. 21–24.
9. Jefferson to BHL, 2 November 1802, Jefferson Papers, LC.
10. BHL to Jefferson, 22 November 1802 and 28 November 1802, Jefferson Papers, LC.

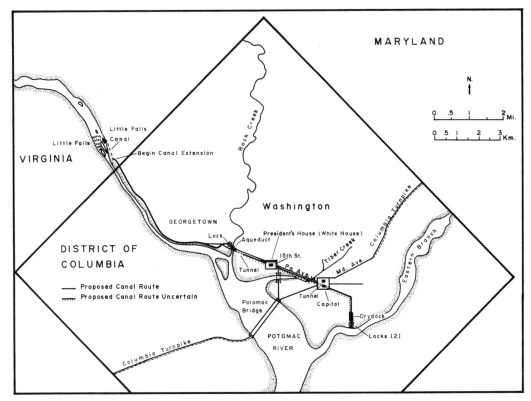

Fig. 3. Map of BHL's proposals for the Naval Drydock and Potomac Canal Extension in 1802. Also included on this map are the Columbia Turnpikes which BHL helped to lay out in 1810–11.

Stephen F. Lintner

town for carriage down the Potomac River and up the Eastern Branch.[11] Latrobe believed that such a canal could be built by a private company, possibly the Potomac Company, which had built the canals and locks around the Great and Little Falls of the Potomac River, or the Washington Canal Company, for which a charter had recently been granted.

Latrobe's plan for the execution of Jefferson's idea began with an eight-mile canal traversing the city of Washington. The canal would have started from the Potomac Company's Little Falls canal, completed in 1795,[12] which lay to the northwest of Washington along the Potomac River. The canal then followed the river to Rock Creek, crossing it by an aqueduct, and entered a tunnel through the hill on the opposite side.[13] Latrobe did not leave drawings of his plans for the rest of the route, but comments he made in letters to

11. [Jefferson], *Message from the President*, p. 10.
12. Corra Bacon-Foster, *Early Chapters in the Development of the Patowmac Route to the West* (Washington: Columbia Historical Society, 1912), p. 90.
13. See pp. 116–24 below for selections from the canal drawings.

Jefferson suggest that he hoped to put the canal on or near Pennsylvania Avenue from the tunnel to the Capitol, then tunnel through Capitol Hill, and turn south to continue to the Navy Yard.[14] Latrobe was certainly aware of Pierre L'Enfant's design for the city which showed a canal running from west to east on the north side of the Capitol axis, and bending southeast in front of the Capitol to eventually intersect the Eastern Branch.[15] He undoubtedly intended the canal to the Navy Yard to be harmonious with that part of L'Enfant's plan.

The prism, or cross section, of the canal was 12 feet on the bottom, and 22 feet at the waterline with 3 feet of water.[16] Precisely how Latrobe derived this dimension is unknown, but perhaps his knowledge of British feeder canals (that is, canals whose major purpose was to carry water to another canal) suggested it. The only feeder canal he ever built, which was for the abortive Chesapeake and Delaware Canal, had virtually the same dimensions.[17]

The culverts, bridges, locks, aqueducts, and tunnels which Latrobe planned for the canals are shown on his drawings. They are characteristic of contemporary British canal designs, particularly the striking aqueduct over Rock Creek which is similar to one designed by Benjamin Outram for the Peak Forest Canal.[18] All the structures are masonry, with no hint of timber (except in the lock gates), indicating not only Latrobe's expectation that the canal would be a great and permanent public work, but also his adherence to the masonry construction tradition of great British canal engineers like Brindley, Smeaton, Jessop, and Rennie.

There was a more direct source for Latrobe's design of the drydock. Jefferson specified in his letter of 2 November 1802 that (1) there should be two basins between the drydock and tidewater, (2) the drydock should lie above the tide, (3) the drydock itself was to be 175 feet wide and 800 feet long, and (4) the drydock was to be covered by a roof like that of the *Halle au Blé* (grain market) completed in 1783 in Paris.[19] Latrobe accommodated these requirements by proposing a combination of structures which would have rivaled in magnitude any naval edifice in Europe.[20]

14. BHL to Jefferson, 22 November 1802 and 28 November 1802, Jefferson Papers, LC.
15. Elizabeth S. Kite, ed., *L'Enfant and Washington* (Baltimore: Johns Hopkins Press, 1929), pp. 55, 69.
16. See p. 121 below.
17. BHL to Strickland, 29 May 1804, Letterbooks; notebook #3, p. 135, Robert Brooke Papers, New York Public Library, microfilm copy in Eleutherian Mills.
18. Robert Harris, *Canals and Their Architecture* (New York: Frederick A. Praeger, 1969), pp. 90–91.
19. Jefferson to BHL, 2 November 1802, Jefferson Papers, LC; Howard C. Rice, *Thomas Jefferson's Paris* (Princeton, N.J.: Princeton University Press, 1976), pp. 18–19.
20. The drawings are reproduced on pp. 119–20.

Locating the site of the drydock on what promised to be firm ground at the Navy Yard on the Eastern Branch, he designed two large masonry locks to carry the vessels from tidewater up to the level of the drydock. Each was to be about 110 feet long and 50 feet wide, about large enough to accommodate one frigate. After ascending both locks the vessel would enter a short canal communicating with the drydock. Latrobe made the canal long enough so that it could form the axis of a turning basin to be excavated in the future. With the turning basin, vessels leaving the drydock could be rotated so that they could enter the Eastern Branch bow first.

Latrobe laid out the drydock according to Jefferson's dimensions and architectural suggestions. It was to have masonry walls on all sides and a sheet iron roof with no intermediary supports. It would cover an area of more than three acres. This daring structure had no American precedents. Nonetheless, Latrobe estimated that it could be built in a year at a cost of about $225,000. The locks, he thought, would add another $170,000.[21]

The last thing that the new Jeffersonian Congress wanted in the budget was a $395,000 monument to the President's imagination and the skill of an architect immigrated from England. The House of Representatives politely received Jefferson's message on the drydock and Latrobe's plans, sent them to committee, debated them briefly on the floor in January 1803, and then tabled the whole idea.[22] Jefferson did not easily give up the scheme, and told Latrobe in March that he thought the long-term savings possible from the drydock plan for preserving the fleet would soon persuade Congress to accept it.[23] But the drydock idea was not revived.

Latrobe's participation in the preparation and presentation of the drydock project was a transitional period between his works in Pennsylvania (the Philadelphia Waterworks, the Bank of Pennsylvania, and the Susquehanna navigation) and his commitment to the Capitol and the Chesapeake and Delaware Canal. In his report he stated that he was dependent on his recent experience in Philadelphia for the accuracy of his estimates,[24] and he hoped that if the work was undertaken he would be able to employ "the very numerous skilful and experienced workmen, who have been collected and in part educated in the execution of large and difficult works at the Pennsylvania bank and the works for supplying the city of Philadelphia with water."[25]

The latter statement in particular, but also the entire plan, was seriously questioned by another of Thomas Jefferson's technically minded acquaintances, Robert Leslie, a Philadelphia watchmaker. Leslie, who was a member of the American Philosophical Society with Latrobe and Jefferson, had read about the drydock proposal in the newspaper and discussed it with Latrobe. He had little respect for Latrobe as a practicing architect, claiming that collusion between Latrobe and the Philadelphia workmen had increased the cost of the bank and the waterworks. Leslie stated that Latrobe benefited from such an arrangement because he was paid by a percentage commission on the final cost.[26] Moreover, Leslie charged that Latrobe's prized workmen were "the most lazey, indolent set of men I ever saw, and as to skill I have seen nothing like it in any of them."[27]

The drydock itself Leslie felt had the great defect of requiring the entire area to be flooded each time one ship had to be moved. Presumably that would be difficult to do when side planking was partially removed from the ships to promote circulation and drying. He described and sketched a plan of his own suggestion which had twelve individual drydocks flanking a common basin, a feature that he thought remedied the defect in Latrobe's plan.[28] Jefferson, who provided Latrobe with the basic ideas for the drydock, was apparently unimpressed by Leslie's objections, and they made little difference as no appropriation for the Jefferson–Latrobe plan was ever made. But they indicate that Latrobe had to answer to an aware technical community in America.

Chesapeake and Delaware Canal

Soon after the seventeenth-century Swedish and Dutch settlement of the

---

21. [Jefferson], *Message from the President*, pp. 7–10.
22. Eugene S. Ferguson, "Mr. Jefferson's Dry Docks," *The American Neptune* 11 (1951): 112–14; *Report of the Committee Appointed on the 17th ultimo, to Whom was referred So Much of the Message of the President of the United States as relates to our Navy Yards and the Building of Docks* ([Washington], 1803), pp. 3–4.
23. Jefferson to BHL, 6 March 1803, Mrs. Gamble Latrobe Collection, MdHS.
24. [Jefferson], *Message from the President*, pp. 14–15. In his estimate for the cofferdam needed to construct the locks, BHL referred to a "similar dam at Philadelphia," which was probably that used to build the west wall of the basin of the waterworks. See pp. 146–51 below. Perhaps, however, he had observed and taken notes on the cofferdam for the Schuylkill bridge piers. See [Richard Peters], *A Statistical Account of the Schuylkill Permanent Bridge, Communicated to the Philadelphia Society of Agriculture, 1806* (Philadelphia, 1807), pp. 57–75.
25. [Jefferson], *Message from the President*, p. 13.
26. Leslie to Jefferson, 10 January 1803, Jefferson Papers, LC. A complete biographical sketch of Robert Leslie is lacking, but see Ophia D. Smith, "Charles and Eliza Leslie," *Pennsylvania Magazine of History and Biography* 74 (1950): 512; Tom Taylor, ed., *Autobiographical Recollections, By the Late Charles Robert Leslie, R.A.* (Boston, 1860), pp. 1–2, 13–14. BHL's salary in each case was 5% of the estimated cost.
27. Leslie to Jefferson, 10 January 1803, Jefferson Papers, LC.
28. Leslie to Jefferson, 26 January 1803, Jefferson Papers, LC.

peninsula between the Delaware and Chesapeake bays, visionary residents began to see it would be possible to build a canal across it. Geographical conditions were favorable, as there was no central range of hills and the tide moved far inland along the creeks which emptied into the bays. A canal would allow water travel on protected inland waters from the falls of the Delaware River (the site of Trenton, New Jersey) to the capes of the Chesapeake, possibly to great economic benefit. It was not until the later eighteenth century that Philadelphia merchants realized that the growing trade of the interior of Pennsylvania had found its natural route down the Susquehanna River to the Chesapeake rather than to their city. In 1769 a surveying party sponsored by the American Philosophical Society and financed by Philadelphia merchants examined several possible routes across the peninsula and estimated the cost of digging a canal along each.

After the Revolutionary War, interest in the canal continued, but political geography hindered action: the project concerned Philadelphians principally, yet all the routes lay in Delaware and Maryland. Moreover, the legislatures of the three states were reluctant to make complementary laws without mutual concessions. Finally, in 1799 the Chesapeake and Delaware Canal Company was incorporated in Maryland, and over a year later in both Pennsylvania and Delaware. Part of the impetus for passage by Maryland was Pennsylvania's stipulation that the improvement of its part of the lower Susquehanna depended on Maryland's incorporation of the canal company.

The company was organized in 1802 and began selling shares to residents of the three states. Public response was unenthusiastic but adequate, and the official organization took place in May 1803, when the stockholders met and elected a board of directors. Among several distinguished men chosen was Joshua Gilpin, a Philadelphia merchant whose father had led the survey of 1769, and who himself had recently examined canals while in England. Gilpin was appointed to the Committee of Survey with five other directors for the purpose of determining the best of the several routes which the canal might take. One of the committee's first acts was to select Benjamin Henry Latrobe to conduct surveys.[29]

Latrobe was a logical choice because he was respected by the Philadelphians for his design and erection of the waterworks, and for his recent direction of the Susquehanna River improvement. Several prominent Philadelphia merchants knew him personally.[30] Moreover, Latrobe had taken an interest in the canal since the company's incorporation, and had expressed his belief in "the easy practicality of the project" in a lengthy letter to President Jefferson.[31]

At first Latrobe shared the duties of engineer with Cornelius Howard of Maryland, but Howard soon resigned, perhaps because Latrobe was shocked by his poor surveying technique.[32] Assistants included surveyors John Thomson of Pennsylvania and Daniel Blaney of Delaware, both of whom Latrobe valued highly.[33] From July 1803 to April 1804 this team was directed to make many surveys, profiles, and maps. At first they examined the possible terminals for the canal, considering particularly the depths of the channels on both bays. Next, they explored the various routes across the peninsula, concentrating on the northernmost possibility from either the Christina River or New Castle on the east to either Frenchtown or Welch Point on the west, and the shorter route from Port Penn on the Delaware Bay to Back Creek on the Chesapeake. Finally, Latrobe carefully looked at the advantages of two sources of water for the summit level of the canal, the Elk River and the White Clay Creek.[34] In November 1803, after the most active phase of this work, Latrobe listed thirty surveys which he or members of the team had completed.[35]

At a directors' meeting in January 1804, Latrobe was officially appointed engineer of the Chesapeake and Delaware Canal Company at an annual salary of $3,500. At the same meeting Latrobe recommended that the Elk River supply the feeder and convinced the directors that work on the feeder could begin before the route of the main canal was chosen.[36] Not until June

---

29. Ralph D. Gray, "The Early History of the Chesapeake and Delaware Canal," *Delaware History* 8 (1959): 223–39; Ralph D. Gray, "Philadelphia and the Chesapeake and Delaware Canal," *Pennsylvania Magazine of History and Biography* 84 (1960): 402–7; 5 July 1803, Minutes, Comm. of Survey, Hist. Soc. Del.

30. Among them, Isaac Hazlehurst, his father-in-law; Samuel Fox, president of the Bank of Pennsylvania; John Miller, chairman of the Watering Committee. Note the comment in Joshua Gilpin, *A Memoir on the Rise, Progress, and Present State of the Chesapeake and Delaware Canal* (Wilmington, Del., 1821), p. 7.

31. BHL to Jefferson, 27 March–24 October 1802, Jefferson Papers, LC.

32. 5 July 1803, 19 September 1803, Minutes, Comm. of Survey, Hist. Soc. Del.; BHL to Gilpin, 19 October 1803, Letterbooks.

33. 26 July 1803, Minutes, Comm. of Survey, Hist. Soc. Del.; BHL to Gilpin, 19 October 1803, Letterbooks; Daniel Blaney daybook, 1803–4, Hist. Soc. Del. Thomson (d. 1842) was the father of J. Edgar Thomson (1809–74), the engineer and president of the Pennsylvania Railroad. Henry Graham Ashmead, *History of Delaware County, Pennsylvania* (Philadelphia, 1884), pp. 727–28.

34. 21 October 1803, 21 November 1803 (two reports), 26 January 1804, 24 April 1804, C & D Engineer's Reports, Hist. Soc. Del. BHL reviewed all of the surveys on 2 June 1804, C & D Engineer's Reports, Hist. Soc. Del.

35. BHL to Gilpin, 24 November 1803, referred to in Gilpin, *Chesapeake and Delaware Canal*, pp. 16–18.

36. BHL to Hazlehurst, 4 February 1804, Letterbooks; 26 January 1804, C & D Engineer's Reports, Hist. Soc. Del. See also BHL to Johns, 24 December 1803, and BHL to Gilpin, 26 December 1803, Letterbooks.

Transportation

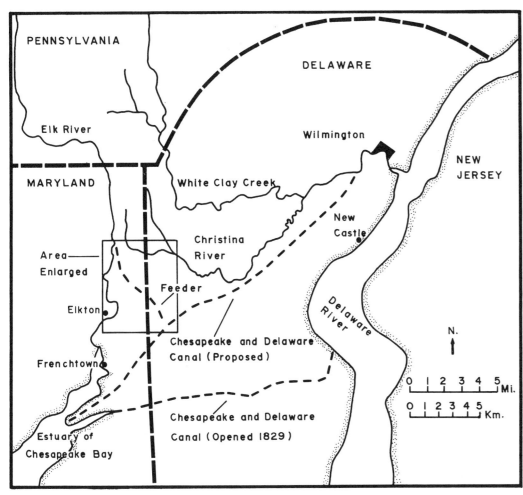

Fig. 4. The directors of the Chesapeake and Delaware Canal Company selected the routes for the canal and feeder in 1804 after considering BHL's survey reports. Neither came into commercial use. For enlarged area see fig. 5, p. 15.

Stephen F. Lintner

did the directors decide in favor of the Christina River–Welch Point route for the main canal.[37]

It was to have a twenty-mile main line beginning at a basin at Welch Point on the Elk River estuary of the Chesapeake Bay, where a flight of locks would raise the canal to a summit level about 74 feet above the tide. The canal maintained that level as it ran east-northeast to the Christina River between Newport and Wilmington, where two flights of locks and a basin brought it back to tidewater on the river. To keep the summit level of the canal filled with water there was a five-mile navigable feeder beginning on the Elk River at the Upper Forge and meeting the canal west of Aitkentown (now Glasgow), Delaware.[38] At Latrobe's urging the directors resolved to begin work on the feeder first, since without the water it provided, the summit of the main canal would be dry.[39] But it was also a gesture to the Pennsylvania investors, since the feeder began just south of the state line, and was itself expected to carry goods from Chester and Lancaster counties when the main line was completed.[40]

The company had been negotiating for the rights to the water of the Elk River since the fall of 1803, and finally made the purchase in April 1804.[41] Latrobe had already begun to organize for the coming work season.

Of the greatest importance to Latrobe in his preparation was the gathering of a competent group of men to work under him. He already had three pupils: Lewis DeMun, a young Frenchman who was Latrobe's most constant assistant on the feeder; William Strickland, troublesome but brilliant, and later one of the United States's most prominent architects and engineers; and Robert Mills, who assisted on the feeder only briefly.[42] The company made some small payments to DeMun for his work, but Latrobe apparently supported Strickland and Mills.[43] All of them assisted in surveying and mapping, and DeMun occasionally had the responsibility for overseeing work when Latrobe was in Washington.[44]

For clerk of the works, an office which had the responsibilities of keeping all the records of daily operations on the feeder as well as assisting him, Latrobe

37. The directors chose the "northern" route over the "southern" route in April, but waited until June to decide that the Christina River rather than New Castle should be the eastern terminal. BHL to Hazlehurst, 10 June 1804, Letterbooks.

38. See description and estimates, 24 April 1804, C & D Engineer's Reports, Hist. Soc. Del.; Gilpin, *Chesapeake and Delaware Canal*, frontispiece map.

39. BHL to Hazlehurst, 4 February 1804, Letterbooks.

40. *Facts and Observations Respecting the Chesapeake and Delaware Canal, and its Extension into Pennsylvania* (n.p., [1805]), p. 4; BHL to McKean, 6 January 1805, Letterbooks.

41. 21 November 1803, 26 January 1804, 7 April 1804, 23 April 1804, Minutes, Comm. of Survey, Hist. Soc. Del.

42. Hamlin, *Latrobe*, pp. 214–20. DeMun's name appears frequently in letters and reports referring to the feeder, and Strickland's less so. Although BHL later stated that Mills "accompanied me during all the operations on the Chesapeake and Delaware Canal" (BHL to Madison, 8 April 1816, Letterbooks), it appears that Mills worked on the feeder only briefly in the winters of 1804 and 1805/6. BHL to Mills, 3 November 1804, Letterbooks; H. M. Pierce Gallagher, *Robert Mills: Architect of the Washington Monument, 1781–1855* (New York: Columbia University Press, 1935), p. 127.

43. 22 September 1804, 21 November 1804, 4 June 1805, Journal, Comm. of Works, Salem Co. Hist. Soc.

44. See payments to Lewis DeMun for supervising operations in BHL's absence, 5 December 1804–24 December 1804, and 4 May 1805–20 May 1805, Journal, Comm. of Works, Salem Co. Hist. Soc.

13

chose Robert Brooke, a Philadelphia surveyor. Brooke had been "Assistant Engineer" to William Weston on the Delaware and Schuylkill Canal in the 1790s and then directed the construction of the Germantown and Perkiomen Turnpike.[45] How Brooke came to be recommended for the position is uncertain, but Latrobe was certainly pleased with his accounting and his surveying skills.[46] Several years later he said of Brooke: "As a Mathematical assistant and calculator, I have never known his equal here or in Europe, and were I employed in projecting or executing any great navigation there is not inducement within my power that I would not hold out to him to obtain his assistance."[47]

Also directly in the employ of the company were three master craftsmen. Thomas Vickers, the chief mason, executed all works in stone and supervised the journeymen masons. Vickers had worked under Weston, had built the brick tunnel of the Philadelphia Waterworks, and had supervised construction of both piers of the Schuylkill Permanent Bridge at Philadelphia. During his career he had demonstrated that he knew how to make and use hydraulic cement, an art which was known to the Romans but was only beginning to be learned again in the Western world.[48] Latrobe had carefully studied John Smeaton's pioneering experiments with hydraulic cements, and highly valued Vickers's talents.[49]

Latrobe hired John Strickland, the father of his pupil William Strickland, to direct the carpentry for the feeder, including construction of workers' housing, sheds for equipment, and items needed for excavation and construction, such as ladders.[50] John Bindley, whom Latrobe described as a "thorough bred Canal Carpenter," worked under Strickland making and repairing wheelbarrows.[51] Peter Bath, a blacksmith, also came to the feeder to make and repair tools, especially those used in the quarry.[52] Early in 1805 another blacksmith repaired the iron parts of the company's wheelbarrows.[53]

Twelve contractors conducted the excavation and embankment of the trough of the feeder, each employing up to eighty laborers.[54] Of these twelve, four had worked for Latrobe on the Philadelphia Waterworks,[55] two were apparently English gangsmen with experience on the Grand Junction Canal,[56] and three had been employed under Weston;[57] only three seem to have had little experience.[58] Obviously Latrobe tended to hire only those he knew personally or whose background he trusted. After a year's work he congratulated himself on his selections, reporting to the company that "the Contractors have performed their duty with fidelity and the work which has been executed may vie with almost any similar work in Europe or America in this respect...."[59]

Latrobe followed British practice in organizing his staff. The office of clerk of the works was common there, and in at least one instance there were "superintendents of masonry" and a carpenter regularly employed on a British canal. British contractors constructed the works, and were allotted sections to complete just as contractors were on the Chesapeake and Delaware Canal feeder.[60] There does not seem to have been any difficulty in adopting this sort of organization in America.

Latrobe's men started construction of the feeder in May 1804, at the

45. BHL to Madison, 8 April 1816, Letterbooks; draft of letter [Robert Brooke to President and Managers of the Delaware and Schuylkill Navigation Company], n.d., Delaware and Schuylkill Canal Papers, Hist. Soc. Pa.; Robert Brooke, "Map and Report on the Germantown and Perkiomen Turnpike," 1802, Map Collection, Hist. Soc. Pa.

46. 7 April 1804, 23 April 1804, 10 May 1804, 1 June 1804, Minutes, Comm. of Survey, Hist. Soc. Del.; BHL to Gilpin, 9 September 1804, Letterbooks; BHL to Helmsley, 29 October 1804, Letterbooks; BHL to Tingey, 6 April 1804, Letterbooks; BHL to Gilpin, 14 December 1805, Letterbooks.

47. BHL to Smith, 25 April 1816, Letterbooks.

48. BHL to Gallatin, 3 August 1805, Letterbooks; Watering Comm., *Report* (1799), p. 14; [Peters], *Schuylkill Permanent Bridge*, pp. 27–28, 41; 7 April 1804, 23 April 1804, Minutes, Comm. of Survey, Hist. Soc. Del.; 23 October 1804, 24 January 1805, Journal, Comm. of Works, Salem Co. Hist. Soc.; BHL to Helmsley, 29 October 1804, Letterbooks.

49. BHL notebook, c. 1793–94, Osmun Latrobe Papers, LC.

50. BHL to Strickland, 23 April 1804, Letterbooks; BHL to Gilpin, 3 May 1804, Letterbooks; BHL to Helmsley, 29 October 1804, Letterbooks; 10 May 1804, Minutes, Comm. of Survey, Hist. Soc. Del.

51. BHL to Bindley, 14 May 1804, Letterbooks; 20 November 1804, C & D Engineer's Reports, Hist. Soc. Del.; 10 May 1804, Minutes, Comm. of Survey, Hist. Soc. Del.

52. BHL to Bath, 1 May 1804, Letterbooks; BHL to Bath, 22 May 1804, Letterbooks; 10 May 1804, 1 June 1804, Minutes, Comm. of Survey, Hist. Soc. Del.

53. Thomas Rutter account, January–March 1805, Salem Co. Hist. Soc.

54. 1 October 1804, C & D Engineer's Reports, Hist. Soc. Del.

55. James Cochran, John Grimes, Joseph Pollock, and William Watson. Ibid.; BHL to Madison, 8 April 1816, Letterbooks; 20 November 1804, C & D Engineer's Reports, Hist. Soc. Del.; BHL to Gilpin, 19 August 1805, Letterbooks; *Report of the Committee Appointed by the Common Council to Enquire into the State of the Water Works* (Philadelphia, 1802), pp. 60, 75.

56. Charles Randall and John Cocksey. BHL to Gilpin, 28 March 1804, Letterbooks; BHL to Gilpin, 2 September 1804, Letterbooks.

57. Clegg, Binnerman, Hugh Sands. BHL to Gilpin, 28 May 1804, Letterbooks; BHL to Cooch, 5 August 1804, Letterbooks; 1 October 1804, C & D Engineer's Reports, Hist. Soc. Del.

58. Bentley, Boyd, and Pettigrew. 5 August 1805, C & D Engineer's Reports, Hist. Soc. Del.

59. 30 May 1805, C & D Engineer's Reports, Hist. Soc. Del.

60. Cyril T. G. Boucher, *James Brindley, Engineer, 1716–1772* (Norwich: Goose and Son, 1968), pp. 69–71; Boucher, *John Rennie*, pp. 62, 72–74; Charles Hadfield, *British Canals*, 4th ed. (New York: Augustus M. Kelley, 1969), pp. 41–42. See contracts, 1804, in Chesapeake and Delaware Canal Company Papers, Hist. Soc. Del.

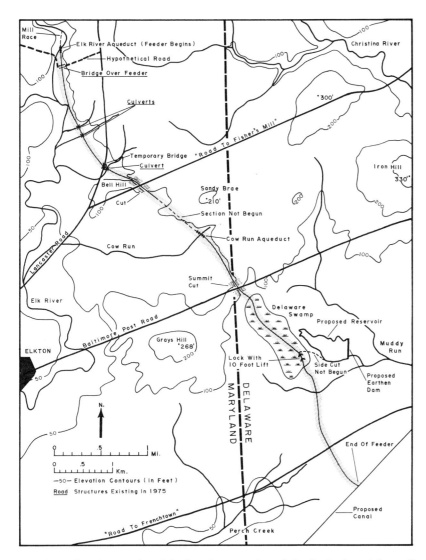

Fig. 5. Chesapeake and Delaware Canal feeder. Construction of the feeder lasted from 1804 to 1805, but for lack of funds it was not completed.

Stephen F. Lintner

beginning of the route near Elk Mills, Maryland.[61] For more than a year, construction progressed steadily southward to where the feeder was to meet the main line, although some difficult points received almost constant attention.

The feeder was to begin with an aqueduct carrying water from the Upper

61. BHL to Gilpin, 3 May 1804, Letterbooks; 10 May 1804, Minutes, Comm. of Survey, Hist. Soc. Del.

Fig. 6. Proposed swing bridge on Chesapeake and Delaware Canal feeder, 1804. (BHL to Gilpin, 9 September 1804, Letterbooks.)

Forge millrace on the west side of the Elk River to the east side. Latrobe designed a masonry aqueduct of three segmental arches, a form common on British canals.[62] He located stone for the aqueduct about a quarter mile to the south of the crossing and purchased quarrying tools and gunpowder to work it.[63] When the feeder was completed from the river bank to the quarry he had the feeder filled with water and had cut and dressed stone boated to the aqueduct site.[64] In the fall of 1804 the foundation of the eastern abutment was excavated and the first stones were laid 5 feet below water level, but no further work was accomplished.[65]

A few hundred feet south of the aqueduct site Vickers built a masonry road bridge across the feeder which is standing today (see figure 9a). It is made of roughly hewn stone cemented together to make a segmental arch of 15-foot

62. See drawings and description of the aqueduct, pp. 125–28 below.

63. 7 April 1804, 10 May 1804, 1 June 1804, Minutes, Comm. of Survey, Hist. Soc. Del.; BHL to Vickers, 14 May 1804, Letterbooks; BHL to Tatem, 14 May 1804, Letterbooks; BHL to Vickers, 20 May 1804, Letterbooks. Although the first quantity of powder was from another source, during the summer and fall of 1804 the Chesapeake and Delaware Canal Company purchased 900 pounds of powder from the new Eleutherian Mills of E. I. du Pont de Nemours & Co., near Wilmington, Delaware. At the time of its final purchase on 17 October 1804, the Chesapeake and Delaware Canal Company had accounted for 12 percent of the sales of du Pont powder. Powder Sales Book, 9 July 1804, 11 September 1804, 24 September 1804, 17 October 1804, item 1500, series Y, part II, Accession 500, Eleutherian Mills; 5 November 1804, Journal, Comm. of Works, Hist. Soc. Del.

64. BHL to Gilpin, 26 August 1804, Letterbooks; 20 November 1804, 30 May 1805, C & D Engineer's Reports, Hist. Soc. Del.

65. BHL to Johns, 1 November 1804, Letterbooks; 20 November 1804, 30 May 1805, C & D Engineer's Reports, Hist. Soc. Del.; 7 September 1805, Journal, Comm. of Works, Hist. Soc. Del.

span across the canal with stone abutments on either side. There was originally a roadway of packed earth, but little remains now.[66] Preparations for bridges were also made at the Lancaster Road, where Vickers laid the masonry abutments for a swing bridge over the feeder, and at the Baltimore Post Road, where some embanking was completed.[67] No traces of the latter bridges are visible.

The feeder required several culverts to carry tributaries of the Elk River safely under it, and Vickers and his masons constructed seven of them in the first mile and a half of the canal. Four of them are demolished or buried, but three remain and are still carrying water under sections of the feeder (see figure 9b). They are made of rough stone, probably from the quarry opened for the aqueduct, and are cemented with a hydraulic mortar composed of lime, sand, and trass (or tarras), a volcanic earth purchased in Philadelphia.[68] They have semicircular ceilings and vertical side walls, and originally may have had flooring stones.[69] The largest culvert is about 6 feet wide and 70 feet long.[70]

By far the largest amount of labor was expended simply in forming the trough of the canal along the line surveyed by Latrobe (see figure 9c). After Latrobe or a member of his staff marked each section with stakes delimiting the width of the canal and towpath,[71] the contractors and their gangs began the arduous job of digging in places where the canal lay below the surface, or embanking where it lay above. The men usually loosened the earth and stones with picks and shovels, although black powder was occasionally necessary.[72] Then they shoveled the debris into wheelbarrows and, following a way laid with planks to keep the wheels from sinking into the soil, carted it to or from the feeder.[73]

Fig. 7. Tools for canal excavation sketched in BHL's letters (*left*: wheelbarrow, rear view; *right*: shovel handle and blade). (BHL to Bindley, 22 May 1804, Letterbooks; BHL to Horner & Son, 2 September 1804, Letterbooks.)

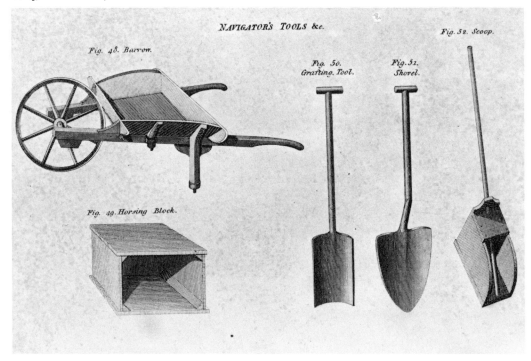

Fig. 8. English tools for excavating canals, c. 1800. Compare with fig. 7. Canal laborers in England were called "navigators." (Rees, *Cyclopaedia*, plates, vol. 2.)

Courtesy of Eleutherian Mills Historical Library

---

66. BHL to Cooch, 5 August 1804, Letterbooks; BHL to Gilpin, 26 August 1804, Letterbooks.

67. BHL to Johns, 1 November 1804, Letterbooks; 20 November 1804, 30 May 1805, C & D Engineer's Reports, Hist. Soc. Del.

68. BHL to Vickers, 20 May 1804, Letterbooks; BHL to Helmsley, 29 October 1804, Letterbooks; 30 May 1805, C & D Engineer's Reports, Hist. Soc. Del.; 26 January 1804, 23 April 1804, Minutes, Comm. of Survey, Hist. Soc. Del.; 23 October 1804, Journal, Comm. of Works, Hist. Soc. Del. See pp. 64–65 below for further discussion of hydraulic cement.

69. BHL to Vickers, 10 February 1804, Letterbooks.

70. Field measurement by author, 27 September 1974.

71. BHL to Strickland, 29 May 1804, Letterbooks; notebooks, 1804–5, Robert Brooke Papers, Eleutherian Mills.

72. 7 April 1804, Minutes, Comm. of Survey, Hist. Soc. Del.; BHL to Cochran, 3 May 1804, Letterbooks; BHL to Strickland, 8 September 1804, Letterbooks; BHL to Horner & Son, 14 May 1804, Letterbooks; BHL to I. and W. Paxson, 14 May 1804, Letterbooks; 20 November 1804, C & D Engineer's Reports, Hist. Soc. Del.

73. BHL to Brindley, 22 May 1804, Letterbooks; BHL to Davis, 10 June 1804, Letterbooks; "List of prices given for making & repairing of Wheel-Barrows at the Ches. & Del. Canal—1804," notebook #3, Robert Brooke Papers, Eleutherian Mills; BHL to Strickland, 23 April 1804, Letterbooks; BHL to Gilpin, 3 May 1804, Letterbooks.

There were exceptions to this general pattern at the deeper excavations and in the Delaware swamp. At Bell Hill, several hundred feet had to be cut from a tongue of land rising 26 feet above the bottom of the feeder,[74] and at the Baltimore Post Road a lesser cut was necessary.[75] Both of these were first attacked in the conventional manner, but in the summer of 1805 Latrobe introduced simple railways to aid in moving earth from the face of the cutting. The railways were probably about 100 feet long, and consisted of wooden rails plated with wrought iron. Special wagons with cast-iron wheels carried the earth over the railways.[76] Latrobe claimed that his railways "answer most perfectly & do very great execution."[77] South of the Post Road in Delaware there was a swampy area over a mile long which first had to be drained, then have a large number of tree stumps removed before excavation could begin.[78]

The prism, or cross section, of the feeder was 12 feet at the bottom and about 22 feet at the waterline, which was $3\frac{1}{2}$ feet from the bottom.[79] These dimensions were almost the same as those which Latrobe had planned in 1802 for the Potomac Canal extension which was to supply water to the proposed drydock at the Washington Navy Yard.[80] Whether these dimensions were typical of British feeders is uncertain, but they appear to be about right for the narrow boats for which most British canals were built.[81] Latrobe built most of the towpath to the standard British width, 9 feet, but in some places made it wider to compensate for "the *washing* rain of [the American] climate, & the effect of [its] frost in winter."[82] A considerable portion of the feeder trough and towpath remains today.

Even though the company completed most of the feeder, the subscribing stockholders were unimpressed by the future prospects of the canal and many

Fig. 9a. Features of the Chesapeake and Delaware Canal feeder, 1976. Road bridge over feeder, built by Thomas Vickers.

Louise Carter

were delinquent on share payments.[83] From the start of work the company was plagued with inadequate capital, and by the spring of 1805 the situation became so difficult that about three-fourths of the laborers were dismissed.[84] As work progressed at a snail's pace Latrobe remained on the feeder attempting to complete the remaining parts. But week by week it became more difficult to pay the contractors, and Latrobe himself received his pay in promissory notes. Finally all debts were settled at a fraction of their face value, and Latrobe dismissed the remaining workers in December 1805.[85]

Latrobe was active in attempts to persuade the states, especially Pennsylvania, and the federal government to aid the company. The directors made strong arguments for the value of the canal in interstate commerce and for its strategic importance during wartime when, it was thought, the entrances to

74. 30 May 1805, C & D Engineer's Reports, Hist. Soc. Del.
75. BHL to Johns, 1 November 1804, Letterbooks.
76. BHL to C. I. Latrobe, 5 June 1805, Letterbooks; BHL to Gilpin, 19 August 1805, Letterbooks; 31 May 1805, 18 June 1805, 6 July 1805, 21 November 1805, 14 February 1806, Journal, Comm. of Works, Salem Co. Hist. Soc.
77. BHL to Gilpin, 19 August 1805, Letterbooks.
78. 20 November 1804, 30 November 1805, C & D Engineer's Reports, Hist. Soc. Del.; BHL to [Cochran], 29 March 1805, Letterbooks.
79. BHL to Strickland, 29 May 1804, Letterbooks; notebook #3, folio 135, Robert Brooke Papers, Eleutherian Mills.
80. See p. 121 below.
81. Hadfield, *British Canals*, p. 56, says that narrow boats were "70 ft. to 72 ft. long, and about 6 ft. 10 in. wide, carrying some 25 to 30 tons." The width of the feeder appears sufficient for such boats, and BHL stated that it was "capable of carrying barges of 40 tons." BHL to McKean, 6 January 1805, Letterbooks.
82. BHL to Gilpin, 19 August 1805, Letterbooks.

83. Gilpin, *Chesapeake and Delaware Canal*, pp. 44–45.
84. BHL to Gilpin, 3 May 1804, Letterbooks; BHL to Hazlehurst, 4 June 1805, Letterbooks.
85. BHL to Johns, 9 June [1805], Letterbooks; BHL to Tatnall, 5 August 1805, Letterbooks; BHL to Tingey, 19 September 1805, Letterbooks; BHL to Hazlehurst, 19 November 1805, Letterbooks; BHL to Roosevelt, 13 December 1805, Letterbooks.

Fig. 9b. Features of the Chesapeake and Delaware Canal feeder, 1976. Culvert with person standing on towpath.

Louise Carter

the Chesapeake and Delaware bays would probably be blockaded. Latrobe accompanied the directors to a meeting of the Pennsylvania legislature in the winter of 1805, and made maps to accompany petitions during the next winter. He officially took leave of the company in 1806, but wrote a strong letter to Secretary of the Treasury Albert Gallatin in 1808 supporting the company's appeal for federal help. In 1809 and 1810 he participated in agitation for an omnibus internal improvements bill which would have

Fig. 9c. Features of the Chesapeake and Delaware Canal feeder, 1976. Canal trough.

Louise Carter

assisted the Chesapeake and Delaware Canal.[86] As late as 1816 he hoped the company would be enabled to renew its efforts, and reported that he had "carefully preserved every paper belonging to the Survey & could start a gang of 1000 men in 10 days from the arrangement of the finances."[87]

But the canal company was not revived until 1822, two years after Latrobe's death. Significantly, William Strickland, one of Latrobe's pupils who had assisted him on the feeder, was employed by the company for the initial re-survey of the main line. Benjamin Wright, later appointed chief engineer, was from the first "disposed to give full credit to Latrobe for his skill and accuracy" in considering the route which the canal should follow, but later

86. BHL to Gilpin, 3 February 1805, Letterbooks; BHL to Tatnall, 21 February 1805, Letterbooks; BHL to Gilpin, 20 November 1805, Letterbooks; BHL to Gallatin, 16 March 1808, in Thomas C. Cochran, ed., *The New American State Papers. Transportation*, 7 vols. (Wilmington, Del.: Scholarly Resources, 1972), 1:368–72; Hamlin, *Latrobe*, pp. 267–69; Lee W. Formwalt, "Benjamin Henry Latrobe and the Development of Internal Improvements in the New Republic, 1796–1820" (Ph.D. dissertation, The Catholic University of America, 1977), pp. 164–85.

87. BHL to Johns, 5 January 1816, Letterbooks.

concurred in the opinion of a board of engineers that a different route was better.[88] He also agreed that the canal could be dug deeply enough that its summit level could be supplied with water from local sources, and Latrobe's feeder, which had long been used by the company as a symbol of how well their money had been spent in the first effort, became useless.[89]

Washington Canal

Pierre L'Enfant, who laid out the new federal city of Washington in 1791, planned a canal to cross the city, connecting the Potomac River and the Eastern Branch. He believed it provided a refined and aesthetically pleasing element in the urban landscape, but he also expected the canal to have practical functions. First, he believed that the canal would assist in the commercial development of the city, and to that end he located the public market on it. Second, he felt that the canal would be an integral part of the works necessary to support the construction of the President's House and the Capitol. Especially stone, but also timber, sand, and lime could be brought cheaply and quickly to the construction sites by the canal. L'Enfant probably anticipated that stone quarried at some distance from the city was likely to be shipped on the Potomac River, and that the canal would eliminate the problems of transshipment to land carriage.[90]

The commissioners of the city were at first very interested in the canal, and expended over $5,000 excavating a section of it before abandoning the project in 1795.[91] In that year a group of men were granted the right to conduct a lottery with the proceeds to be applied to the canal, but little money was forthcoming.[92] Then in 1802 Congress incorporated a private company to undertake the canal.[93] Thomas Law, a real estate developer and a stock commissioner of the new company, attempted to raise capital for it in England during his visit there in 1803, but was unsuccessful.[94] However, during the summer of that year the officials of the company employed Benjamin Henry Latrobe to survey the route, and he furnished a report with plans early in 1804. Although only parts of the report and plans are extant, it is clear that Latrobe did not suggest any significant deviation from the route that L'Enfant planned. He wanted substantial stone locks and stone bridges to make the canal a permanent improvement, and he echoed L'Enfant's belief that the canal would have commercial value to the city. When the stock subscription books of the company were finally opened in Washington in June 1804, Latrobe's report was exhibited as proof of the feasibility of the canal.[95] Interest in the project was minimal, however, and the company did not fulfill its charter requirement of establishing navigation within a five-year period.

Congress enacted another charter for a Washington Canal Company in February 1809, and it was under that act that the promoters had some measure of success. Within two months subscription books were opened, and about half the stock was taken up, of which Latrobe subscribed thirty shares.[96] The first meeting of the stockholders was on 10 January 1810, and several prominent men of the city were elected as directors.[97] A committee was at once appointed to speak to Latrobe about his estimate of construction costs, but he refused to provide any information until he was regularly appointed engineer to the company.[98] The directors formally tendered the appointment within a few days, and Latrobe then stated his terms for accepting the position, which included taking his salary in shares of canal stock rather than cash. With one change the company accepted his proposition, and Latrobe entered

---

88. Ralph D. Gray, "The Early History of the Chesapeake and Delaware Canal—Part II," *Delaware History* 8 (1959): 383, 386–92; Henry Gilpin to Joshua Gilpin, 28 May 1823, Gilpin Collection, Hist. Soc. Del.

89. Gilpin, *Chesapeake and Delaware Canal*, pp. 30–31; 6 May 1822, Minutes, Comm. of Survey, Hist. Soc. Del.

90. Kite, ed., *L'Enfant and Washington*, pp. 55–57, 69, 117–20, 127, map facing p. 62.

91. Saul K. Padover, ed., *Thomas Jefferson and the National Capitol* (Washington: Government Printing Office, 1946), pp. 97, 105, 109, 112, 129, 141–42; Wilhelmus Bogart Bryan, *A History of the National Capitol*, 2 vols. (New York: Macmillan Company, 1914–16), 1:242; "Rough Account of Expenditures in the City of Washington from its Commencement in 1791 to the 15th of December 1806," [1806], District of Columbia Papers, LC. BHL twice mentions "the old cut": BHL to President and Directors of the Washington Canal Company, 2 June 1810, Letterbooks; BHL to Law, 17 July 1810, Letterbooks.

92. "An Act to Authorise two Lotteries in the City of Washington," n.d., District of Columbia Papers, LC; Gustavus Scott et al., to Managers of Canal Lottery, 18 April 1799, District of Columbia Papers, LC; Bryan, *National Capitol*, 1:493.

93. *ASP. Miscellaneous*, 1:258–59; Bryan, *National Capitol*, 1:493–94.

94. *Federal Gazette* (Baltimore, Md.), 28 April 1803, quoting an unidentified Alexandria, Va. newspaper; *National Intelligencer* (Washington), 5 September 1803.

95. 2 July 1803, 5 July 1803, 10 July 1803, Robert Mills diary, Special Collections Division, Tulane; BHL to Law and Brent, 13 December 1803, Letterbooks; *National Intelligencer* (Washington), 26 May 1809, quoting BHL's report of 1804; BHL's Washington Canal Plans, 5 February 1804, Geography and Map Division, LC; *National Intelligencer* (Washington), 15 June 1804; Thomas Law, "Observations on the Intended Canal in Washington City," *Records of the Columbia Historical Society*, 8 (1905): 159–68.

96. Bryan, *National Capitol*, 1:499; BHL to Tingey, 16 July 1810, Letterbooks; May to Crawford, 2 December 1818, copy in Minutes, Wash. Canal Co., NA.

97. 10 January 1810, Minutes, Wash. Canal Co., NA.

98. BHL to Caldwell, 17 January 1810, Letterbooks.

upon his second canal venture, which came to be almost as frustrating as his experience with the Chesapeake and Delaware Canal.[99]

It appears that Latrobe depended upon his survey of 1804 in preparing for the construction of the canal, although he made further studies such as determining the quality of the soil at various sites.[100] He wrote a long letter to the canal directors in which he agreed to design some elements of the canal according to their ideas, such as locating the locks as far inland as possible so that ships could use much of the canal without going to the trouble of locking through. He also agreed to make the canal "on the most oeconomical scale, & in the most oeconomical manner which is consistent with its utility & with the provisions of the law." That meant using wood rather than stone or brick for locks and bridges.[101] The directors later had reason to regret that provision.

Having reached an agreement with the directors, Latrobe turned to organizing a staff and gathering materials for the coming work season. He chose his son Henry as clerk of the works, although Henry received no pay from the company. Henry worked only one season, leaving for New Orleans late in 1810. No new clerk of the works was appointed.[102] Latrobe also had at least one other assistant, Eugene Leitensdorfer, to help lay out the canal.[103] For a contractor he turned to his old friend from the Philadelphia Waterworks and the Chesapeake and Delaware Canal, James Cochran. Cochran and his gang worked on the canal during the summers of 1810, 1811, and 1812.[104] When Cochran gave up his contract Latrobe called on another of his frequent associates, Charles Randall, to work on the canal in the summer of 1813.[105] To direct some work not covered by the digging contracts, such as building the locks, Latrobe hired Simeon Meade, a foreman at the Capitol, and John White, another local man. A few craftsmen were also employed to build bridges and scows.[106]

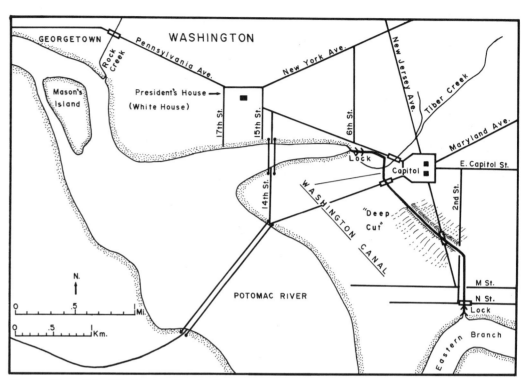

Fig. 10. Map of the Washington Canal, 1810–17.
Stephen F. Lintner

One of Latrobe's earliest transactions with the directors of the canal was to ask for the purchase of equipment. He wanted "100 wheelbarrows, 150 shovels (American), 100 spades (English long handles), 1 Archimedean screw with its apparatus, & to these I should add, post & rail fencing sufficient to fence the work on both sides"; Latrobe busied himself acquiring these items in the spring of 1810.[107] In addition he constructed a pile driver ("piling engine") for the company.[108]

With preparations complete, the official beginning of work on the canal was 2 May 1810, exactly six years after the first sod of the Chesapeake and Delaware Canal feeder was cut. President James Madison, Elias B. Caldwell, president of the Washington Canal Company, and the mayor of Washington participated in turning the first shovelful of earth, and James Cochran personally plowed open the first strip of grass. Latrobe was on hand and wrote a

---

99. BHL to Caldwell, 19 January 1810, Letterbooks; BHL to Caldwell, 23 January 1810, Letterbooks; 24 January 1810, Minutes, Wash. Canal Co., NA.

100. BHL to Cochran, 30 March 1810, Letterbooks.

101. BHL to President and Directors of the Washington Canal Company, 6 February 1810, Letterbooks.

102. BHL to Caldwell, 19 January 1810, Letterbooks; BHL to Caldwell, 31 July 1810, Letterbooks. The company did hire a clerk, but he apparently had only office duties: 27 April 1810, Minutes, Wash. Canal Co., NA.

103. BHL to Bradley, 30 January 1811, Letterbooks.

104. BHL to Caldwell, 19 April 1810, Letterbooks; BHL to Cochran, 8 July 1811, Letterbooks; BHL to President and Directors of the Washington Canal Company, 20 July 1812, Letterbooks; BHL to Henry Latrobe, 24 April 1813, Letterbooks; 19 April 1810, Minutes, Wash. Canal Co., NA.

105. BHL to Randall, 2 May 1813, Letterbooks; BHL to Randall, 19 June 1813, Letterbooks; BHL to May, 10 February 1814, Letterbooks.

106. BHL to Law, 27 June 1811, Letterbooks; BHL to Caldwell et al., 13 June 1810, Letterbooks; BHL to Caldwell, 23 July 1810 and 26 January 1811, Letterbooks; BHL to President of the Washington Canal Company, 29 June 1812, Letterbooks; BHL to May, 14 August 1812 and 21 July 1812, Letterbooks; BHL to Caldwell, [10 July 1811], Letterbooks; 7 August 1809, Latrobe Journals, PBHL, MdHS.

107. BHL to President and Directors of the Washington Canal Company, 6 February 1810, Letterbooks; BHL to Caldwell, 14 March 1810, Letterbooks; 26 May 1810, Minutes, Wash. Canal Co., NA.

108. BHL to Caldwell et al., 13 June 1810, Letterbooks.

description of the event which was published in a Washington newspaper.¹⁰⁹

The canal as planned and constructed was about a mile and a half long. The wide tidal mouth of Tiber Creek provided entrance to the canal from the Potomac River, and the shoreline at 7th Street, N.W. was wharfed so that vessels could land goods for the nearby public market. At 6th Street the canal proper began with the Tiber or Western lock, which raised the water level four feet above low tide, about equal to common high tide. The canal then ran east to between 3rd and 2nd streets, and turned south along the base of Capitol Hill until it met Canal Street, which was part of L'Enfant's original plan. The canal followed Canal Street southeast until near 2nd Street, S.W., where it turned south toward the Eastern Branch. The eastern lock was just south of N Street, and below it there was another wharf provided by the company.¹¹⁰

The dimensions of the canal were nearly the same as those of Latrobe's proposed Potomac Canal Extension and the Chesapeake and Delaware Canal feeder: 12 feet wide at bottom, and 21 feet wide at the water line, which was 3 feet from the bottom. Latrobe provided another 2 feet of height for the canal's banks, at the top of which the canal was 27 feet wide; and he planned towpaths from 6 to 12 feet wide.¹¹¹

The major difficulty was an excavation south of the Capitol through a hill which at its height was about 35 feet above the planned bottom of the canal, and which averaged more than 20 feet above the bottom. It took two seasons to complete that section (although it was just over 200 yards long), and the treacherousness of the banks required extensive piling to hold the soil.¹¹² Apparently scows helped carry earth away from this excavation by being floated to and from the worksite on the completed part of the canal.¹¹³ When the canal company dredged part of the Tiber Creek to improve access to the canal, scows removed the mud.¹¹⁴

109. *National Intelligencer* (Washington), 4 May 1810; BHL to Smith, 3 May 1810, Letterbooks.
110. A contemporary map of the entire canal has not been located, but see BHL, "Map exhibiting the property of the U.S. in the vicinity of the Capitol," 3 December 1815, Geography and Map Division, LC; W. Howard and F. Harrison, Jr., "Washington Canal," April 1827, Canals 32-2, RG 77, NA.
111. BHL to President and Directors of the Washington Canal Company, 6 February 1810, Letterbooks; BHL to Caldwell, 19 April 1810, Letterbooks.
112. BHL to Caldwell, 29 January 1811, Letterbooks; BHL to Fulton, 8 July 1811, Letterbooks; BHL, "Report on the Improvement of Jones' Falls, Baltimore," 31 May 1818, PBHL, MdHS; BHL to Cochran, 8 July 1811, Letterbooks.
113. BHL to President and Directors of the Washington Canal Company, 2 June 1810, Letterbooks; BHL to Caldwell et al., 13 June 1810, Letterbooks; BHL to President and Directors of the Washington Canal Company, 1 April 1813, Letterbooks.
114. BHL to Hamilton, 20 October 1810, RG 45, NA.

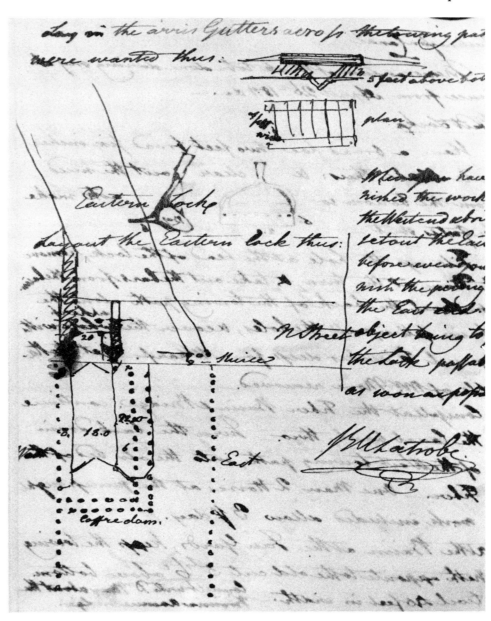

Fig. 11. Construction of Eastern Lock, Washington Canal, 1811. BHL's sketch shows the extensive piling needed for the lock's foundation. The piles surround the lockpit on the left side of the sketch (the lock is "18.0" feet wide), and there is a double row of piles at its southern end for the protective coffer dam. To the north of the lock is the N Street bridge. (memorandum, [26 April 1811], Letterbooks.)

Latrobe constructed the canal's locks of wood as the directors of the company originally stipulated, although he was never enthusiastic about using that material. He thought it would prove more expensive in the long run to replace rotten wood than to build with "permanent" stone.[115] There were, however, ample precedents in Britain and the United States for locks of wooden construction; and Latrobe's beautifully detailed plans for the Washington Canal locks suggest he was familiar with them.[116]

The lock on the Eastern Branch gave Latrobe a great deal of difficulty: it was located on swampy soil which, although drained by the Archimedean screw pump, twice surged into the lockpit and destroyed the partially completed work.[117] And although the Eastern and Tiber locks were in operable condition in the summer of 1812, within a few years both were decayed. In 1816 Latrobe supervised the construction of a stone lock costing $8,000 to replace the Tiber lock, and it was an enduring improvement. All the locks had the generous dimensions of 90 feet length and 18 feet width—possibly large enough to take two boats at once.[118]

From the beginning Latrobe considered it poor engineering to locate the locks as far inland as possible, the way the directors wanted. Although vessels could reach the company's wharves without going through the locks, the long approaches to the locks were subject to the tides and sedimentation. Latrobe argued that if the locks were located closer to the Potomac River and the Eastern Branch, or even if single lock gates were placed closer to the rivers, the water level of the canal could be maintained to those points. Then the entrance channels would not have to be dredged so deeply at such great expense. But the expected additional costs of canalizing the entrances and the anticipated revenues from the wharves kept the directors from taking Latrobe's advice. Unfortunately, the traffic to the wharves was never substantial, yet the tide-level entrances silted up at an increasing rate as rainwater runoff and sewage entered the canal, the Tiber, and the Potomac from the growing city of Washington.[119] The canal was never easily navigable.

In addition to directing the excavation and building of the locks, Latrobe was responsible for a number of other features on the Washington Canal. Timber piling was necessary to support certain portions of the walls of the canal and the banks of the Tiber Creek, as well as to provide foundations for the locks. The wharves located near the Tiber lock at 6th and 12th streets, N.W., and at the Eastern Branch entrance to the canal also required piles. A floating pile driver was used for the work, and most of the piles were pine logs rafted down the Potomac River from the interior.[120]

Since the Washington Canal went through the heart of the city, intersecting several major streets, the company probably built some road bridges and the approaches to them. Presumably the bridges were wooden, in keeping with the directors' preferences for cheap construction, although years later one of Latrobe's sons claimed that his father had built a brick bridge over the canal at one location. The city erected at least two bridges over the Tiber.[121]

In spite of spending $85,000 by the end of 1816 on digging the canal, building locks and bridges, and piling the banks, the Washington Canal projectors never realized their great hopes. Although some traffic did develop, it was minimal and irregular, and the tolls and wharf rental were inadequate to cover maintenance costs, to say nothing of improvements. The company was never able to collect all the installments on its shares of stock, and resorted to a lottery and borrowing from banks for capital within six years. The canal company lingered on the verge of insolvency through the 1820s, and in 1831 the city purchased the canal in order to improve it. Shortly thereafter the Chesapeake and Ohio Canal Company built an extension of its line from Georgetown to connect with the Washington Canal, and for some time it was possible for large canal boats to travel through the city. In spite of frequent attempts to keep the canal in operation, however, the problems of silting and sewage

---

115. BHL to President and Directors of the Washington Canal Company, 6 February 1810, Letterbooks; BHL to Fulton, 31 July 1811, Letterbooks.

116. Hadfield, *British Canals*, pp. 185, 189; Christopher Roberts, *The Middlesex Canal, 1793–1860*. Harvard Economic Studies, no. 61 (Cambridge, Mass.: Harvard University Press, 1938), pp. 110–11; see pp. 130–36 below.

117. BHL to Law, 27 June 1811, Letterbooks; BHL to Cochran, 8 July 1811, Letterbooks; BHL to Fulton, 31 July 1811, Letterbooks. For a drawing of an Archimedean screw, see p. 153 below.

118. BHL to President and Directors of the Washington Canal Company, 20 July 1812, Letterbooks; BHL to Stuart, 1 July 1816, Letterbooks; BHL to Van Ness, 16 October 1816, Letterbooks; *National Intelligencer* (Washington), 6 September 1817; May to Crawford, 2 December 1818, copy in Minutes, Wash. Canal Co., NA; BHL to Randall, 19 August 1815, Letterbooks; BHL to Ingle, 13 September 1816, Letterbooks.

119. BHL to President and Directors of the Washington Canal Company, 6 February 1810, 20 July 1812, and 5 April 1817, Letterbooks; BHL, memorandum, "Washington Canal. Aug. 9, 1811," Letter-

books; Cornelius W. Heine, "The Washington City Canal," *Records of the Columbia Historical Society*, 53 (1953–54): 10; BHL to Randall, 19 August 1815, Letterbooks.

120. BHL to Caldwell et al., 13 June 1810, Letterbooks; BHL to President and Directors of the Washington Canal Company, 20 July 1812, Letterbooks; BHL to Law, 17 July 1810, Letterbooks; BHL to Caldwell, 31 July 1810, Letterbooks; BHL to Caldwell, 29 January 1811, Letterbooks; BHL to Coombe, 1 July 1810, Letterbooks; 29 March 1811, Minutes, Wash. Canal Co., NA.

121. BHL to May, 8 January 1811, Letterbooks; BHL, "Washington Canal. Memorandum of Work to be done during my absence from April 26th 1811 &c.," Letterbooks; BHL to May, memorandum, 15 May 1811, Letterbooks; BHL to Law, 27 June 1811, Letterbooks; BHL to May, 14 August 1812, Letterbooks; BHL to President and Directors of the Washington Canal Company, 1 April 1813, Letterbooks; Benjamin H. Latrobe, Jr. to McAlpine, 31 August 1876, Mrs. Gamble Latrobe Collection, MdHS; Donald Beekman Myer, *Bridges and the City of Washington* (Washington: U.S. Commission of Fine Arts, 1974), p. 85; *National Intelligencer* (Washington), 29 October 1810, 8 December 1810.

accumulation continued to plague it, and in the 1870s the canal was finally filled in.[122]

Latrobe lost money by his association with the Washington Canal. He paid some of the installments on the original thirty shares which he purchased, and when he was pressed for cash in 1813 found it impossible to sell the seventeen shares he had received in lieu of salary, nor were they acceptable as collateral for a loan. By retaining the stock, Latrobe remained one of the largest shareholders of the company. He was appointed to the board of directors in 1815 to fill a vacancy, and was elected for regular terms in 1816 and 1817. The shares apparently were reassigned before Latrobe's bankruptcy of December 1817.[123]

Other Waterways

Throughout his American career Latrobe maintained a wide interest in canals and navigations in addition to those of which he was the engineer. On several occasions he hoped to obtain commissions to survey or direct construction of new waterways, particularly in the latter years when his personal finances were straitened. Although he entertained hopes of becoming connected with the James River Canal, a Niagara canal, a Patapsco canal, the Virginia Public Works, and a canal at the falls of the Ohio,[124] it was an appointment to survey the New York Western Navigation which came the closest to fruition. Latrobe had an early interest in the project, which involved connecting the Hudson River and Lake Erie by an all-water route, and he discussed it with New York congressmen.[125] His friend Robert Fulton was a member of the state's commission to develop a plan for the consideration of the state legislature. Early in 1811 first Governor Clinton and then the commissioners asked Latrobe if he would conduct a survey of the region to determine a practical route for the improvement. Late in July, after considerable hesitation, Latrobe agreed to do the survey if he was advanced the money for some of his expenses. He prepared to go, but the commissioners refused his condition and the arrangement fell through.[126] Latrobe continued his interest in the project for a few months, at one point writing a report for Governor Clinton concerning "ideas on the best mode of raising a fund by which any great national, or state undertaking might be carried into effect."[127] But of the later start of construction of the Erie Canal in 1817 and its progress until Latrobe's death in 1820, there is no comment in his papers.

Occasionally Latrobe gave advice by letter about a fledgling canal or navigation. In 1809, at the request of the directors of the Santee Canal in South Carolina, he wrote a report regarding the canal's water supply, and he was extremely annoyed when the directors neither acknowledged his report nor paid his fee.[128] The commissioners of the proposed Roanoke and James Canal asked Latrobe to survey their line in 1812 and in declining the appointment he wrote a letter advising the commissioners how a good survey could be done.[129] Similarly, when he refused a position as engineer of the North Carolina state works in 1818 he gave extensive information on the preliminary steps a good engineer would take to make the Yadkin River navigable, which he understood to be the first project of the state.[130]

More frequently Latrobe had advice to give which was expressed through unofficial or indirect means. Shortly after his arrival in the United States in March 1796, he ventured into the Dismal Swamp near Norfolk, Virginia, and probably looked at the Dismal Swamp Canal which had been under construction for three years.[131] In the summer of the following year he briefly examined the property of the Dismal Swamp Land Company, partly to determine the damage done to its mills by the diversion of water to the canal.[132] Several years later Latrobe was still familiar enough with the Dismal Swamp Canal to

---

122. May to House of Representatives, 30 January 1817, copy in Minutes, Wash. Canal Co., NA; 20 November 1815, Minutes, Wash. Canal Co., NA; Heine, "Washington City Canal," pp. 7–27; BHL to Tingey, 16 July 1810, Letterbooks.

123. BHL to Law, 17 July 1810, Letterbooks; BHL to May, 10 September 1813, Letterbooks; BHL to President and Directors of the Washington Canal Company, 26 August 1813, Letterbooks; 1 September 1813, 10 September 1815, 10 January 1816, 10 January 1817, Minutes, Wash. Canal Co., NA. The shares are not listed as assets in BHL's bankruptcy petition of 19 December 1817, RG 21, NA.

124. BHL to Hazlehurst, 17 December 1809, Letterbooks; BHL to Caton, 21 May 1817, Letterbooks; BHL to Jefferson, 20 November 1817, Letterbooks; BHL to Burr, 7 April 1805, Letterbooks; BHL, *Journals*, 1:80.

125. BHL to Eppes, 20 December 1809, Letterbooks.

126. BHL to Fulton and Eddy, 23 May 1811, Letterbooks; BHL to Mark, memorandum, 29 May 1811, Letterbooks; BHL to Fulton, 31 July 1811, Letterbooks; BHL to Fulton, 2 August 1811 [misdated 20 August 1811], Letterbooks; BHL to Eddy, 12 August 1811, Letterbooks; BHL to Fulton, 12 August 1811, Letterbooks.

127. BHL to Clinton, 20 January 1812, Letterbooks.

128. BHL to Gilchrist, 7 December 1809, Letterbooks; BHL to Smith, 7 December 1809 and 13 February 1810, Letterbooks.

129. BHL to Newton, 29 October 1812, Letterbooks.

130. BHL to Gales, 16 April 1818, Archibald D. Murphey Papers, North Carolina State Archives, Raleigh, N.C.

131. BHL, *Journals*, 1:80; Alexander Crosby Brown, *The Dismal Swamp Canal* (Chesapeake, Va.: Norfolk County Historical Society, 1970).

132. BHL to Wood, 14 February 1798, Executive Papers of Governor James Wood, Virginia State Library, Richmond, Va., printed in BHL, *Journals*, 2:362–65.

discuss its faulty engineering and what could be done to correct it.[133] Because of his two years' residence in Richmond he knew of the James River Canal, and in 1803 and 1808 criticized the existing and proposed portions of that work.[134] During the last years of his work on the Capitol in Washington Latrobe attended stockholders' meetings of the Potomac [River] Company, and took an interest in the physical condition of that enterprise.[135]

Latrobe also saw in the canal and navigation projects the opportunity to provide employment for his pupils. In 1804 Frederick Graff was appointed engineer to the Catawba Canal in South Carolina on his recommendation, and Graff directed the company's operations for a short time before returning to Philadelphia to become superintendent of the waterworks.[136] When the Salem Creek Canal Company of southern New Jersey wanted a surveyor in 1805 Latrobe declined the appointment but sent Lewis DeMun to do the work.[137]

Latrobe's association with and interest in American waterway projects is only what could be expected of an engineer trained in Britain during its canal age and working in the United States during the first stirrings of internal improvement activity. Latrobe was the engineer of three waterways, and had a professional connection with several others. He was the primary technical advisor in the early congressional attempts to plan a national system of internal improvements.[138] Had he been alive after the Erie Canal proved a success and internal improvement fever spread rapidly, he undoubtedly would have had more commissions. But the tentative and ill-funded projects of his time provided few opportunities for him, and those opportunities which he did have were the source of many disappointments.

Roads

Latrobe's interest and involvement in road construction was relatively slight compared to his career as a waterway engineer. He believed that laying out and building good roads did not require the services of an engineer if accurate surveyors and experienced contractors were employed, and there is no evidence that he ever sought the supervision of a road.[139] Nonetheless, Latrobe's American career coincided with the first attempts to establish a network of good highways in the United States, and he occasionally took an interest in new roads.

Latrobe followed the planning and construction of the National, or Cumberland, Road by the federal government more closely than any other.[140] Although the road was authorized and first surveyed from Cumberland, Maryland, to Wheeling, Virginia, in 1806, preparations for construction did not begin for another four years. In the summer of 1810 Albert Gallatin, who as secretary of the Treasury had overall responsibility for the road, asked Latrobe to find a competent man to make an estimate of the cost of constructing the first 72 miles. After one unsuccessful inquiry he acquired the services of Charles Randall, who had been a contractor under Latrobe on the Philadelphia Waterworks and the Chesapeake and Delaware Canal. Randall made a detailed report of how the road might be constructed, and also suggested changes in the route which he thought would be improvements.[141]

Latrobe later looked into the first contracts which were let for the road. He objected to setting the end of February 1811 as the deadline for bids on the road, and told Gallatin in December 1810 that "the contractors whom I know to be the best Men cannot at this season examine the ground & know what they are about. The first of April would be a better day."[142] Presumably the soil was frozen in winter and could not be sampled to determine the difficulty of excavation and embankment. Interestingly, two of the first three contracts on the National Road were later awarded to Latrobe's friends Charles Randall and James Cochran, the latter already the contractor for the Washington Canal. There is no evidence that Gallatin showed favoritism, but Latrobe certainly brought the men to his attention.[143]

---

133. BHL to Wheeler, 16 December 1803, South Carolina Historical Society, Charleston, S.C. See also p. 138 below for BHL's 1803 sketch of the Dismal Swamp Canal.

134. BHL to Gallatin, 16 March 1808, in Cochran, *New ASP. Transportation*, 1:263–65. See also p. 140 below for BHL's 1803 sketch of the James River Canal.

135. BHL to Harper, 16 November 1816, Letterbooks; BHL to Carroll, 12 December 1816, Letterbooks; BHL to Orr, 5 August 1817, Letterbooks; BHL to Seaton, 3 September 1817, Letterbooks.

136. BHL to Smith, 21 March 1804, Letterbooks; BHL to Graff, 26 April 1804, Letterbooks; BHL to Editor of the Philadelphia *Gazette*, 9 June 1804, Letterbooks; BHL to Traquair, 24 October 1804, Letterbooks; Cochran, *New ASP. Transportation*, 1:105, 106.

137. BHL to President and Directors of the Salem Creek Canal, 27 May 1805, Letterbooks; BHL to DeMun, 21 June 1809, Letterbooks; survey map SM-95, n.d., Salem Co. Hist. Soc.; Thomas Cushing and Charles Sheppard, *History of the Counties of Gloucester, Salem, and Cumberland, New Jersey* (Philadelphia, 1883), pp. 333, 487.

138. Formwalt, "Benjamin Henry Latrobe and the Development of Internal Improvements in the New Republic, 1796–1820," chap. 5.

139. BHL to Madison, 8 April 1816, Letterbooks.

140. See pp. 141–43 below.

141. Thomas B. Searight, *The Old Pike: A History of the National Road* (Uniontown, Pa., 1894), pp. 25, 28, 29; BHL to Jessop, 17 July 1810, Letterbooks; BHL to Randall, 31 July 1810, Letterbooks; BHL to Jenkins, 9 January 1811, Letterbooks; BHL to Madison, 8 April 1816, Letterbooks.

142. BHL to Gallatin, 19 December 1810, Letterbooks.

143. Ibid.; BHL to Cochran, 8 July 1811, Letterbooks; Cochran, *New ASP. Transportation*, 2:128.

There is no evidence that BHL was a "Commissioner of the National Road" as reported by Hamlin, *Latrobe*, pp. 522, 548. Hamlin may have been misled by references to BHL as a commissioner of the Columbia Turnpikes.

After construction of the road was underway Latrobe apparently expressed to Gallatin misgivings about the permanence of the types of drainage and surfacing required in the contracts. Gallatin asked him to put his objections in writing, which he did in October 1813. Later Latrobe published the letter in the *Emporium of the Arts and Sciences* with a drawing. Little official notice was taken of his opinion, and in January 1820 he learned that the road was in disrepair, as he had predicted it would be.[144]

The only roads with which Latrobe had a direct connection were the turnpikes in the District of Columbia. He was a signer of a petition to Congress in 1809 which asked for the incorporation of a company to establish turnpikes between the limits of the city of Washington and the district boundaries. In the spring of 1810 Congress passed and President Madison signed an act creating the Columbia Turnpikes Company with the authority to sell $60,000 of stock. The company was empowered to lay out three turnpikes within the District on which it could charge tolls: one towards Baltimore, one towards Montgomery Court House (later Rockville, Maryland), and one on the west bank of the Potomac toward the Little River Turnpike in Virginia.[145]

The act required that the roads be laid out by three disinterested commissioners appointed by the District court. In June 1810 Benjamin Henry Latrobe, Griffith Coombe, and Joseph Forrest were appointed commissioners, under an oath that they had no interest in the Columbia Turnpikes Company nor in any other turnpike company. Within a few days they published an announcement in a Washington newspaper that the three turnpike routes would be surveyed and laid out within the next two months.[146] In spite of the intention to perform the work so quickly, Latrobe's affiliation with the Columbia Turnpikes dragged on until late in 1811. Apparently there were difficulties finding routes which satisfied all of the persons whose lands were intersected by the new turnpikes.[147]

Latrobe was the chairman of the board of three commissioners and bore the brunt of the work.[148] He remarked to his father-in-law that his position with the Columbia Turnpikes was "the most troublesome & the least advantageous of all my engagements."[149] In the field he had two surveyors to assist him: his son Henry, who helped before he went to New Orleans, and Eugene Leitensdorfer, who was in his office during most of 1810.[150]

Although Latrobe surveyed all three turnpikes, only two were actually constructed. One began at the intersection of Maryland Avenue and Florida Avenue northeast of the Capitol and led north to Bladensburg; today it is known as the Bladensburg Road. The other began at the west end of the causeway connecting Alexander's Island to the western shore of the Potomac and ran west in the direction of the Little River Turnpike; it is still called the Columbia Pike. It appears that the company never attempted to use its rights to a route leading toward Montgomery Court House.[151] (See figure 3, p. 10.)

Latrobe had no connection with the construction of the Columbia Turnpikes during 1811 and 1812. They were substantially complete by the end of the latter year and were already collecting tolls. From 1812 to 1815 they yielded from 6 percent to 9 percent on the invested capital. For just the latter half of 1813 the company declared a dividend of $8.14 per $100 share.[152] There is no evidence that Latrobe benefited from his association with the company other than by collecting the hourly salary due him as a commissioner.[153] He did, however, have the pleasure of contributing to the creation of a transportation mode which was heavily used from its inception.

Bridges

Bridges were an important phase of civil engineering in Latrobe's time, both as adjuncts to canals and turnpikes, and as important works in themselves. Although Latrobe designed and built few bridges, they are a revealing record of his engineering attitudes.

While in England Latrobe built a stone bridge, possibly in connection with

---

144. BHL, "To the Editor of the Emporium," *Emporium of the Arts and Sciences*, new series, 3 (1814): 284–97; BHL to Harper, 22 January 1820, Mrs. Gamble Latrobe Collection, MdHS.
145. "Petition of sundry inhabitants of the District of Columbia praying for the establishment of sundry turnpike roads," 19 December 1809, RG 233, NA; *Journal of the House of Representatives of the United States, at the First Session of the Eleventh Congress* (Washington, 1826), pp. 276, 365, 383; *National Intelligencer* (Washington), 2 May 1810.
146. Carroll to Judges of the Circuit Court of the District of Columbia, 31 May 1810, RG 21, NA; notarized statement of BHL, 5 June 1810, RG 21, NA; *National Intelligencer* (Washington), 11 June 1810.
147. BHL to Queen, 5 September 1810, Letterbooks; BHL to Teakle, 5 September 1810, Letterbooks; BHL, Notice of meeting of turnpike commissioners, 28 November 1810 and 1 December 1810, Letterbooks; BHL to Carroll, 28 December 1810, Letterbooks; BHL to Madison, 19 July 1811, Letterbooks; BHL to Mark, 7 November 1811, Letterbooks; *National Intelligencer* (Washington), 28 February 1811.

148. BHL to Carroll, 27 August 1810, Letterbooks; BHL to Mark, 7 November 1811, Letterbooks.
149. BHL to Hazlehurst, 1 September 1810, Letterbooks.
150. Ibid.; BHL to Bradley, 30 January 1811, Letterbooks.
151. *National Intelligencer* (Washington), 11 June 1810; Bryan, *National Capitol*, 1:532–33.
152. Treasurer's reports for the Columbia Turnpikes, 1 January 1813, 1 January 1814, 17 January 1815, 6 January 1816, RG 21, NA; *National Intelligencer* (Washington), 10 January 1814.
153. BHL to President and Directors of the Columbia Turnpikes, 11 February 1811, in BHL to Coombe, 10 February 1811, Letterbooks.

the Basingstoke Canal.[154] His first American bridge crossed the feeder of the Chesapeake and Delaware Canal, and was a stone segmental arch of 15-foot span.[155] He also planned a swing bridge on another part of the feeder, but, although the masonry abutments were completed, only a temporary wooden deck was laid.[156] Latrobe also may have directed the construction of wooden bridges over the Washington Canal.[157] All of these bridges crossed narrow waterways and did not have any special problems requiring the supervision of an engineer.

In the early years of the 1800s, however, private capital was just beginning to finance the construction of bridges across major rivers such as the Merrimack, Hudson, Delaware, Schuylkill, Susquehanna, and Potomac.[158] Latrobe was informally consulted about the Potomac Bridge at Washington, a mile-long wooden structure erected on timber piles and built in 1808 and 1809.[159] Although his advice was unofficial, Latrobe stated that "from the formation of the Company, & during its whole progress I have been consulted, by Dr. May on all the details respecting building of the bridge, especially on the Icebreakers by Mr. Carol."[160] Later the company retained him as a professional witness in a suit to prevent the construction of wing dams which threatened to alter the river's current and undermine the piles.[161]

Latrobe also designed bridges for two other major crossings. In 1799 or 1800 he submitted a plan for a masonry bridge to the directors of the Schuylkill Permanent Bridge, but they thought his "elegant" design too expensive and built one with stone piers and a timber superstructure instead.[162] At the request of his friend and future son-in-law, Nicholas Roosevelt, Latrobe calculated the cost of bridging the East River between Manhattan and Long Island at Blackwell's Island. His estimate of nearly $1,000,000 for a masonry bridge (he thought that a wooden bridge would cost almost as much) was well beyond the resources of private investors and ended consideration of the scheme for some time.[163]

Latrobe designed three smaller bridges for urban locations. In behalf of a Mr. Farrell, who had worked as a contractor on public projects in Washington, Latrobe made a proposal for a bridge over the Rock Creek at Georgetown in the District of Columbia in 1808. Apparently his bridge was not built.[164] Two years later Latrobe planned a bridge at an unknown location for Daniel Brent, one of the major real estate developers in Washington, but Brent did not pursue the matter.[165] Much later, in connection with his Jones' Falls Improvement report of 1818 for the city of Baltimore, Latrobe designed a bridge to carry Wilkes Street over Harford Run. Although there are two differing drawings of the Harford Run bridge, the major features are the same. Timber piles form the foundation, and the bridge itself is masonry. The arch is segmental, in one drawing with a 16-foot span and 8-foot rise, and in the other with a 30-foot span and $10\frac{1}{2}$-foot rise. The Harford Run bridge, although never built, would certainly have been similar in appearance to some late eighteenth-century bridges over waterways in England.[166]

Latrobe's bridge building principles favored masonry construction, and he summarized them in 1818.

> Of all Bridges ... stone Bridges are assuredly the best and handsomest, and in the end the cheapest. They may be as easily and as cheaply built of one arch as of more, Although the span be eighty feet, and their clear elevation only ten feet. Cast iron Bridges have their great defects, and cannot be recommended.... of wooden Bridges, everything that can be said is universally known.[167]

Latrobe did not believe that the usual bulk of the classic masonry bridge was necessary to give it its strength, and had a design for reducing the amount of stone or brick incorporated in the structure above the piers. In an application for an ungranted American patent in 1806, he proposed making up the

---

154. BHL to Thornton, 21 February 1806, Letterbooks.
155. BHL to Gilpin, 26 August 1804, Letterbooks. See fig. 9a, p. 17.
156. 20 November 1804, C & D Engineer's Reports, Hist. Soc. Del. Swing bridges were common on part of the Basingstoke Canal, where BHL may have become familiar with them. Harris, *Canals and Their Architecture*, p. 45. See fig. 6, p. 15 above.
157. BHL to May, 14 August 1812, Letterbooks.
158. Carl W. Condit, *American Building* (Chicago: University of Chicago Press, 1968), pp. 53–56.
159. *National Intelligencer* (Washington), 13 June 1808, 17 June 1808; Myer, *Bridges and the City of Washington*, p. 19; Thomas Pope, *A Treatise on Bridge Architecture* (New York, 1811), p. 173.
160. BHL to Caldwell, 17 January 1810 [misdated 1809], Letterbooks.
161. BHL, *Opinion*.
162. [Peters], *Schuylkill Permanent Bridge*, p. 25.

163. BHL to Roosevelt, 10 November 1804, Letterbooks.
164. *National Intelligencer* (Washington), 17 June 1808; BHL to Mason, 22 June 1808 and 24 June 1808, Letterbooks; Myer, *Bridges and the City of Washington*, p. 60.
165. BHL to Brent, 26 November 1810, Letterbooks.
166. BHL, "Design for a bridge proposed to be built over Hartford [Harford] Run, on Wilkes Street," 14 March 1818, and "Design of a Tunnel proposed to be built in Wilkes," 16 March 1818, Peale Museum, Baltimore, Md.; cf. Harris, *Canals and Their Architecture*, pp. 40–42.
167. BHL, "Report on the Improvement of Jones' Falls, Baltimore," 31 May 1818, PBHL, MdHS.

arches of several ribs tied together with iron bars, single blocks of stone, or even timber. The spandrels raised on the arches could either be solid or pierced with openings. A continuous masonry arch would be sprung from the arches to carry the roadway. He believed that a bridge of this construction saved nine-tenths of the cost of the superstructure. Latrobe followed this design for a bridge he built in England, and based his proposals for the Schuylkill River bridge and the Rock Creek bridge on it. According to his son Benjamin, he built a bridge over the Washington Canal on this plan.[168]

Latrobe believed that masonry bridges if well executed were permanent improvements. There were, after all, many Roman bridges still used in Europe. In 1820 Latrobe approved of the stone bridges on the National Road near Wheeling, Virginia, even though he did not think that they had a practical and inexpensive design.[169] He did not admire timber bridges, such as the Potomac Bridge at Washington, which he thought were not objects for the attentions of a professional engineer, but could be erected by any "New England bridge builder."[170] Latrobe's attitude about materials for bridge construction is symbolic of the conflict between the ideals of the European or European-trained engineer and American society, which tended to accept cheapness rather than permanence as the major aim of public works.

Railroads

There is no direct evidence that Benjamin Henry Latrobe observed or was in any way associated with railroads before he came to the United States, but there seems little doubt that he was familiar with them. He was born and raised in the north of England, where railroads with horses were in common use in connection with coal mining, and he returned there on occasion after his years in Germany. He also personally visited coal mines at some point.[171] Railroads were also intimately intertwined with the growing canal network in England. They frequently served as feeder lines to canals, but sometimes were used as temporary connections when unfinished tunnels or flights of locks prevented opening a canal to through traffic.[172] As a canal engineer, Latrobe had a working knowledge of railroads, and he employed them for excavations on the Chesapeake and Delaware canal feeder in 1805. He also proposed using a railway at the Aquia quarries in 1806.[173]

In the "Postscript" of his report to Gallatin in 1808 Latrobe revealed his fundamental understanding of railroads and their function in transportation. He reviewed the materials which were then used for constructing rails and track, and concluded that although "cast iron *rails* let down on stone foundations . . . will last for ages," iron rails on timber ties were less expensive and suitable for the United States.

The rails of which Latrobe spoke were not the ancestor of the modern T-rail, but the L-shaped "plate rail" common in Britain in the latter eighteenth and early nineteenth centuries. The plate rail did not require a flange on the wheels of the vehicles using the railroad, since its raised side guided them. Latrobe stated that the rails had a tread 3 inches wide and a flange 2 inches high, could be cast in 6-foot lengths, and weighed about 56 pounds each.

The rails were "pinned" to flattened timber ties at each end, the holes for the pins being cast into the rail. Latrobe thought that the gauge of the rails could be from $3\frac{1}{2}$ to 5 feet, and that the "perfection of the road consists in the parallel rails being laid perfectly level with each other across the road, and perfectly jointed." He estimated that such a railroad could be built in the United States for $10,000 per mile.

Latrobe stated that he did not expect railroads to come into early use in this country, because "to render a rail road sufficiently saving of the expense of common carriage, to justify the cost of its erection, there must be a very great demand for its use." The only locations which Latrobe foresaw capable of immediate benefit from a railroad were mines or quarries which sent heavy loads over short distances for transshipment to water carriage. He also thought that once long lines of canals were constructed, railroads might be used, as in England, to overcome "difficult parts."[174]

It is true, as Talbot Hamlin has pointed out, that Latrobe's papers indicate no suspicion of the great revolution that steam locomotives were about to bring to railway transportation.[175] Yet few British engineers were advocates

---

168. BHL to Thornton, 21 February 1806, Letterbooks; BHL, "Specifications of a new method of building Bridges of Stone or Brick," [21 February 1806], Letterbooks; BHL to Mason, 22 June 1808, Letterbooks; Benjamin Henry Latrobe, Jr. to McAlpine, 31 August 1876, Mrs. Gamble Latrobe Collection, MdHS.
169. Latrobe Journals, 1 February 1820, PBHL, MdHS.
170. BHL to Caldwell, 17 June 1810 [misdated 1809], Letterbooks. According to the *National Intelligencer* (Washington), 17 June 1808, the Potomac Bridge was under the direction of "Mr. Mills . . . [who] has directed the erection of several of the most extensive bridges in New-England."
171. BHL to Ogden, 3 March 1814, Letterbooks.

172. Hadfield, *British Canals*, pp. 110–14.
173. BHL to C. I. Latrobe, 5 June 1805, Letterbooks; BHL to Gilpin, 19 August 1805, Letterbooks; BHL, *A Private Letter to the Individual Members of Congress* (Washington, 1806), p. 8.
174. Cochran, *New ASP. Transportation*, 1:269–70.
175. Hamlin, *Latrobe*, p. 555.

of steam locomotion before 1820, and the matter was not settled decisively until the Rainhill Trials of the Liverpool and Manchester Railroad in 1829. Even if he had been fully informed of the development of the steam locomotive beginning with Trevithick's demonstration of it in 1802, it seems unlikely that in his generally mature and conservative judgment he would have expected it to be a useful innovation under American conditions.

Latrobe believed that the only useful sort of railroad in America would be horse-drawn with a generally downhill grade for freight. For some months in 1809 he promoted a railway from the Chesterfield coal mines (near Richmond, Virginia) to the James River on the premise that it could be constructed with a constantly descending grade from the mines to docks on the river at Ampthill below Richmond.[176] When his prospective partner, a man who owned a sizeable portion of the coal field, suggested instead that the railroad terminate at Manchester, across the James from Richmond, Latrobe remained adamantly in favor of his original plan. He said:

> If the rail road is to terminate at Manchester the idea is not worth pursuing. Nature has said that it shall terminate nowhere but at Ampthill a point to which there is a regular descent from all the coal country.... One horse to Ampthill will take 40 tons. One horse to Manchester up & down could only take say 8 tons, making in 32 tons a difference of 4 horses....[177]

Although he thought current technology limited the usefulness of railroads in America, Latrobe obviously had firm ideas about their capabilities.

3. WATERWORKS

Philadelphia Waterworks

When Benjamin Henry Latrobe came to Philadelphia in the late 1790s the city had a population of nearly 50,000. The water supply for this city, perhaps the largest metropolis in the United States, was taken exclusively from wells which lined the streets at intervals of sixty to seventy feet.[1] Philadelphians were increasingly aware that their dependence on wells entailed several problems, including pollution by nearby privies and an insufficient quantity of water for fighting fires and cleaning streets.[2] Many citizens had also come to believe that a larger and purer water supply could prevent the city's annual yellow fever epidemics, which had been particularly severe in the summers of 1793, 1797, and 1798. With water to cleanse the streets, purify the air (with fountains), and provide wholesome drink for citizens who formerly depended on the wells, it was thought that yellow fever would not grow to epidemic proportions.[3]

Although the yellow fever made a new water supply appear to be a pressing need in 1798, since 1792 the Delaware and Schuylkill Canal Company had had a clause in its charter allowing it to provide water from the canal it intended to dig across the northern portion of the city. Construction of the canal went on sporadically after incorporation, and was suspended in 1795. In 1797 the city councils negotiated with the company concerning terms on which the canal could supply water, but no agreement was concluded. Then, after the yellow fever epidemic in the summer of 1798, the Common Council appointed a Joint Committee on Supplying the City with Water (later known as the Watering Committee). The committee was empowered to employ an engineer to find sources of pure water and ways of bringing it to the city.[4]

Latrobe came to Philadelphia in December of 1798 after his appointment as architect of the Bank of Pennsylvania. Late that month he was asked by John Miller, Jr., chairman of the Joint Committee, to investigate Spring Mills, located north of Philadelphia, for its suitability as a water source.[5] Undoubtedly Latrobe had come to the attention of Miller and other members of the committee because of his knowledge of European waterworks, especially those in London. (In Latrobe's writings there is evidence that he had observed the operations of several waterworks while he lived in London, and that he

---

176. BHL to Bollman, 21 May 1809 and 30 May 1809, Letterbooks; BHL to Heth, 3 July 1809 and 25 September 1809, Letterbooks; BHL to Bollman, 8 December 1809, Letterbooks.

177. BHL to Heth, 21 November 1809, Letterbooks.

1. BHL, *Journals*, 2:379–81. For the population of Philadelphia: John K. Alexander, "The Philadelphia Numbers Game," *Pennsylvania Magazine of History and Biography* 98 (1974): 314–24; Gary B. Nash and Billy G. Smith, "The Population of Eighteenth Century Philadelphia," *Pennsylvania Magazine of History and Biography* 99 (1975): 362–68. In 1801 only seven British cities had populations exceeding 50,000: B. R. Mitchell, *Abstract of British Historical Statistics* (Cambridge, England: Cambridge University Press, 1962), pp. 20, 24.

2. BHL, *Journals*, 2:379–81; Blake, *Water for the Cities*, pp. 5, 10–13; BHL, *View of the Practicability and Means of Supplying the City of Philadelphia with Wholesome Water* (Philadelphia, 1799), p. 7n.

3. Blake, *Water for the Cities*, pp. 6, 8–11; Thomas Condie and Richard Folwell, *History of the Pestilence, Commonly Called Yellow Fever, Which Almost Desolated Philadelphia, in the Months of August, September & October 1798* (Philadelphia, [1799]), pp. 7–8, 27.

4. Blake, *Water for the Cities*, pp. 18–22.

5. Watering Committee, *Report to the Select and Common Councils, on the Progress and State of the Water Works, on the 24th of November, 1799* (Philadelphia, 1799), p. 6.

knew of others in Europe.⁶) A long letter to Miller, dated 29 December 1798, presented his ideas for a comprehensive water supply system for Philadelphia.

Published early in 1799 as *View of the Practicability and Means of Supplying the City of Philadelphia with Wholesome Water*, Latrobe's letter revolutionized thinking about the possibilities for a water supply. Although he did consider Spring Mills as a probable future source for drinking water for the city, Latrobe thought it insufficient for the purposes of washing the streets and cooling the air, which he considered to be immediate needs. For those purposes he believed that only the Schuylkill River provided water both clean and copious enough. The water could be collected in a reservoir on the river bank, pumped up to just below street level by a steam engine, allowed to flow through a tunnel to Centre Square (where Broad and High, or Market, streets met), and there pumped into an elevated reservoir by a second steam engine. From that reservoir the water could be distributed by gravity flow through wooden pipes under the city streets. Latrobe believed that this system could be in operation in about seven months—by the end of July 1799.⁷ Obviously he was thinking in terms of the yellow fever season, which was usually at its height in August, September, and October.

The most daring element of the plan was the use of steam engines. There were at this time only three steam engines in the United States, and no established steam engine builders.⁸ In his letter to Miller, Latrobe attempted to minimize the problem of procuring engines, commenting: "I have no doubt but that this city can produce Smiths capable of constructing very efficient Engines, under proper direction."⁹ Apparently he thought that he would give the "proper direction."

Faced with the boldness of Latrobe's plan and the need to combat the yellow fever, the Joint Committee reported Latrobe's plan to the city councils for consideration. Shortly thereafter the committee commissioned Latrobe to visit the Soho Works in northern New Jersey, where a steam engine was under construction for the Stevens-Livingston steamboat. The committee wanted Latrobe's opinion on whether engines could be rapidly and "perfectly" constructed at Soho.¹⁰ He reported favorably.

With momentum increasing in favor of Latrobe's plan, and no alternative water supply in view, councils took two important steps early in February 1799: first, they contracted with Nicholas J. Roosevelt, Soho's proprietor, for two steam engines; and second, they asked the citizens for subscriptions to a $150,000 loan, the proceeds to be applied to the waterworks.¹¹

On 2 March 1799 Latrobe filed a more elaborate report with the Joint Committee.¹² He no longer discussed bringing water from Spring Mills (apparently the Joint Committee had dropped that project), but detailed his plan for taking water from the Schuylkill and distributing it throughout the city. He stated that he expected the waterworks to supply water free to citizens at public hydrants, and to provide an "inexhaustible quantity" to fire engines when needed. He also took steps to assuage fears that the steam engines would be subject to frequent breakdowns or could do damage through explosions or fires. In a memorable passage Latrobe stated that "soon after the invention, steam engines were justly considered as dangerous, man had not yet learned to controul the immense power of steam, and now and then they did a little mischief. A steam engine is, at present, as tame and innocent as a clock."¹³

In an appendix to the report, Latrobe described European waterworks using steam engines as he remembered them, appealing to "fellow-citizens [who] have been in Europe" to verify his account. He began by describing the Chelsea waterworks in London, which had essentially the same elements in its system as he was proposing for Philadelphia: water pumped from a tidal river by an engine, with part of that water pumped into a higher reservoir by another engine. He proceeded to list five other waterworks in London, several private systems in Britain, and Wilkinson's waterworks at Paris.¹⁴ Latrobe obviously believed that Philadelphia's waterworks could be just another instance of the successful use of steam engines for water supply.

---

6. BHL, *View*, pp. 12, 13, 15, 17; BHL, *Remarks on the Address of the Committee of the Delaware and Schuylkill Canal Company . . .* (Philadelphia, 1799), p. 16; BHL, *Answer to the Joint Committee of the Select and Common Council of Philadelphia, on the subject of a plan for supplying the city with water* ([Philadelphia, 1799]), pp. 5–7; BHL, "The Waterworks," unidentified Philadelphia newspaper, 15 March 1801, Peter Force Papers, LC; BHL to Fulton, 31 July 1811, Letterbooks. For a brief description of contemporary London waterworks, see Charles Singer et al., eds., *A History of Technology*, 5 vols. (Oxford: Clarendon Press, 1954–58), 4:491–94.

7. BHL, *View*, pp. 1–20.
8. Pursell, *Early Stationary Steam Engines in America*, pp. 32, 40.
9. BHL, *View*, p. 20.
10. Watering Comm., *Report* (1799), pp. 7–8; Pursell, *Early Stationary Steam Engines in America*, pp. 30–31.
11. Blake, *Water for the Cities*, p. 30; 2 February 1799, Minutes of the Select Council of Philadelphia, Phil. Archives.
12. BHL, *Answer*.
13. Ibid., p. 7.
14. Ibid., pp. 5–6.

On 2 March 1799, the same day as Latrobe filed his second report, the city councils took the momentous step of authorizing the Joint Committee to proceed with the implementation of Latrobe's plan.[15] The committee also began negotiations with Latrobe concerning his fee as engineer of the project (he was already retained as "consulting engineer"), although an agreement was not signed until August.[16] Latrobe had begun the most important engineering commission of his career.

His function as engineer of the Philadelphia Waterworks was varied and extensive. Having supplied the original plan, he was responsible for all the details of its construction, and in the course of operations his office must have provided many drawings, most of which are lost. Latrobe was responsible for making contracts for all phases of the work from the steam engines to the logs used for timber and pipe. He directed construction at each site, and he reported frequently to the Watering Committee on its progress. To the dismay of some members of the committee, he occasionally took the opportunity to instruct them how to proceed with waterworks business. Ultimately Latrobe hoped to become the permanent engineer of the waterworks, but he only succeeded in associating himself with Roosevelt in the ill-fated lease of the "extra power" of the Schuylkill engine for a rolling mill.[17]

The complete control of a construction project as complex as the waterworks was beyond the ability of one man, however, and Latrobe soon found others to assist him. The two most important were John Davis and Frederick Graff. Davis, with architectural and engineering experience in his native England, had emigrated to the United States in 1793. A member of the Watering Committee, Samuel Fox, brought Davis to an interview with Latrobe, who immediately recognized Davis's value and hired him. Davis was actually on the city payroll as clerk of the works, a position which required him to keep accurate records of hours worked, materials purchased, and all financial accounts. Frederick Graff was the son of a bricklayer and became a carpenter's apprentice. At an early stage of the construction of the waterworks he entered Latrobe's office and worked as his personal assistant, drafting plans and supervising some phases of construction. Davis and Graff became, successively, superintendents of the waterworks, and Davis later went to Baltimore to design and construct that city's water system.[18]

John Barber was also employed by Latrobe as a clerk until, in the summer of 1800, he abused his privileged position and absconded with a large sum of money and important papers. Thomas Breillat, a draftsman in Latrobe's office, was apparently an accomplice.[19] Adam Traquair, the son of a marble mason, joined Latrobe's office as a clerk and draftsman sometime during the erection of the waterworks.[20]

Construction of the waterworks began in the spring of 1799 after segments were let out to various contractors. John Lewis, for example, was in charge of excavating the tunnel from the intake on the Schuylkill to the pump well at the Schuylkill Engine House.[21] Latrobe was pleased with his contractors' performance, and later employed four of them on the Chesapeake and Delaware Canal feeder.[22] Latrobe also relied on local craftsmen for their skills in brickmaking, log boring, stone quarrying, carpentry, and masonry. Peter Brady and James Traquair, father of Adam, the draftsman, did the fine marble work for the Centre Square Engine House.[23] Thomas Vickers constructed the brick tunnel between the two engine houses; he had the "skill and experience ... necessary in forming a good water cement." The Watering Committee regarded Vickers's work as excellent.[24]

The construction of the Philadelphia Waterworks, although conducted by a skilled and capable group of men under the most highly trained engineer in the United States, was a complex and difficult task. Perhaps the most unusual and exacting segment of the work was the basin for collecting Schuylkill water. Latrobe designed it to project about 250 feet into the river from the

15. Watering Comm., *Report* (1799), p. 12.
16. Ibid.; *Report of the Committee Appointed by the Common Council to Enquire into the State of the Water Works* (Philadelphia, 1802), pp. 36–44.
17. 11 August 1800, 1 December 1800, 28 January 1801, 4 February 1801, 15 March 1801, Cope Diaries, Haverford College; "List of Reports & Communications from B. H. Latrobe [1799–1801]," Thomas Pym Cope Papers, Haverford College (not in microfiche ed.); Watering Comm., *Report* (1799), pp. 12–13; 23 January 1799, Minutes of the Select Council, Phil. Archives.
18. Watering Comm., *Report* (1799), p. 15; Watering Committee, *Report of the Committee for the Introduction of Wholesome Water into the City of Philadelphia* (Philadelphia, 1801), p. 11; [John Davis], "Autobiography of John Davis, 1770–1864," *Maryland Historical Magazine* 30 (1935): 12–15; "Frederick Graff," in Henry Simpson, *The Lives of Eminent Philadelphians, Now Deceased* (Philadelphia, 1859), pp. 431–36; BHL to Smith, 21 March 1804, Letterbooks. Robert Mills, another BHL pupil, succeeded John Davis in 1816 as engineer of the Baltimore Waterworks.
19. Hamlin, *Latrobe*, pp. 134–35; 13 August 1800, Cope Diaries, Haverford College.
20. Hamlin, *Latrobe*, pp. 134–35.
21. Watering Comm., *Report* (1799), p. 14; BHL to Madison, 8 April 1816, Letterbooks; BHL to Cope, 3 February 1801, Cope Papers, Haverford College (not in microfiche ed.).
22. See n. 55, p. 14 above.
23. Padover, ed., *Thomas Jefferson and the National Capitol*, pp. 163–64; Traquair to Jefferson, 30 May 1801, Massachusetts Historical Society, Boston, Mass.; Abraham Ritter, *Philadelphia and Her Merchants* (Philadelphia, 1869), pp. 199–200.
24. Watering Comm., *Report* (1799), pp. 14–15.

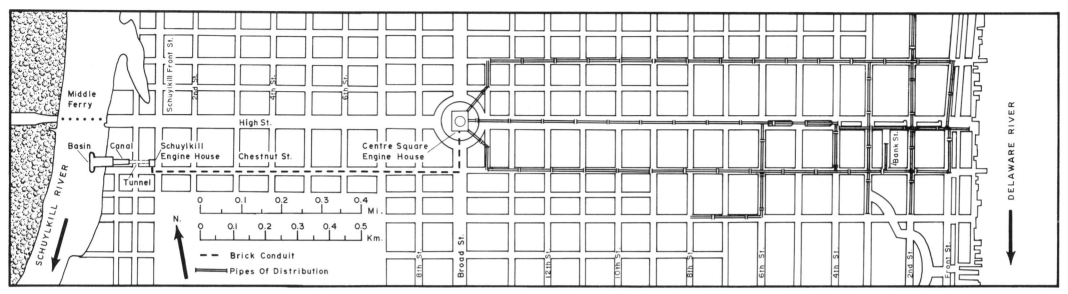

Fig. 12. Map of the Philadelphia Waterworks, 1801. Water entered the system at the basin in the Schuylkill River, then flowed through the canal and tunnel to the well of the Schuylkill Engine House. There a steam engine pumped it up nearly 50 feet to the beginning of the brick conduit which ran almost a mile to Centre Square. Next, a second engine in the Centre Square Engine House raised the water to reservoirs in the dome. The water descended by gravity to a distributing chest where the principal pipes of the system were connected. At the end of 1801 the city had laid 30,000 feet of pipe and had erected 42 hydrants for the public. For an annual fee of $5 a resident could have water piped into his home. Stephen F. Lintner

east bank, and to be about 150 feet wide. The west wall of the basin, which had a "flood gate" for admitting water, was masonry, while the north and south sides were earthen dikes.[25]

To allow the construction of the west wall Latrobe erected a cofferdam consisting of piling joined with sheet timber on the inside and of earth fill on the outside. Pumping kept the work site dry. The masonry wall rested on bedrock as deep as 11 feet below the bottom of the river, and rose about 5 feet above high water mark.[26] The mortar was probably a hydraulic (i.e., waterproof) cement made with tarras, and it seems likely that the mason was Thomas Vickers, although he is nowhere mentioned as the builder.[27] Erecting the basin directly in the current of a river subject to frequent floods was regarded as a daring feat at the time, and at every rise in the river level it was anxiously watched. Yet it stood undamaged for years.[28]

Between the basin and the Schuylkill Engine House to the east were a canal, connected to the basin by a scouring sluice, and a tunnel. The canal, about 200 feet long and 40 feet wide, was partly blasted out of a rocky surface with black powder. The tunnel, which carried the water about 360 feet to the bottom of the Schuylkill Engine House well, was created partly by open-cutting, arching over with brick and recovering with earth, and partly by tunnelling with the aid of black powder.[29] At the end of the tunnel was the well which rose nearly 50 feet to the Schuylkill Engine House. For a considerable distance the well proceeded through loose gravel which caved in once during excavation and had to be held back with a brick casing 18 inches thick.[30]

25. For drawings of the basin, see pp. 146–51 below; cf. drawings concerning development of city property on the Schuylkill, c. 1820, Frederick Graff Collection, Franklin Inst.

26. BHL to Roosevelt, 10 November 1804, Letterbooks; BHL to Poulson, in *Poulson's American Daily Advertiser*, 27 October 1801.

27. BHL later stated that Vickers was "employed by me at the Waterworks of Philadelphia, and executed all the difficult & heavy work in stone belonging to it." BHL to Gallatin, 3 August 1805, Letterbooks. With hydraulic cement in use for the brick tunnel, it is unlikely that the wall would have been attempted with ordinary mortar. Shortly after completion of the waterworks, Vickers built the masonry piers of the Schuylkill bridge with hydraulic mortar. [Peters], *Schuylkill Permanent Bridge*, pp. 27, 30, 41.

28. Watering Comm., *Report* (1799), p. 25; Watering Comm., *Report* (1801), p. 4; 20 October 1800, Cope Diaries, Haverford College; Benjamin Henry Latrobe, Jr. to McAlpine, 31 August 1876, Mrs. Gamble Latrobe Collection, MdHS.

29. [BHL], [Report to the Select and Common Councils of Philadelphia], 12 July 1799, *Gazette* (Philadelphia), 20 July 1799; Watering Comm., *Report* (1799), pp. 25–26.

30. BHL, "Designs of Buildings...in Philadelphia," 1799, Hist. Soc. Pa.; Watering Comm., *Report* (1799), p. 27.

Fig. 13. Schuylkill Engine House, c. 1830. View to northwest. At this time the building contained a porcelain factory. (John F. Watson, *Annals of Philadelphia and Pennsylvania*, 2 vols. [Philadelphia, 1856–57], 2: facing p. 456.)
Courtesy of Eleutherian Mills Historical Library

Fig. 14. Brick conduit of the Philadelphia Waterworks, 1975. An inside view of the remaining segment of the 6-foot diameter conduit under Chestnut Street between Broad and 15th Streets.
Karin E. Peterson

The Schuylkill Engine House was a square, stone building designed to contain both the engine for pumping water up the well and the rolling and slitting mill which used the "extra power" of the engine. Latrobe's knowledge of English steam-powered rolling mills may have influenced the design of the building.[31] The engine house was in constant use until September 1815, when the new works at Fairmount started to supply the city with water. The Schuylkill engine was first advertised for sale in 1816, but apparently found no purchaser until after 1818. From about 1825 to 1831 William Tucker rented the building from the city and manufactured porcelain there (see figure 13), and later in the 1830s the structure was torn down.[32]

A brick conduit six feet in diameter connected the Schuylkill and Centre Square engine houses. The mason, Thomas Vickers, built it in 1799 with hard bricks laid in the pattern of American common bond and united by a hydraulic cement. From the pump well of the Schuylkill Engine House the conduit ran a few feet south to Chestnut Street, then turned east toward Broad Street. Between 21st and 22nd streets it crossed a gulley on an aqueduct of three arches, but otherwise was three feet below ground level. At Broad the conduit turned north to the pump well of the Centre Square Engine House.[33]

The brick conduit was so excellently constructed that there is no record that it had any repairs during its first fifteen years of service. After the waterworks were moved to Fairmount the conduit may have been abandoned, but by 1905 sections of it were in use again as the Chestnut Street sewer.[34] By 1975 all of the conduit had been destroyed except for a segment under Chestnut between 16th and Broad streets. A photograph of the interior shows that at that date the brickwork was still in excellent condition, and that the perfect curve of the wall was interrupted only by an accumulation of debris on the bottom (see figure 14).

31. For a description of contemporary English steam rolling mills, see John Farey, *A Treatise on the Steam Engine, Historical, Practical, and Descriptive* (London, 1827), p. 507.
32. Watering Committee, *Report of the Watering Committee to the Select and Common Councils, 14 January 1819* (Philadelphia, 1819), p. 3; Greville Bathe and Dorothy Bathe, *Oliver Evans* (Philadelphia: Historical Society of Pennsylvania, 1935), "additional notes," following p. 282. For information on the fate of the engine houses after 1820, I am indebted to the research of Karin E. Peterson.
33. [BHL], [Report to the Select and Common Councils]; Watering Comm., *Report* (1799), p. 27; BHL, "The Waterworks"; Harley J. McKee, *Introduction to Early American Masonry* (Washington: National Trust for Historic Preservation, 1973), pp. 50, 52.
34. John C. Trautwine, Jr., "The Water Supply of Philadelphia," *Journal of American Water Works* 22 (1905): 422–23.

Fig. 15. Centre Square Engine House, 1812. ("The Water Works," *Port Folio* 8 [1812]:30–31.)

The Centre Square Engine House, which received water from the brick conduit, was located on one of the original five public squares planned for the city of Philadelphia. Standing at the intersection of the city's two major streets, the building was appropriately a grand edifice of marble sixty-two feet high. Latrobe designed it to have a domed central cylinder containing the steam engine and pump, surrounded at its base by a rectangular block containing the waterworks offices. Architecturally the engine house was a strong expression of Latrobe's style.[35]

When the new waterworks was erected at Fairmount, Centre Square remained important for water supply only because it was the originating point for the network of pipes which carried water to the customers. The steam engine was sold in 1818, and the building became a watch house for the western part of the city, although for a nominal sum the American Philosophical Society rented the roof and two rooms "for astronomical purposes." In 1827 the engine house was demolished.[36]

Within the Centre Square Engine House, water from the brick conduit was pumped up to wooden reservoirs inside the dome, and from there it flowed down to an iron distributing chest in the ground just outside the eastern portico. The chest had a capacity of about 250 gallons of water and brass cocks which admitted water to several six-inch wooden pipes. Each pipe served as a main for a major east–west street in the city. Virtually all of the pipe used during the first twenty years of the Philadelphia Waterworks' operation was wooden pipe bored from logs brought down the Schuylkill and Delaware rivers.[37]

Part of Latrobe's original proposal for a water system was the free distribution of water to the citizens through street pumps, and from the beginning of operation the waterworks provided that service. The resulting problem of waste inevitably associated with "free" water made the city's Watering Committee search constantly for a pump which would conserve water and encourage citizens to install water taps in their homes for an annual fee. Commercial and business establishments had been important paying customers from the first.[38]

Within a few years of the introduction of Schuylkill water there was little doubt that everyone preferred it to well water, and it became a valued part of Philadelphia's amenities.[39] Contaminated wells no longer had to be relied on for drinking water. Fire companies came to depend upon the system of hydrants lining the streets. The decline of yellow fever epidemics, the last of which occurred in 1822, was credited to the introduction of Schuylkill water.[40]

The waterworks was also instrumental in establishing Philadelphia as the early steam capital of the United States. Steam engineering in America had

---

35. See fig. 15 and pp. 173–80 below and the appropriate sections of Charles E. Brownell, ed., *The Architectural Drawings of Benjamin Henry Latrobe* (New Haven: Yale University Press, forthcoming).

36. Watering Comm., *Report* (1819), p. 3; Thomas Wilson, *Picture of Philadelphia, for 1824....* (Philadelphia, 1823), p. 26.

37. 7 July 1800, Cope Diaries, Haverford College; BHL, "The Waterworks." There are many references to wooden pipes in Philadelphia Waterworks papers and publications. The first trial of iron pipe was in 1801, and some was laid in Water Street in 1804, yet it was not widely used until the 1820s.

38. BHL, *View*, p. 12; *Report of the Watering Committee to the Select & Common Councils, November 1, 1803* (Philadelphia, 1803), p. 4; Watering Comm., *Report* (1801), p. 7; *Report of the Watering Committee to the Select and Common Councils, October 7th, 1805* (Philadelphia, 1806), p. 5. Also see "The First Subscribers for the City Water from Centre Square Water Works," 1801–6, Am 3055, Hist. Soc. Pa.; and pp. 196–97 below.

39. See comments in *Aurora, General Advertiser* (Philadelphia), 24 September 1805; and 1 June 1807, Cope Diaries, Haverford College.

40. Eugene S. Ferguson, ed., *Early Engineering Reminiscences (1815–1840) of George Escol Sellers* (Washington: Smithsonian Institution, 1965), pp. 4, 5; Mease, *Philadelphia*, p. 150. The introduction of piped water could not have had a major effect on the severity of yellow fever epidemics, although it may have reduced the number of rainwater cisterns and shallow wells, which were breeding grounds for mosquitoes, the yellow fever vector. See Latrobe Journals, 29 September 1819, PBHL, MdHS.

been in a rudimentary state.[41] When Latrobe went to Nicholas Roosevelt's Soho Works early in 1799 it was believed to be the only establishment in America capable of manufacturing the waterworks engines.[42]

Three years earlier Roosevelt had taken a lease on a nearby copper mine which had an old imported Newcomen engine. In order to smelt and refine the copper, and to repair the engine, Roosevelt erected a foundry and machine shop which he named *Soho* after Boulton and Watt's famous engine works in Birmingham, England.[43] The principal buildings of the American Soho were a large smith's shop, several furnaces, three machine shops, and two boring mills.[44]

Since Roosevelt was not skilled in the foundry or machinist trades, he hired a number of men with European experience to erect and operate his works. James Smallman, who had personally erected a Boulton and Watt engine in Holland, was the most important of the European technicians since he was the only practical steam engineer.[45] Others were John Hewitt, a pattern-maker who had worked for Boulton and Watt, Frederick Rhode, a "German master miner & Smelter," and Charles Stoudinger, who before emigrating from Germany must have had experience in drawing since he became the chief draftsman under Smallman. A mixed group of European and American miners, smelters, and craftsmen filled the lesser positions at Soho.[46]

At what point Roosevelt decided to engage in steam engine manufacture is unknown, but as early as 1797 he made plans for supplying an engine to a coal mine in Virginia and discussed the construction of a steamboat with Robert Livingston and John Stevens. Soon the latter project was under way. In 1798 Roosevelt and Smallman secured an American patent for a "double steam engine" (one deriving power from both strokes of the piston) essentially of the Boulton and Watt design.[47] When Latrobe went to Soho in January 1799 he found there a group of workmen who were prepared for the challenge of making the waterworks' engines.

Latrobe reported favorably to the Philadelphia city councils on what he had found at Soho and presented them with a draft of a conditional contract for two engines from Roosevelt. Negotiations for a final agreement began almost immediately, and the manufacturing of engine parts was soon under way.[48]

The engines designed and built by the Soho group were of the contemporary Boulton and Watt type. They had separate condensers, parallel motions linking the piston and working beam, and double-acting steam cylinders. However, they also had some features not commonly associated with Boulton and Watt pumping engines, including double-acting air pumps, wooden boilers, flywheels, and the substitution of wood for cast iron in many parts.[49]

The inexperience in engine construction of most of the Soho workmen and the lack of adequate machinery to conduct some operations caused delays in the manufacture of the engines for the Philadelphia Waterworks. The problem of boring the cylinder for the Schuylkill engine is an example. Thomas P. Cope, a member of the Watering Committee who visited Soho in July 1800 to report on the progress of the engines' manufacture, described the operation in detail:

> The parts of the Engines which are unfinished, & which it is of importance to have as early as possible I will now endeavor to mention—The first that presents itself to my mind is the *Great Cylinder* for the lower engine. This cylinder I discovered to be cast in 2 pieces, united by copper & secured externally by a strong iron band of cast iron 18 inches broad & weighing, as I am told, 1200 lbs. 7,500 lbs. of metal was used in casting the cylinder—when bored the workmen suppose it will weigh about 7000 lbs. exclusive of the band. It is rather more than $6\frac{1}{2}$ feet long, & will finish about $38\frac{1}{4}$ inches bore. It was cast hollow, & required but about $\frac{3}{4}$ of an inch to be bored to smooth it off for service. Only $\frac{1}{2}$ inch

---

41. Pursell, *Early Stationary Steam Engines in America*, chaps. 1, 2.
42. 23 January 1799, Minutes of the Select Council, Phil. Archives.
43. Pursell, *Early Stationary Steam Engines in America*, pp. 28–30.
44. 7 July 1800, Cope Diaries, Haverford College.
45. Eric H. Robinson, "The Early Diffusion of Steam Power," *Journal of Economic History* 34 (1974): 98, 99n; Boulton and Watt to James Smallman, 20 March 1794 and 1 May 1794, Foundry Letter Books, Birmingham Reference Library, Birmingham, England. This evidence solidly establishes Smallman's connection with Boulton and Watt. Cf. Pursell, *Early Stationary Steam Engines in America*, p. 38n.
46. Pursell, *Early Stationary Steam Engines in America*, pp. 29–30; Nicholas Roosevelt to Boulton & Watt, précis, 5 February 1800, Boulton Papers, Birmingham Reference Library, Birmingham, England; "Testimony of Richard Newsham & James Smallman," n.d., Woods Collection, Franklin Inst.; 7 June 1800, Cope Diaries, Haverford College; BHL, *American Copper Mines* (n.p., [1800]), p. 5.
47. Drawings in the Stoudinger Collection, New Jersey Historical Society, Newark, N.J.; Pursell, *Early Stationary Steam Engines in America*, pp. 30–31; *Letter from the Secretary of State Transmitting a List of All Patents Granted by the United States*, 21st Congress, 2nd Session, House Doc. 50 (Washington, 1831), pp. 360, 398.
48. 23 January 1799, 2 February 1799, Minutes of the Select Council, Phil. Archives; BHL, *Answer*, p. 3.
49. See the drawings and descriptions of the engines, pp. 157–64 and 181–88 below.

has been cut throughout its whole length. The process is excessively tardy. The cylinder rests on a movable frame & is bored by a bit fixed to the shaft of a large water wheel. The bit may be said to consist of several parts—in the first place a rod of iron is connected with the shaft of the wheel. This rod is about 8 inches diameter, at the extremity is a rim 6 inches thick, & a little less in circumference than the bore of the cylinder. This rim or head piece has 12 or 14 mortices on the edge, into which the cutters or steelings are inserted, & secured by wedges. No more than 3 of these cutters were inserted or used. They have an edge of about $2\frac{1}{2}$ inches. Two men attend the operation. One of them is employ[e]d in adjusting the cutters & almost lives in the Cylinder. The other attends the frame. These 2 are relieved by [an]-other 2 at night, so that the business goes on day and night. One other hand is constantly employed in the day time in grinding the cutters. When the men dine the boring stops, which they suppose is no disadvantage as it allows time for the cylinder to cool—constant boring makes it very hot, altho cold water is continually dripping on it. The boring ceases also when one set of cutters is removing & another fixing. So that of 24 hours not more than 11 are actually taken up in boring. About 10 minutes are consumed in replacing the cutters, & 10 minutes more dull them. I examined several & found that in that time, some parts of them were worn nearly an eighth of an inch by the operation. They are made of the best steel. The workmen say the boring will be compleated in 3 or 4 weeks—[I] very much doubt it, from present appearances.[50]

It is not surprising that the engines which Latrobe predicted would be ready in the summer of 1799 were not completely delivered to Philadelphia until the fall of the next year. In December 1800 the Schuylkill engine was tested, and on 27 January 1801 the Centre Square engine was operated with the mayor and members of the city councils present. Regular operation of the engines commenced in February.[51]

During the nearly fifteen years of their operation the original waterworks engines had many problems. Some parts were poorly made and had to be replaced at an early date; wooden elements rotted; the boilers consumed tremendous quantities of fuel.[52] The plan of the waterworks itself was criticized for depending upon two engines in series, so that if either engine did not work, water might be unavailable. A report of the Watering Committee in 1812 stated: "So sensible indeed are your committee of the radical defects of the present works, that they believe no alteration or improvement can render them efficient."[53] In that year plans were made for a new waterworks at Fairmount on the Schuylkill, and a contract was made with a Philadelphia foundry for a new Boulton and Watt type engine. It was with that engine that the Fairmount Works began operations on 7 September 1815. In 1817, an Oliver Evans high-pressure engine was installed along with the first engine, but neither was entirely satisfactory. In 1822 waterwheels were installed to take over all the pumping.[54]

In spite of the short life of the waterworks steam engines, they had a significant impact on Philadelphia's technological development. The operation and repair of the engines brought several men from Soho to the city,[55] including James Smallman, who later went into the engine-making business on his own.[56] Such activity, along with the establishment of at least two other engine-makers in the city, led to the assertion in 1808 that "there is no part of the world in proportion to its population, where a greater number of ingenious mechanicks may be found than in Philadelphia and its immediate vicinity. Steam engines with all their various improvements are built and

---

50. This is the original passage from Thomas P. Cope's "Notes taken on a journey to Soho & N York 1800," 8 July 1800, Cope Diaries, Haverford College. Frederic Graff (the son of Latrobe's pupil) apparently saw this diary prior to writing his article, "Notice of the Earliest Steam Engines used in the United States," *Journal of the Franklin Institute*, 3d series, 25 (1853):269-71. At that time he made the error of dating the passage four days earlier than it should have been. The Cope description as it appeared in Graff's article has since been quoted or paraphrased several times, apparently without knowledge that the original source existed.

51. 29 September 1800, 28 October 1800, 15 December 1800, 27 January 1801, 4 February 1801, Cope Diaries, Haverford College; *Gazette* (Philadelphia), 28 January 1801.

52. The printed annual reports of the Watering Committee, 1803-15, contain details of engine problems.

53. *Report of the Watering Committee to the Select and Common Councils, May 2, 1812* (Philadelphia, 1812), p. 3.

54. *Report of the Watering Committee to the Select and Common Councils, November 5th, 1812* (Philadelphia, 1812), p. 4; contract for steam engine from Eagle Foundry, 3 October 1812, Watering Committee Papers, Phil. Archives; *Report of the Watering Committee to the Select and Common Councils Read January 25, 1816* (Philadelphia, 1816), p. 3; Blake, *Water for the Cities*, pp. 78-87.

55. Richard Newsham (a pattern-maker), F. I. Schneider, John Brown, and John Vanreiper were part of this group. See "Testimony of Richard Newsham & James Smallman," n.d., Woods Collection, Franklin Inst.; receipt, Newsham to Waterworks, 3 April 1804, Watering Committee bills received, Phil. Archives; BHL to Smallman, 6 February 1804, Letterbooks; "N. J. Roosevelt account with James Smallman settled the 4 of April 1806," Watering Committee Papers, Phil. Archives.

56. *Report of the Watering Committee to the Select & Common Councils, November 13th, 1807* (Philadelphia, 1807), accounts; *Report of the Watering Committee to the Select & Common Councils, November 2d, 1809* (Philadelphia, 1809), accounts; Pursell, *Early Stationary Steam Engines in America*, pp. 42-44.

applied, beneficially, to the most useful purposes."[57] Philadelphia was soon joined by Pittsburgh and New York as centers of steam engine manufacturing, but it remained important in that field for many years.

Latrobe's Philadelphia Waterworks had far-reaching effects not only on the health of one city but on the course of technological development in the United States.

New Orleans Waterworks

After the United States took possession of the Louisiana Territory in 1803, the city of New Orleans was for the first time open without restriction to American commerce. As the natural emporium for all Mississippi River goods needing to be transferred to ocean craft, New Orleans immediately experienced growth, and its economic advantages buoyed speculation about its future. The census of 1810 enumerated 17,000 residents, making it one of the largest southern cities.

However, like most other American urban centers of the time, New Orleans lacked a water supply system. Wells in the mucky soil yielded clear but unpalatable water, and all drinking water was obtained from the Mississippi. Wealthy families had their servants procure it, while most residents purchased it from street vendors. Although sufficient supplies of water for drinking were available by this means, it was inconvenient and expensive. Moreover, neither the wells nor the water carriers were capable of providing sufficient water for fire fighting or washing the streets (the latter service highly desirable in an era when horse manure and garbage paved American streets more commonly than gravel or cobblestone).[58]

It seemed unlikely that the extremely conservative New Orleans community would erect a municipal water system, especially since there was no sense of urgency regarding the almost annual yellow fever epidemics.[59] Yet the high cost of water and the city's rapid expansion created a vision of private profit. In 1807, only four years after Louisiana had become part of the United States, several persons discussed the likelihood that a franchise for a New Orleans waterworks would yield a fortune. One promoter, the governor of Louisiana, W. C. C. Claiborne, mentioned the idea to President Jefferson, who passed it along to Latrobe. Latrobe did not act on the suggestion for two years, although meanwhile he learned a great deal about the city from Robert Alexander, who went to New Orleans to erect the United States Customs House designed by Latrobe.[60]

Sometime in the late summer or fall of 1809 Latrobe and Alexander combined to petition the territorial legislature of Louisiana for a waterworks franchise. Latrobe told his father-in-law that the enterprise might be both safe and lucrative.

> ... in conjunction with a Mr. Alexander I have petitioned the Legislature of N. Orleans for an exclusive right in supplying the city with Water. Nothing is there so much wanted, & at the same time nothing is so easily & cheaply accomplished. The first expense, before we begin to receive any compensation I estimate at 30.000$,—but to cover errors Say 50.000. There are now 2.000 Houses which would take the Water,—as Mr. Alexander says, at 20$ a Year, but we will say,—10$ P annum, or 20.000$ a Year. Mr. Poydras the representative from N. O. an old French miser, worth 75.000$ P Annum, to whom I have had occasion to be useful, says that we shall assuredly succeed, & be the richest Men in America. But I hope for nothing, & therefore shall not feel disappointed.[61]

In the spring of 1810 his application was rejected, but he reapplied almost immediately. To support his new effort he sent to New Orleans his son Henry, who was only eighteen years old but had already begun an engineering education under his father and spoke French fluently. Latrobe believed that Henry would have greater success with the Creole legislators and city council members than Alexander, who spoke only English.[62]

Henry arrived in New Orleans early in 1811 and quickly became acquainted with the important members of the New Orleans city council. His father afterward related Henry's experience.

---

57. Thomas Green Fessenden, *The Register of Arts, or A Compendium View of Some of the Most Useful Discoveries and Inventions*, quoted in Pursell, *Early Stationary Steam Engines in America*, pp. 40, 42.
58. BHL to Fulton, 31 July 1811, Letterbooks; John Duffy, ed., *The Rudolph Matas History of Medicine in Louisiana* (Baton Rouge, La.: Louisiana State University Press, 1958), 1:406–8.
59. John Duffy, "Nineteenth Century Public Health in New York and New Orleans," *Louisiana History* 15 (1974): 331–33; Constance McLaughlin Green, *American Cities in the Growth of the Nation* (New York: Harper & Row, 1965), pp. 73–74.

60. R. Claiborne to Evans, 24 August 1807, Woods Collection, Franklin Inst.; BHL to Bollman, 11 August 1811, Letterbooks; Samuel Wilson, Jr., ed., *Impressions Respecting New Orleans by Benjamin Henry Boneval Latrobe: Diary & Sketches, 1818–1820* (New York: Columbia University Press, 1951), p. xiv; Hamlin, *Latrobe*, pp. 295–96, 354.
61. BHL to Hazlehurst, 17 December 1809, Letterbooks.
62. BHL to Alexander, 14 October 1810, Letterbooks.

The City councils consist of three Americans to thirteen French. The Men of the most influence were Blanque, Fortier, Boret and others of the same *french manners & principles*. With these it was necessary to be on good terms, to frequent their clubs, & in every innocent respect to enter into their habits. The facility with which Henry speaks french rendered all this easy & agreeable to him.[63]

The city council responded to this attention on 27 April 1811 by granting Latrobe the exclusive privilege of supplying New Orleans with water. To comply with the ordinance he had to commence distribution to residences by 1 May 1813, at which time 2,000 feet of pipe had to be laid. Water for the hospital, for public buildings, for fires, and for watering the streets had to be supplied free. In return Latrobe received the franchise for fifteen years with an optional five-year extension, as well as the unrestricted right to determine his own price for water. Moreover, the city agreed to donate the land necessary for the engine house site.[64]

Latrobe's plan for the New Orleans Waterworks was in conception similar to the Philadelphia Waterworks, but in detail somewhat different. The unstable soil of the area would not allow a tunnel or even a canal as a means of bringing river water to an engine house, so Latrobe planned for an iron suction pipe to extend into the river. A steam engine of moderate size would draw water from the river and force it into huge wooden casks or reservoirs elevated above the streets. Both the engine and the reservoirs would be in an engine house close to the river. From the reservoirs the water would flow by gravity through a system of wooden pipes to the consumers.[65]

Having obtained the charter, Henry returned to Washington to discuss the situation with his father and inform him of the arrangements he had made. He remained there throughout the busy fall of 1811 while his father sought financial support and arranged for the manufacture of equipment. Henry returned to New Orleans in late February 1812.[66] Within three weeks the city council had granted a lot for the engine house, which Latrobe promised would be "an ornament to Your city, as well as the means of salubrity & pleasure."[67]

Henry immediately set to work on its construction, and finished it within the year. His father at first complained that Henry had built it stronger than was necessary, but in 1813 he proudly related that "it was the only building in the city which did not suffer by the [recent] Hurricane, while everything around it was leveled with the Ground."[68] Unfortunately, lack of capital and the War of 1812 kept Henry from completing any other part of the waterworks before he died of yellow fever at New Orleans on 3 September 1817.[69]

The burden of providing an engine and accessories such as the suction pipe and pump had always been with the elder Latrobe in Washington. At first he had asked his old friend James Smallman, who was engineer of the Philadelphia Waterworks and Washington Navy Yard engines, to take a preliminary order for one engine. But on 10 August 1811, within a month of making the first arrangement, Latrobe cancelled the order due to lack of funds.[70] He decided to manufacture much of the engine himself.

In a building in Washington which he owned he set up a small foundry and hired several craftsmen.[71] He employed two smiths unsuccessfully before he found one who satisfied him,[72] and not until the summer of 1812 did he find a competent machinist, William Robinson, to be his foreman. Latrobe told Henry how valuable an acquisition Robinson was:

> He is a good Smith, an excellent Turner in Brass, Iron & wood, and perfectly master of all the English improvements of the last 4 or 5 years, which are many & important. He is very Silent, keeps the men at a proper distance, & sees that they do their duty. I give him $2.50 P day, & have promised him 100$ when the Engine starts & performs well.[73]

---

63. BHL to Mark, 29 August 1811, Letterbooks. The club, or coffeehouse, long remained a prominent feature of male society in New Orleans. See [Joseph H. Ingraham], *The South-West By a Yankee*, quoted in Charles N. Glaab, ed., *The American City: A Documentary History* (Homewood, Ill.: The Dorsey Press, 1963), pp. 101–3.
64. "Ordonnance pour la fourniture de l'eau dans la Ville et les Faubourgs," 27 April 1811, BHL Papers, Tulane; BHL to Bollman, 29 July 1811, Letterbooks. Apparently the charter from the city was sufficient authority for establishing a franchise, since no act of the state legislature confirmed the grant to BHL. BHL to Graf, 21 August 1812, Letterbooks (this is Baltimore merchant Frederick C. Graf, who should be distinguished from the engineer Frederick Graff, BHL's pupil).
65. As no descriptions of the planned or completed waterworks are extant, only the general features of the design are clear. See p. 200 below for Henry's drawing of the original plan, c. 1811.
66. BHL to Bollman, 2 February 1812, Letterbooks; Mather to President and Members of City Council, 29 February 1812, Louisiana Division, New Orleans Pub. Lib.
67. Mather to President and Members of City Council, 14 March 1812, Louisiana Division, New Orleans Pub. Lib.; BHL to Mather, 17 September 1811, BHL Papers, Tulane.
68. BHL to Graf, 8 April 1812, Letterbooks; BHL to Henry Latrobe, 1 August 1812, Letterbooks; BHL to Graf, 6 March 1813, Letterbooks.
69. For brief sketches of Henry Latrobe's activities in the interval, see Hamlin, *Latrobe*, pp. 429, 472–73, and Wilson, ed., *Impressions Respecting New Orleans*, pp. xxi–xxii.
70. BHL to Smallman, 6 July 1811, 22 July 1811, and 10 August 1811, Letterbooks.
71. BHL to Mark, 14 September 1811, 20 August 1811, and 14 November 1811, Letterbooks.
72. BHL to Patton & Butcher, 10 November 1811, Letterbooks; BHL to Henry Latrobe, 4 August 1812 and 6 September 1812, Letterbooks.
73. BHL to Henry Latrobe, 6 September 1812, Letterbooks.

Another important workman was Jonathan Criddle, a pattern-maker whom Latrobe borrowed from the Navy Yard.[74]

With this group of workmen Latrobe set about making a standard Boulton and Watt style low-pressure steam engine with a 24-inch cylinder.[75] From his drawings the men cast and forged all of the smaller parts of the engine, including the working gear, pillow blocks, sun-and-planet gears, and nuts and bolts. Twenty-three cases of such items were sent to New Orleans in June 1812.[76]

Latrobe made contracts with several iron manufacturers for larger and more difficult parts. The Bishop and Malin rolling mill near Chester, Pennsylvania, made the wrought-iron plate for two boilers. Half the plate for each boiler came by land, but the rest was lost when the ship carrying it sank near Cape Henry, Virginia.[77] A number of castings, including the flywheel, were made at the Antietam Furnace in Maryland.[78] John Davis, formerly Latrobe's clerk of works for the Philadelphia Waterworks, made a pattern for the iron suction pipes and a casting for the flywheel gudgeon at the Baltimore Waterworks foundry.[79] Latrobe was disappointed when the Principio Furnace in Maryland attempted but was unable to cast the 16- and 18-inch pipe he wanted.[80] The owner of Principio also declined to cast the steam cylinder, air pump, condenser, and water pump, so Latrobe let the work to Henry Foxall who formerly operated the Eagle Works in Philadelphia but had now located his foundry along the Potomac north of Georgetown.[81] Foxall's cylinders were finished after the War of 1812 began and Latrobe decided not to risk sending them by sea.[82]

Thus, for the considerable effort which Latrobe expended in finding men and contractors who could make the parts for the waterworks engine, Latrobe actually received little but frustration. In fact, the war ended Latrobe's hope for manufacturing the engine in Washington and its vicinity, since a British blockade severely limited coastwise shipping, and overland transportation was prohibitive for items the size and weight of engine parts.

Latrobe nonetheless continued the operations of the forge and machine shop he had erected. He worked on a small engine (14-inch cylinder) which Henry had ordered for a friend's sugar mill in Louisiana. Although the war made it impossible to deliver the engine, Latrobe completed it in hope of finding another purchaser. He sold it in the spring of 1813 for a Baltimore gristmill owned by William Hartshorne, but the engine did not work properly and he received only partial payment for it. Over three years later he was still trying to arrange resale of the engine to cover his losses.[83]

Latrobe accepted orders for many smaller items as well, including equipment for the powder mill which Thomas Ewell built near Washington.[84] This source of income and the termination of his appointment as surveyor of the public buildings in 1812 put Latrobe in an ironic situation which he explained to his father-in-law.

> There is one discovery that I have made in the course of my leisure from the cares & honors of my public situation. It is this, that altho' a life spent in the study & practise of my art will not make my fortune, I could easily make it as a blacksmith. There are few days [which] pass in which the honor of shoeing horses by the Year or Quarter is not offered to me, and my neighbors the printers find me a much cleverer fellow at mending their presses, than I was considered by congress in building the Capitol. Even the tavernkeeper near me, who never could tell why they should employ a man at a large Salary to build for them who never had worked at the bench, has found out that I can be very useful at mending his pokers & gridirons. Thus by keeping a Blacksmith shop, I have actually maintained my family for some Months.[85]

It was not until the fall of 1813, after receiving an extension of his franchise from the New Orleans city council, that Latrobe was able to return to the

---

74. BHL to Henry Latrobe, 30 January 1812, Letterbooks.
75. BHL to Hughes, 12 September 1811, Letterbooks.
76. BHL to Henry Latrobe, 17 May 1812, Letterbooks; BHL to Fulton, 5 July 1812, Letterbooks; BHL to Mark, 15 August 1812, Letterbooks.
77. BHL to Bishop & Malin, 15 September 1811 and 12 January 1812, Letterbooks; BHL to Henry Latrobe, 22 January 1812, Letterbooks; Ashmead, *Delaware County*, p. 678.
78. BHL to McPherson, 3 December 1811, Letterbooks; BHL to Henry Latrobe, 22 March 1812, Letterbooks; BHL to McPherson, 18 June 1812, Letterbooks.
79. BHL to Davis, 19 August 1811, Letterbooks; BHL to Hughes, 1 October 1811, Letterbooks; BHL to Davis, 8 February 1812, Letterbooks.
80. BHL to Hughes, 21 December 1811, Letterbooks.
81. BHL to Hughes, 20 September 1811, Letterbooks; BHL to Fulton, 5 July 1812, Letterbooks.
82. BHL to Fulton, 5 July 1812, Letterbooks. There is no record that BHL made any shipment from Washington to New Orleans after the war began.

83. BHL to Davis, 10 June 1814, Letterbooks; BHL to Graf, 24 November 1816, Letterbooks. There is an extensive correspondence concerning the Hartshorne engine in the Latrobe Letterbooks.
84. BHL to Hamilton, 12 June 1812, Letterbooks; BHL to Henry Latrobe, 30 December 1812, Letterbooks.
85. BHL to Hazlehurst, 13 December 1812, Letterbooks. Congress voted no appropriation for BHL's salary in 1812.

problem of providing a steam engine for the New Orleans Waterworks. By then he had transferred ownership of his Washington shop to a local millwright and had decided to join with Robert Fulton to superintend the construction of boats in Pittsburgh for the Ohio Steamboat Company.[86] In Pittsburgh he established an extensive machine shop in his shipyard, and reported to Henry that he was making new patterns for the waterworks engine.[87] The work went slowly. Latrobe encountered delays because of financial trouble, although in August 1814 he was cautiously optimistic about completing the engine. He had already gathered sheet iron for the boiler, and a local founder would cast the cylinder.[88]

Then, however, the Ohio Steamboat Company collapsed, and Latrobe lacked the capital and credit necessary to complete the waterworks engine.[89] In desperate straits, Latrobe even attempted to make an agreement with George Evans (the son of Philadelphia millwright Oliver Evans) and his partners, whose Pittsburgh Steam Engine Company manufactured high-pressure engines. Latrobe offered the company a combination of cash and stock for an engine, but the British presence in the New Orleans vicinity at the end of 1814 deterred them from investing "till the fate of that section of the Union is decided."[90] Latrobe still hoped to get an engine from Evans, especially once the news of the Peace of Ghent reached Pittsburgh,[91] but his recall to Washington in March 1815 made him look in other directions.

In August, when he knew that his application to the New Orleans city council for a second extension of time (to 1 August 1816) was granted, Latrobe wrote to Robert McQueen, an engine manufacturer in New York, and contracted for a new engine.[92] The work went slowly, partly because Latrobe could not supply McQueen with adequate cash advances; in late winter the work was cancelled.[93] Nicholas Roosevelt then located a small engine (with an 18-inch cylinder) in New York which Latrobe purchased even though it was not particularly appropriate for pumping water. It was shipped to New Orleans on 2 June 1816.[94]

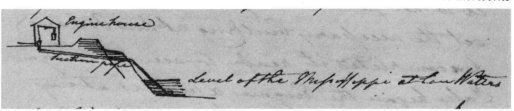

Fig. 16. Proposed design of the New Orleans Waterworks, 1815. The city of New Orleans is to the left of the engine house, and the levee is to the right. (BHL to McQueen, 13 August 1815, Letterbooks.)

Latrobe wanted to send an experienced steam engineer to New Orleans with the engine, and arranged through his New York business associate Jacob Mark for a man named Noble to make the trip. Noble, however, refused to go when news of a May flood in New Orleans reached New York, and he became convinced that the city would be unhealthy through the rest of the summer. Two other engineers whom Latrobe contacted also declined his offers, and all hope of meeting the August deadline faded.[95]

The New Orleans city council threatened to revoke the waterworks charter, and Henry despaired of any possibility of continued official support. Then, on 28 September 1816, a fire gutted a large building for which Henry had a commission, and he lost the hope of a substantial profit which might have been applied to the waterworks.[96] For a year the Latrobes waited for better fortunes, each day expecting the city council to revoke their charter. Then came an even worse tragedy: in the first serious yellow fever epidemic since his arrival in New Orleans, Henry Latrobe died.[97]

Instead of ending the New Orleans affair, however, Henry's death only temporarily checked Latrobe's interest in New Orleans. With his renewed position as architect of the U.S. Capitol again in jeopardy, and personal insolvency imminent, Latrobe turned his personal attentions to the waterworks as his best hope for his family's security. In mid-October he wrote to the city council asking for a new extension of his privilege and promised that

---

86. Hamlin, *Latrobe*, pp. 377–79, 381.
87. BHL to Henry Latrobe, 8 January 1814, Letterbooks; BHL to Stackpole, 24 May 1814, Letterbooks.
88. BHL to Henry Latrobe, 6 August 1814 and 24 September 1814, Letterbooks.
89. Hamlin, *Latrobe*, pp. 429–32.
90. BHL to Henry Latrobe, 13 November 1814, Letterbooks; "Proposition to the Pittsburg St. En. Company on the Subject of the New Orleans Waterworks," 23 November 1814, Letterbooks; Evans to BHL, 15 December 1814, private collection.
91. BHL to Henry Latrobe, 23 February 1815, Letterbooks.
92. BHL to Graf, 2 August 1816, Letterbooks; BHL to McQueen, 13 August 1816, Letterbooks.
93. BHL to Mark, 1 December 1815 and 27 January 1816, Letterbooks.
94. BHL to Mark, 31 March 1816, Letterbooks; BHL to Harper, 9 May 1816, Mrs. Gamble Latrobe Collection, MdHS; BHL to Henry Latrobe, 6 June 1816, Letterbooks.
95. Noble to BHL, 11 June 1816, Letterbooks; BHL to Henry Latrobe, 16 June 1816 and 1 July 1816, Letterbooks.
96. 20 July 1816, New Orleans City Council Proceedings, Louisiana Division, New Orleans Pub. Lib.; BHL to Henry Latrobe, 5 August 1816, Letterbooks; Hamlin, *Latrobe*, pp. 472–73.
97. BHL to Henry Latrobe, 22 July 1817 and 7 August 1817, Letterbooks; BHL to Mayor and Council of the City of New Orleans, 13 October 1817, Letterbooks; Duffy, *Medicine in Louisiana*, 1:346.

he would come to the scene and personally direct operations. In spite of the reservations of the mayor, the council once again granted an extension, this time until 1 January 1819.[98]

During 1818, from early January when he heard of the extension until mid-December when he left for New Orleans, Latrobe sought to make final arrangements for the waterworks from his new home in Baltimore. He approached various potential supporters and found that only one, Colonel Benjamin G. Orr, would provide ready cash.[99] With that money he purchased and sent three separate shipments of equipment to New Orleans, including a pipe-boring mill, fire bricks (for the boiler foundation), cast-iron pipes, and other articles. He also finally sent on an engineer, Andrew Coulter, who when he arrived in New Orleans found that the engine sent two years before had seriously deteriorated. Rather than expend time and money repairing that engine, Latrobe bought in Baltimore a used engine with "double the power" of the one in New Orleans.[100]

Latrobe left Baltimore for New Orleans in mid-December, and as soon as he arrived he petitioned city council for another (the fourth) extension of time to complete the waterworks. A special committee of council met with him for almost two months to draft a new ordinance, and on 6 March 1819 a detailed agreement became law.[101] In the meantime Latrobe was busy making preparations for further work by setting up the boring mill behind the engine house and experimenting with a method of paving the streets. The spongy soil of the city, which instantly filled trenches with water to within a few inches of the surface, must have made Latrobe think seriously about the problem of restoring the streets once they had been dug up for pipes.[102]

As soon as the new agreement was made, Latrobe became a whirlwind of activity. He asked city council to complete its payments on stock it had begun buying several years before, and advertised the sale of more stock to the public at $500 a share. He also announced that he wanted contracts for 35,000 feet of round pine or cypress logs in diameters varying from 12 to 18 inches. Apparently he had his boring mill in operation by July because he then began laying pipe in Levee Street, the major artery of the city paralleling the waterfront.[103]

The new steam engine had arrived on 20 March and a gang of slaves owned by the city carried it from the ship to the engine house. Immediately Latrobe set about erecting the engine, assisted by Coulter, a group of smiths who came from Baltimore at Latrobe's urging, and some local craftsmen. The erection went painfully slowly, but Latrobe reported in mid-June that little further work remained, and it seems likely that the engine first operated in July to drive the boring mill.[104]

However, the yellow fever season of 1819 was severe and after early July very little was done toward completion of the waterworks. On 3 July Latrobe employed thirty-two men, but by 17 September only five of them were still working. Ten men died, six were sick or recovering, and eleven had left the city or quit working for Latrobe. He hired only two new men as replacements.[105] Coincidentally, Latrobe himself sustained a debilitating attack of a disease (not yellow fever) and spent most of August and early September in bed. Since it was unlikely that the city and waterworks operations could return to normal for some months, Latrobe returned to Baltimore in October to arrange for his family to move to New Orleans. In his absence Coulter took charge of

---

98. Hamlin, *Latrobe*, pp. 473–78; BHL to Mayor and Council of the City of New Orleans, 13 October 1817, BHL Papers, Tulane; Rogers to Mayor and Council of the City of New Orleans, 15 November 1817, BHL Papers, Tulane; 26 November 1817, New Orleans City Council Proceedings, Louisiana Division, New Orleans Pub. Lib.; Macarty to [City Council], 29 November 1817, Louisiana Division, New Orleans Pub. Lib.

99. BHL, "No. II. Statement of the funds from which the Waterworks of New Orleans have been executed," 16 June 1819, BHL Papers, Tulane.

100. BHL to Mayor and Council of the City of New Orleans, 6 June 1818 and 9 October 1818, private collection. It seems likely that at this time he shipped the cast-iron pipes purchased at New York in 1816. He used iron pipes for the suction pipe laid in 1820, but their origin is not mentioned. BHL to McQueen, 12 June 1816, Letterbooks.

101. BHL to Mayor and Council of New Orleans, 14 January 1819, private collection; 16 January 1819, 30 January 1819, 9 February 1819, New Orleans City Council Proceedings, Louisiana Division, New Orleans Pub. Lib.; Macarty to [City Council], 4 March 1819, Louisiana Division, New Orleans Pub. Lib.; BHL, "No. II. Statement of the funds from which the Waterworks of New Orleans have been executed," 16 June 1819, BHL Papers, Tulane.

102. Latrobe Journals, 8 March 1819, PBHL, MdHS; 30 January 1819, New Orleans City Council Proceedings, Louisiana Division, New Orleans Pub. Lib.

103. BHL to Mayor and Council of the City of New Orleans, 14 January 1819, private collection; *Louisiana Courier* (New Orleans), 24 March 1819, 2 June 1819; 3 July 1819, New Orleans City Council Proceedings, Louisiana Division, New Orleans Pub. Lib.

104. BHL to Mayor and Council of the City of New Orleans, 20 March 1819, Louisiana Division, New Orleans Pub. Lib.; 20 March 1819, New Orleans City Council Proceedings, Louisiana Division, New Orleans Pub. Lib.; BHL to Mary E. Latrobe, 17 April 1819 and 18 May 1819, Mrs. Gamble Latrobe Collection, MdHS; BHL, "State of the Water Works," 16 June 1819, BHL Papers, Tulane.

105. "Statement respecting the Workmen that have been employed at the Waterworks," enclosed with BHL to Mayor and Council of the City of New Orleans, 17 September 1819, BHL Papers, Tulane; Duffy, "Nineteenth Century Public Health," p. 333.

the preparations and the first board of directors of the Water Works Company of New Orleans was elected.[106]

Latrobe spent considerable time preparing for a complete move to New Orleans: he tried to finish all work relating to the Exchange and Cathedral in Baltimore, and took his family on a trip to West Point to see his son John (a new cadet), stopping to see family, friends, and business associates along the way. Finally the family packed their belongings and left for New Orleans in mid-January 1820. They traveled overland to the upper Ohio River and took a steamboat from there to their new home, not arriving until the first week of April.[107]

Latrobe was determined to complete the waterworks by the end of the summer of 1820. As usual, he barely had enough money: only his income from architectural commissions and loans from a bank kept the enterprise going. By July he had bored and laid over 5,000 feet of wooden pipes, and in August he began laying the iron pipe through the levee and into the river. The city's gang of slaves again assisted him, and the work was essentially completed at the end of the month.[108] Then the ultimate tragedy struck: while overseeing the remaining work, Latrobe contracted yellow fever. With the suction pipe laid, the engine in operation, pipes in the streets, and (as his wife later claimed) water actually flowing through the system, Benjamin Henry Latrobe died on 3 September 1820.[109] With its superintendent and guiding spirit gone, all of the creditors of the Water Works Company panicked and sued or put liens on the property. The work stopped and Latrobe's family was left penniless. Fortunately they soon escaped the financial trials by returning to friends in Baltimore.[110]

The Water Works Company struggled unsuccessfully to obtain new financing, and in June 1822 the city purchased all of its assets. In the latter half of 1822 Andrew Coulter restored the engine, which had been neglected for two years, and the waterworks finally began to provide water for the city. Latrobe's decade-long dream of supplying New Orleans with Mississippi water had finally come true after his death. For eighteen years the Latrobe-designed system operated, until in 1840 a new and larger system, planned by Albert Stein, replaced it.[111]

4. RIVER CONTROL

Many of the major achievements of engineering in the early modern period were in the control of rivers to reduce flooding and improve navigation.[1] Some of Latrobe's earliest engineering experiences were in that area, and he had read extensively in the subject. In America, however, there was little call for the services of a professional engineer to control rivers, and Latrobe seldom had the opportunity to exercise his understanding. None of the projects which he planned was carried out, and only in one instance (on the Potomac River) was his advice of some advantage to the public. Yet his approach to river control is interesting because it is an example of the solutions offered by a European-trained engineer facing the problems of a recently exploited continent.

Mud Island Bar

In May 1807 the Chamber of Commerce of Philadelphia requested Latrobe's opinion on how to control or reduce a shoal accumulating north of Mud Island (sometimes known as Fort Island) in the Delaware River.[2] The shoal threatened to close up an important passageway for the larger vessels coming to Philadelphia, and the city's merchants were concerned.

Latrobe's response was to inform the chamber that there was no artificial way to eliminate the shoal because massive and irreversible forces of nature were at work. The shoal, Mud Island, and a whole series of islands in the lower

---

106. BHL to Mayor and Council of the City of New Orleans, 17 September 1819, BHL Papers, Tulane; Macarty to President and Members of the City Council, 4 December 1819, Louisiana Division, New Orleans Pub. Lib.
107. Hamlin, *Latrobe*, pp. 519–25.
108. BHL to President and Directors of the State Bank of Louisiana, 17 July 1820, Department of Archives and Manuscripts, Louisiana State University, Baton Rouge, La.; Roffignac to President and Members of the City Council of New Orleans, 16 August 1820, Louisiana Division, New Orleans Pub. Lib.; 16 August 1820, 26 August 1820, New Orleans City Council Proceedings, Louisiana Division, New Orleans Pub. Lib.; Mary E. Latrobe to Harper, 4 November 1820, Mrs. Gamble Latrobe Collection, MdHS; BHL to Mary E. Latrobe, [30 August 1820], Mrs. Gamble Latrobe Collection, MdHS.
109. The yellow fever epidemic of 1820 was relatively minor. Duffy, *Medicine in Louisiana*, 1:361–62.
110. Mary E. Latrobe to Harper, 4 November 1820, Mrs. Gamble Latrobe Collection, MdHS; 9 September 1820, New Orleans City Council Proceedings, Louisiana Division, New Orleans Pub. Lib.; Roffignac to President and Members of the City Council, 9 June 1821, Louisiana Division, New Orleans Pub. Lib.; Hamlin, *Latrobe*, pp. 528–30.

111. Hamlin, *Latrobe*, p. 529; Duffy, *Medicine in Louisiana*, 1:408; Mary E. Latrobe to Harper, 5 January 1821, Mrs. Gamble Latrobe Collection, MdHS; New Orleans Waterworks records, 1822, BHL Papers, Tulane. Albert Stein had experience in waterworks planning: David Stevenson, *Sketch of the Civil Engineering of North America* (London, 1838), p. 289.

1. William Barclay Parsons, *Engineers and Engineering in the Renaissance* (Cambridge, Mass.: MIT Press, 1968), chaps. 20–21, 25.
2. See Latrobe Journals, 25 June 1800, PBHL, MdHS, concerning Fort Mifflin on Mud Island in the Delaware River.

section of the Delaware existed because of the massive amounts of sediment which the Schuylkill River carried into the Delaware above Mud Island. The sediment resulted from soil erosion caused by the extensive and continuing agricultural development of the Schuylkill watershed. Under such conditions, the development of shoals and islands in the Delaware was a predictable occurrence.

Latrobe offered only one solution to the dilemma. By analyzing the action of the river current and the tides he determined why the river had an eddy, or slowly moving section, which deposited sediment at the point of the shoal. He then suggested that the shoal could be naturally reduced by diverting a greater portion of the ebb tide over it. For that purpose he designed an inexpensive timber barrier to change the route of the ebb tide. Latrobe's advice was considered, but not taken, and it was with some difficulty that he collected his fee for consultation.[3]

Four years later the Philadelphia Chamber of Commerce again asked Latrobe about the bar north of Mud Island, this time inquiring about the efficacy of a certain dredging machine. Considering their estimate that the machine could raise forty-five tons per hour, Latrobe replied that it would take six months to make an adequate ship channel through the bar. Since the action of the currents would be unchanged by dredging there was no guarantee that the channel would remain open. In fact, Latrobe reiterated that diverting the river currents, as he had suggested earlier, was the only means of permanently reducing the bar. He implied that his plan was cheaper as well as more effective than dredging.[4]

Latrobe's advice on this matter was corroborated twenty-five years later by an officer in the United States Army Corps of Engineers who considered the likelihood of improving the navigation of the Delaware by removing sediment and concluded: "I cannot too strongly recommend that, in this instance, nature be left to work for herself, unaided by art."[5] Not until late in the nineteenth century was dredging an effective means of controlling silt accumulation. The equipment used then was far more powerful than what was available in Latrobe's time.[6]

Potomac River

By the end of the first decade of the nineteenth century, Georgetown in the District of Columbia faced the likelihood that the Potomac River would soon be too shallow for large vessels to reach its wharves. In particular, a bar had developed just below the point where the ship channel reached Georgetown. At first the city government had a new channel dredged which bypassed the bar, but the channel quickly filled with new sediment. Next they employed Thomas Moore, a surveyor and engineer active in the Potomac region, to dam up the channel between the western (Virginia) shore and Mason's Island opposite the city. When that failed to sufficiently concentrate the current on the Georgetown side and reduce the bar, Moore suggested constructing wing dams (long obstructions nearly perpendicular to the current) to further direct the current so that it would scour out a new channel. He was, in a sense, taking the same approach which Latrobe had taken concerning the Mud Island bar—using the force which had produced the problem to eliminate it.[7]

The city of Georgetown accepted Moore's suggestion, and hired him to design and construct the wing dams. Almost as soon as Moore set to work early in 1811, however, major landowners in Washington and the Potomac Bridge Company, whose properties were downstream from the proposed wing dams, obtained a preliminary court injunction to restrain Moore from completing the project. In the subsequent legal proceedings the bridge company called Latrobe as a witness and he testified that the wing dams might indeed be injurious to the Potomac Bridge. A permanent injunction against Moore's wing dams was granted, but in 1812 a Congressional committee again investigated the subject and obtained Latrobe's opinion on the matter. Latrobe then collected his official statements and some additional material and published them in a pamphlet, *Opinion on a Project for Removing the Obstructions to a Ship Navigation to Georgetown, Col.*[8]

---

3. BHL to Fitzsimmons, 25 May 1807, 6 June 1807, 19 November 1807, and 7 December 1807, Letterbooks; BHL, *Opinion*, pp. 12–13, 13n, 14n.
4. BHL to Fitzsimmons, 27 March 1811, Letterbooks.
5. Quoted in Frank E. Snyder and Brian H. Guss, *The District: A History of the Philadelphia District U.S. Army Corps of Engineers, 1866–1971* (Philadelphia: U.S. Army Engineer District Philadelphia, 1974), p. 66. This statement refers specifically to the "Mifflin Bar," which was probably the same one which the Chamber of Commerce wanted to remove. See also David M'Clure, *Report on the Survey of a Section of the River Delaware* (Philadelphia, 1820), pp. 5–9.
6. Snyder and Guss, *The District*, pp. 64–70.
7. BHL to Fitzsimmons, 27 March 1811, Letterbooks; BHL, *Opinion*, pp. vi–viii, 1. On Thomas Moore (1760–1822), see Calhoun, *The American Civil Engineer*, pp. 31, 36, 62.
8. BHL to Mason, 8 January 1811, Letterbooks. The remainder of this section is based on BHL's pamphlet, *Opinion*.

River Control

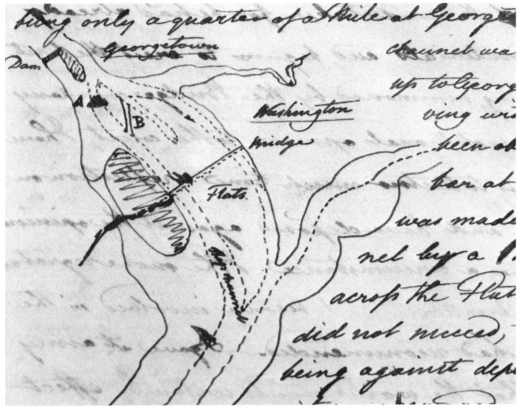

Fig. 17. BHL's sketch of the Potomac River at Washington, 1811. *A* is the Georgetown bar, and *B* is the dredged channel. (BHL to Fitzsimmons, 27 March 1811, Letterbooks.)

Latrobe's analysis of the problem rested on his assumption that the Potomac, in common with other coastal North American rivers, carried a heavy load of sediment because of the continuing agricultural development of the basin which it drained. As a result, the area in which the river current met the retarding influence of the tide would always be a place where considerable sediment was deposited.[9] Such was the case at Georgetown and Washington where the Potomac, having passed its final rapids, met the tide which rolled up the river estuary from the Chesapeake Bay. The situation was aggravated because the width of the Potomac increased from a quarter mile just above Georgetown to a mile or more at Washington (see figure 17), thus making the current slow down even more than it might have under the influence of the tide.[10]

9. Cf. BHL to Fitzsimmons, 25 May 1807, Letterbooks; BHL to Bate, 3 February 1810, Letterbooks; BHL, "Report on the Improvement of Jones' Falls," 31 May 1818, PBHL, MdHS; BHL to Thompson, 24 March 1806, Letterbooks.
10. BHL to Fitzsimmons, 27 March 1811, Letterbooks.

To demonstrate the validity of his assumption concerning the hydrological basis of the phenomenon, Latrobe reminded his readers that as recently as 15 years before the Potomac had carried the largest ships to Georgetown with ease, but that its channel had become progressively shallower. Therefore, as he had pointed out in 1807 to the Philadelphia Chamber of Commerce, the problem of the silting up of rivers was of long duration and would extend well into the foreseeable future. Latrobe believed that the technical skill of his day was capable only of changing the location where some of the sediment was deposited, and could not eliminate the problem. To support his belief he called upon European experience, especially citing the writings of the French engineer Fabre, who had proved the folly of man's attempts to contravene the natural causes of sedimentation in river estuaries.

Referring to Moore's proposals, Latrobe stated that wing dams extending into the river from the Virginia shore would undoubtedly concentrate the current and scour out a new channel. Where the new channel might be he could not predict, but he thought that it might pass under the bridge and along the eastern shore about where the present secondary channel was located. Since the bridge had been designed for deeper water at that point, with a draw for ships, little injury could be done to the bridge or the interests of Washington landowners. If the channel passed elsewhere, however, the shorter piles in shallower water might be endangered.

Latrobe concluded that ultimately the wing dams would merely remove the heavy accumulations of sediment from the Georgetown vicinity and deposit them somewhere further downstream, since at some point the current would become diffuse enough that the tide would again exert its retarding influence. The effect of the wing dam would thus be to cause the bars and mud flats to form further from Georgetown, with the result that ships would be unable to come as close to it as they had prior to the wing dams' erection.

Having destroyed the presumption that Moore's proposal would effect a real improvement of Georgetown's situation, Latrobe declined to offer any hope for a different solution. He thought that the Potomac below Georgetown would become increasingly blocked and the head of navigation would become more and more distant. Somewhat superciliously and mysteriously, he provided only a glimpse of his professional engineer's view of the future.

> There is indeed a means of bringing up to Georgetown ships of the largest dimensions, and only one means. But our present state of population and wealth, so little warrants the undertaking, that to propose it, would

be to expose one's self to ridicule; and yet I believe it will be undertaken and executed by our posterity.[11]

Possibly Latrobe was referring to canalizing the entire river.

New Orleans Levee

Virtually from the founding of New Orleans in the 1720s the city had a levee, or dike, to protect it from the overflow of the Mississippi River. Such protection was needed because much of New Orleans was built on the river's floodplain and any rise of the Mississippi above normal level was a serious matter. Yet even during the second decade of the nineteenth century (when Latrobe had the franchise for the New Orleans Waterworks) there was no organized and compulsory plan for repairing the levee which was so vital to the city's welfare.[12]

The flood of May 1816, possibly the worst since the city's founding, thus found the city unprepared. A crevasse, or break, in the levee occurred about five miles north of the city in the first week of the month. The city council at first appointed Henry Latrobe (Benjamin's son) as one of two engineers responsible for directing a team of two hundred men to close the crevasse. Then a day later the council retracted its action because it learned that the governor had selected another engineer to direct the repair. Unfortunately the state-supported operation failed within a few days, and neither the state nor the city attempted anything further. Water spread into the city and flooded the half closest to the river, causing immense damage. It was a month after the flood began before all of the water receded from New Orleans.[13]

Benjamin Henry Latrobe took a serious interest in the flood of 1816 because the value of his waterworks franchise was dependent on the future growth of the city—growth which was hampered by the spectacle of an uncontrolled, disastrous flood. To help New Orleans and the surrounding area prepare for future crevasses, Latrobe suggested adopting a method of strengthening dikes used for centuries along the rivers of eastern Germany. He had observed it when just beginning to learn the profession of civil engineering.

Drawing on his own knowledge and on the writings of two German engineers, Latrobe wrote a long essay on the usefulness and method of making fascine works. Fascines were bundles of green branches about ten feet long bound tightly by a few other branches. Standing on a river bank, an experienced workman could take a quantity of previously prepared fascines and build layers of them projecting into the water. With stakes holding the fascines to the bank and soil packed into the completed mass so that it offered sufficient resistance to the current, a fascine work was an effective device for diverting a stream or closing a crevasse. It also offered the advantages of being made entirely of local materials. Latrobe wrote that fascines made from aquatic trees and bushes such as willow and poplar could actually take root and sprout shoots so that the fascine bank became a new shoreline.[14]

Latrobe submitted his essay to the mayor of New Orleans in August 1816, and suggested to Henry that he insert a notice in the newspapers informing readers of its availability.[15] There is no record that anyone attempted to make fascine works in New Orleans or its vicinity, probably because no person skilled in the technique was present to make a successful introduction. However, since fascine works were a standard element in flood control engineering in central Europe until the latter part of the nineteenth century,[16] it seems likely they could have been used effectively in New Orleans.

Jones' Falls

The most comprehensive river control project proposed by Latrobe was for Jones' Falls in Baltimore, Maryland. In the second decade of the nineteenth century the falls originated in an agricultural area north of the city and flowed fourteen miles to the harbor in the heart of the city. Until its final segment it had a rapid drop and was used to power several mills and the pumps for the city's waterworks. Within the city limits there were several private residences in the meadows along its banks and four major bridges across its waters. Portions of the stream had stone retaining walls defining its course.[17]

Although there was a serious flood of Jones' Falls in 1795, not until 1817

---

11. BHL, *Opinion*, p. 11.
12. Hodding Carter, *The Lower Mississippi* (New York: Farrar & Rinehart, 1942), pp. 350–51.
13. Samuel Wilson, Jr., *The Vieux Carré New Orleans: Its Plan, Its Growth, Its Architecture* (New Orleans: Bureau of Governmental Research, 1968), pp. 74–75. Apparently the state engineer, Joublanc, attempted to close the crevasse by sinking a ship in it. See BHL to Henry Latrobe, 6 June 1816, Letterbooks.
14. BHL, "Essay on the Means of Preventing, Meeting, and Repairing the Calamities Occasioned by Inundations," 10 June 1816, Mrs. Gamble Latrobe Collection, MdHS. For BHL's drawings of a fascine work and further discussion of his essay, see pp. 208–13 below.
15. BHL to Henry Latrobe, 16 August 1816, Letterbooks; BHL to Mayor of New Orleans, 17 August 1816, Letterbooks.
16. David Stevenson, *The Principles and Practice of Canal and River Engineering*, 2d ed. (Edinburgh, 1872), pp. 152–54.
17. Ralph D. Nurnberger, "The Great Baltimore Deluge of 1817," *Maryland Historical Magazine* 69 (1974): 405–6. The term "falls," meaning river, is generic in the Baltimore region.

(when Baltimore ranked as the third largest port on the Atlantic coast) did the falls demonstrate its potential danger to the city.[18] Heavy rains began on the night of 8 August and continued until midday of 9 August throughout the Baltimore area. The quantity of rain was sufficiently large to strain and break one or more milldams, unleashing a far greater volume of water than the bed and banks of the falls could carry. The surging current carried away farm buildings, fences, timber, and other debris which smashed into and destroyed two wooden bridges within the city limits. The accumulated debris then piled up against the center pier of the stone bridge at Gay Street in Baltimore, forming a dam and spreading water throughout the lower part of the city. The rapidity of the flood and the unprecedented volume of water was responsible for killing several people, completely destroying many homes, and damaging many other buildings. A small stream to the east of Jones' Falls, Harford Run, also flooded and caused extensive damage.[19]

Soon after the flood the Baltimore city council asked Robert Mills, a former pupil of Latrobe's who was then erecting Baltimore's Washington Monument, to present a plan for controlling Jones' Falls and providing new and better bridges. At the desire of the council, Mills rushed his investigations and on 25 September 1817 he reported that the falls might be less liable to flooding if its course were widened and straightened, and if stone bridges without obstructive central piers were erected. Secondary advantages of the Mills plan were that the falls would then be rendered navigable, tolls could be charged by the city, and pleasant walks might be laid out along its banks. Mills thought that the cost of implementing his plan would be $126,757.[20]

City council politely declined to accept Mills's report and merely passed a resolution urging the council elected for the next year to take appropriate action.[21] Their successors again avoided the onus of committing so much of the taxpayers' money to public works, and on 4 March 1818 decided

> to advertise that a premium of five hundred dollars would be paid to the person who should submit a plan for the improvement of Jones' Falls, provided the plan was adopted by the corporation, and also to advertise in the papers of Boston, New York, Philadelphia, Baltimore, and Washington that the corporation of Baltimore is desirous to receive proposals, with plans and estimates of the expense, for the improvement of Jones' Falls, by making a canal navigable for flats and scows, or by securing the waters within their bed so that it shall not overflow.[22]

At the same time the council seems to have invited proposals for designs of flood-proof bridges suitable for Jones' Falls and Harford Run.[23]

Benjamin Henry Latrobe was one of more than a dozen people who took up the challenge of planning for the future safety of Baltimore. He was a resident of the city in March 1818 when the city council offered its premium, and his recent financial problems made the offer tempting. Within two weeks of the council's action he submitted a plan and estimate for a single-span stone bridge over Harford Run, and late in April, assisted by surveyor Joseph Jeffers, he investigated the possible improvement of Jones' Falls. His final report, submitted a month later than the 1 May 1818 deadline, was nonetheless received and considered by city council. It contained a comprehensive examination of current and future problems and benefits which the city faced from Jones' Falls.[24]

Latrobe noted that seaports located on streams of fresh water had the advantages of immediate access to drinkable water and the possibility of a transportation route to the interior, but also the disadvantages of floods and the destruction of the harbor by the stream's sediment. Concerning the seriousness of the threat to the harbor Latrobe referred the council to his testimony during the Potomac River case, a copy of which he appended to his report.[25] He then briefly summarized his views by emphasizing that as long as Jones' Falls flowed into the city's harbor it would continue to carry sediment, and that the annual deposit could be expected to increase with the continuing agricultural development of the uplands.

---

18. Benjamin H. Latrobe, Jr., Issac R. Trimble, John H. Tegmeyer, *Report of the Board of Engineers upon Changing the Course of Jones' Falls, With a View to prevent Inundations, to the Mayor and City of Baltimore* (Baltimore, 1868), p. 4.

19. Nurnberger, "The Great Baltimore Deluge of 1817," pp. 405–8. BHL quickly learned of the flood: see BHL to Jefferson, 12 August 1817, Letterbooks.

20. *Baltimore American*, 3 October 1817.

21. Ibid.

22. Quoted in J. Thomas Scharf, *History of Baltimore City and County*, 2 vols. (Philadelphia, 1881; reprint ed., Baltimore: Regional Publishing Company, 1971), 1:209.

23. Stiles to City Council, 28 May 1818, Baltimore City Archives, Baltimore, Md.; see BHL's designs for the Harford Run bridge, Peale Museum, Baltimore, Md.

24. BHL, "Report on the Improvement of Jones' Falls, Baltimore," 31 May 1818, PBHL, MdHS. Most of what follows is derived from this document, although the surviving copy is not in BHL's hand and has numerous errors of spelling and grammar.

Joseph Jeffers appears in Baltimore directories as a "carpenter." See also McKim to Hamilton, 28 March 1811, Miscellaneous Letters Received (M-124), RG 45, NA.

25. BHL, *Opinion*. See Richard Walsh and William Lloyd Fox, eds., *Maryland: A History, 1632–1974* (Baltimore: Maryland Historical Society, 1974), pp. 171–72.

Latrobe estimated the rate of sedimentation by measuring the amount of silt deposited in a bucket of water he dipped from the falls. He calculated that if the flood of 9 August 1817 carried sediment at the same rate, it must have brought down enough to cover three acres of land three feet deep. He pointed out that most of the sediment undoubtedly washed out to sea, but that perhaps one-third of it remained, and that was an enormous amount. As the city annually experienced several freshes, or minor floods caused by hard rains, Latrobe believed that the city's harbor was rapidly filling up.

Latrobe then turned to the problem of floods and considered the inadequacy of the attempt to control them by channelizing the stream bed. He pointed out that the falls dropped more slowly in its passage through the city than in its upper reaches, and near the harbor met the retarding effect of the tide. By natural law water moving at a slower rate needs a channel of greater capacity than if moving at a faster rate. To provide safety to the city 94,600 cubic yards of soil had to be removed to widen the falls and to provide adequate flood capacity. Yet if that were done without straightening the falls, the current would not be strong enough to prevent new accumulations in the stream bed, thus defeating the purpose. If the falls were both widened and straightened Latrobe thought that floods might be avoided, but then all of the sediment would go directly into the harbor, speeding the harbor's decline.

Having discussed the major problems presented by Jones' Falls, as well as the defects of plans (such as Mills's) for channelizing the falls, Latrobe proposed diverting all or part of the water from its normal course. He thought that from the substantial milldam of Keller and Foreman near the city limits a canal could be cut to Herring Run, a tributary of Chesapeake Bay about four miles to the east. The diversion of water would eliminate flooding and sedimentation in one bold stroke.

The canal needed two substantial engineering feats for its accomplishment. Immediately upon departing from the dam it required a tunnel of five-eighths of a mile through Gallows Hill. In cross section the tunnel would have a semicircle of 50 feet diameter at the top, and a canal 40 feet wide and 5 feet deep on the bottom, or a total section capable of carrying the entire flood of 9 August 1817. Latrobe ascertained through his geological examinations that the roof of the tunnel was located in durable rock, but he still wanted it reinforced with brick laid in "Roman Cement"; the sides he wanted lined with stone masonry.[26] He thought that the tunnel itself was nothing to fear: "The manner of conducting this work cannot now be described: but nothing is more common than the tunneling of hills by canals, in Europe [;] nor is our own country, nor my own practice without experience, in this operation. The Tunnel will commence at both ends at once: and if the work be carried on vigorously, also in the middle by means of shafts."[27]

Another obstacle was Mine Hill, which rose forty-eight feet above the projected canal's water line, and which had to be traversed with either an open cut or tunnel slightly shorter than that needed at Gallows Hill. Latrobe could not choose between the two options because he was not able to examine the site sufficiently, but, citing his experience with the Washington Canal, he suggested that the open cut might be cheaper.[28]

If these two segments of the canal were made as planned, the canal would begin at Keller and Foreman's dam and proceed through Gallows Hill; come out in the valley of Harford Run, proceed north along its valley and intercept its water; proceed through Mine Hill; cross Harris's Creek; cross and run to the south of the Philadelphia Turnpike and the Old Philadelphia Road; and empty into Herring Creek.[29] In its course the canal would drop forty feet.

Latrobe foresaw many advantages of this plan besides preventing floods and the silting of the harbor. The capacious dimensions of the canal would make it navigable, so that with a slight enlargement of the mouth of Herring Creek boats could come into the city by that route. Latrobe thought that Jones' Falls might also be made navigable above Keller and Foreman's dam, thus tapping the stone and marble quarries in the upper valley. He pointed out that property values would rise in the areas formerly threatened by flooding, producing greater tax revenue for the city. The annual expense of dredging the harbor would be considerably reduced. In a visionary spirit, he also predicted that the valley of Harris's Creek south of the canal could be turned into a vast system of drydocks, because the canal provided a sufficient head of water to fill drydocks when ships were floated in.[30] Latrobe did not provide estimates of the financial costs or benefits for any of his plan.

The Baltimore city council considered Latrobe's proposal, but, as it did with all of the others submitted, rejected it as too expensive to implement. Latrobe

---

26. In 1817 BHL found "Roman Cement" from England very useful for the Capitol's masonry. BHL to Dawson, 21 July 1817, Letterbooks.

27. BHL, "Report on the Improvement of Jones' Falls, Baltimore," 31 May 1818, PBHL, MdHS.
28. See the discussion of the cut south of Capitol Hill, p. 21 above.
29. See fig. 18, p. 47.
30. This idea is similar to his plan for the Navy Yard drydock. See pp. 9–11 above.

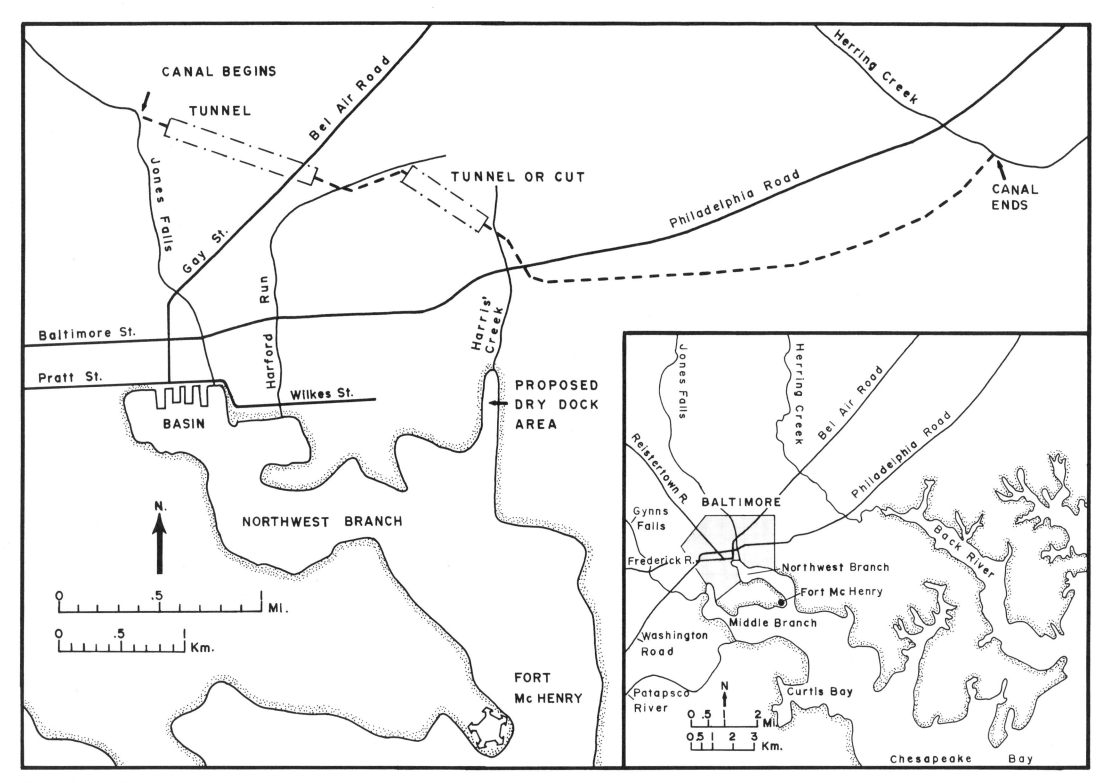

Fig. 18. Map of BHL's proposal for the Jones' Falls Improvement, 1818.

Stephen F. Lintner

did not receive the $500 premium.³¹ City council authorized the Wardens of the Port of Baltimore to expend no more than $25,000 to widen and straighten Jones' Falls, but that sum was far below even the estimate for implementing the modest Mills plan.³² Essentially city council decided to leave the problem to posterity.

There were several minor floods of Jones' Falls in the next fifty years, but the first which exceeded the one of 1817 occurred on 24 July 1868. Shortly thereafter, the city appointed a board of three engineers, including Latrobe's youngest son, Benjamin Henry Latrobe, Jr., not only to examine the cause of the flood, but specifically "to examine the most practicable route for turning the course of Jones' Falls outside the city into Herring Run." After two years' study the board reported that the diversion project was the best means of eliminating floods, and advanced a plan "virtually the same as that proposed by Mr. Latrobe [Sr.]... but modified by the results of experience since his day...." The board also outlined an alternative project of widening and straightening the falls, plus raising nearby streets, noting that it was a cheaper undertaking but would probably not be as effective in preventing floods.³³

The 1868 report, however, was not acted on, and to the present, the Jones' Falls diversion project has never been seriously considered. In 1912 the falls were completely covered over south of the point where Latrobe suggested starting the diversion, making minor floods impossible. The silting of the harbor is dealt with by a continuous program of dredging conducted by the Army Corps of Engineers. ConRail (earlier the Pennsylvania Railroad, then the Penn Central Railroad) comes into Baltimore from the east over a route very similar to that proposed by Latrobe for the diversion canal.

Latrobe's plan for controlling Jones' Falls was undoubtedly ahead of its time in terms of the expense which it would have entailed for the city of Baltimore, yet it was based on an accurate and prescient understanding of the problem. Had the city of Baltimore merely adopted as future policy the engineering principles expressed in Latrobe's proposal, it would have provided an approach to the problem which would have proved a valuable guide to succeeding generations.

---

31. Memorial of the Representatives of B. H. Latrobe to the Mayor and City Council of Baltimore, 15 March 1830, Baltimore City Archives, Baltimore, Md.
32. *Ordinances of the Corporation of the City of Baltimore; Passed at the Extra Session in June and July, 1818 ...* (Baltimore, 1818), pp. 9, 20–21.
33. Latrobe, Jr., Trimble, and Tegmeyer, *Report of the Board*, pp. 3–6, 11, 33.

5. INDUSTRIAL WORKS

Schuylkill Rolling Mill

The Schuylkill rolling mill was the first of Latrobe's industrial works in the United States, and of all his engineering projects was probably the most destructive of his personal fortune. It is a good example of how Latrobe's enthusiasm for novel and dramatic technical enterprises led him blindly and inexorably toward financial failure.

Sheet iron was a standard commodity in the late eighteenth century. For centuries in Europe water-powered forge hammers flattened wrought-iron bars into sheets. By the sixteenth century pairs of iron cylinders powered by a waterwheel appeared, making sheets more rapidly and uniformly than hammers. Rolling mills also had a pair of meshing ribbed rollers, called slitters, which cut sheets into rods for nail and hoop manufacture. But the higher power requirements of rolling mills, as well as complicated gearing which was subject to frequent breakdowns, limited their spread.¹

When James Watt improved the steam engine so that it became practical motive power for industrial purposes, he was aware of its potential for the iron industry. In his patent specification of 28 April 1784 he listed a design for applying steam power to rollers and slitters as one of his innovations.² Within two years Watt's business associate John Wilkinson had a rolling mill operated by steam power,³ and interest in the new combination spread rapidly. A contemporary has left a concise description of these first English steam-powered rolling mills.

> Messrs. Boulton and Watt's 36 horse, and 40 horse rotative-engines became a sort of standard for rolling-mills at iron works, and a great number ... were erected in Staffordshire, Shropshire, and Yorkshire, between the years 1790 and 1800. The engine-house is built with four massive walls, of a proper size to inclose the cylinder and working gear; one of these walls supports the centre of the great lever, but the axes of the fly-wheel, and of the crank and multiplying-wheel, as well as of the other wheel-work, and of the rollers, are supported upon very solid foundation

---

1. H. R. Schubert, *History of the British Iron and Steel Industry from ca. 450 B.C. to A.D. 1775* (London: Routledge & Kegan Paul, 1957), pp. 306–8; Singer, *History of Technology*, 3:32, 342–43, 4:104, 106–7.
2. Eric Robinson and A. E. Musson, *James Watt and the Steam Revolution* (London: Adams and Dart, 1969), pp. 117–19, figs. 15–17.
3. T. S. Ashton, *Iron and Steel in the Industrial Revolution*, 2d ed. (Manchester: Manchester University Press, 1951), p. 74.

walls, which do not rise above the level of the ground, so that the wheel-work is very securely fixed; but as it is not inclosed by the walls, it is more accessible for repairing.

The whole of the space occupied by the mill-work, and by the rollers, and their furnaces, is covered over by a series of large roofs, to form a vast shed. . . .

It is usual to apply two series of rollers, by suitable connexions of wheel-work, at the end of the axis of the great fly-wheel; each set of rollers being adapted for a particular purpose, such as for rolling bars, or for plate, or for round bolts, or for nail rods.[4]

In 1800 in the United States there were many rolling and slitting mills in the areas where iron was produced, but all were water-powered.[5] Men who were familiar with the great change that steam power had introduced into the English industry undoubtedly anticipated that a profitable increase in production was likely to occur when steam rolling mills were erected in the United States, but the infant state of American steam power prior to 1800 made the innovation risky. Nicholas Roosevelt and the workmen at his Soho works near Newark, New Jersey, were prepared to take the risk, however. Roosevelt was the builder of the Philadelphia Waterworks steam engines.

James Smallman, Roosevelt's chief engineer, apparently had acquired first-hand experience in English rolling mill technology by helping to build John Wilkinson's mill at Bradley in Shropshire.[6] Depending upon Smallman's skill, Roosevelt in 1798 contracted with the Navy Department of the United States to supply rolled copper for sheathing the bottoms of new ships in the American fleet.[7] Plans were drawn for a large steam-powered rolling mill in the English style, but for rolling copper instead of iron.[8] Then in January

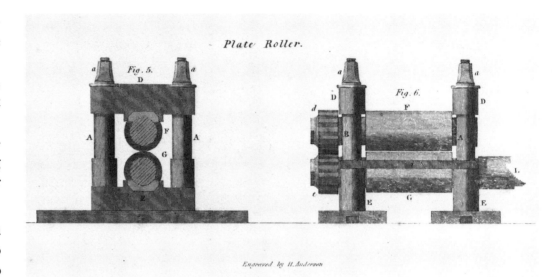

Fig. 19. Side views of rollers in an English rolling mill, c. 1800. (Rees, *Cyclopaedia*, plates, vol. 3.)
Courtesy of Eleutherian Mills Historical Library

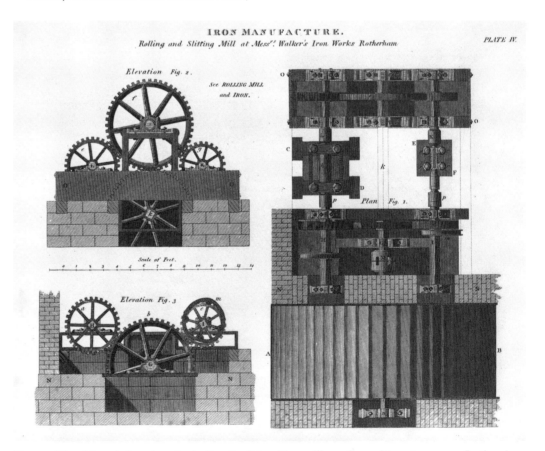

Fig. 20. Plan (1) and elevations (2,3) of an English rolling mill, c. 1800, with water power. In the plan, note the rollers (C, D) and slitters (E, F). (Rees, *Cyclopaedia*, plates, vol. 3.)
Courtesy of Eleutherian Mills Historical Library

4. Farey, *Treatise on the Steam Engine*, p. 507.
5. J. Leander Bishop, *A History of American Manufactures from 1608 to 1860*, 2 vols. (Philadelphia, 1864), 1:491–92, 496–97, 503, 513, 524, 541, 556, 558, 583, 587, 592, 606.
6. *Proposals for Establishing a Company, For the purpose of employing the surplus power of the Steam-Engine Erected near the river Schuylkill: Under the Title and Firm of the Philadelphia Rolling Company* [Philadelphia, 1801], p. 3, Society Collection, Hist. Soc. Pa. See Pursell, *Early Stationary Steam Engines in America*, p. 38n; Ashton, *Iron and Steel*, p. 74.
7. *Proposals for Establishing a Company*, p. 2; Hamlin, *Latrobe*, pp. 168–69; BHL to Ingersoll, 9 December 1811, Letterbooks; BHL to Hamilton, 18 July 1811, Letterbooks.
8. See drawings in Stoudinger Collection and the "Explanation of the drawings of a new improved Steam Engine [Plan, Section and Elevation]," n.d., New Jersey Historical Society, Newark, N.J. These drawings are probably connected with Roosevelt and Smallman's patent for a "double steam engine" in 1798.

1799 Latrobe came to Soho to make a preliminary contract for the Philadelphia Waterworks steam engines and in the negotiations someone realized that the waterworks provided the opportunity to significantly reduce the capital required to build and operate a rolling mill. If the waterworks was to be constructed so as to provide for not only the current, but also the future needs of the growing city of Philadelphia, the steam engines would at first have power far in excess of that needed to pump the requisite number of gallons of water in a day. This "extra power" might be applied to other purposes—in this case, a rolling mill.

Thus when Latrobe returned to Philadelphia he not only had a preliminary contract for the waterworks steam engines, but also a suggestion for a complicated arrangement with Roosevelt. The Soho proprietor would be permitted to lease the "extra power" of the Schuylkill engine (the larger of the two waterworks engines) and the building and the lot on which it was located, in return for promising to keep the engine in repair and to pump the necessary amount of water into the brick conduit leading to Centre Square. Roosevelt would be paid to maintain the engine, but the amount of his rent would be deducted from the payment.[9]

From the earliest stage of planning, then, the Schuylkill Engine House was to be both a pump station and a rolling mill. Its design appears to have been very similar to that of English rolling mills, with large open spaces and the rolling apparatus located at ground level.[10] In designing the Schuylkill Engine House, Latrobe may have drawn upon Smallman's recollections of his English experiences.

Although the waterworks began operation in January 1801, the rolling mill took somewhat longer to get in operation. Undoubtedly Roosevelt's chronic lack of capital caused the delay, since he had barely enough to complete the waterworks engines without forfeiture. He had in fact relied upon Latrobe's guarantee of several loans to keep going.[11] As compensation for that entanglement Latrobe agreed to become a partner in the rolling mill and contribute his engineering talents to it.[12] In the summer of 1800 he drew designs for the planned rollers, slitters, and shears.[13]

The poor financial state of Roosevelt became public in April 1801 when the partners issued a prospectus for a joint-stock company to be titled the "Philadelphia Rolling Company." Roosevelt offered to assign all his rights to the Schuylkill Engine House to the new company for $42,000, which he would in turn use to subscribe $30,000 in shares. The capital of the new enterprise was to be $100,000 in hundred-dollar shares.[14] But nothing came of the projected company.

Sometime in the next few months Roosevelt and Latrobe solved the problem of finding capital by bringing in a new partner, Eric Bollman, then a prosperous Philadelphia merchant. Bollman, a native of Hanover, Germany, had an extensive chemical and technical background before he came to the United States in 1796. At an early date he met Roosevelt, though he resided in Philadelphia where he established a mercantile partnership with his brother Lewis. Their international connections were supplemented by local ones when Eric married a wealthy merchant's daughter, and the Bollmans did quite well during the general prosperity fostered by European warfare. After Eric Bollman became a partner in the rolling mill with Roosevelt and Latrobe, he invested about $30,000 in the concern. However, when the Treaty of Amiens ended the European conflict in March 1802, American neutral trade suffered; the Bollmans were especially vulnerable. Eric turned to the rolling mill as the best hope for solvency and assumed the position of resident director.[15]

By the fall of 1802 the rolling mill was in production. Although the original plans were to roll copper as well as iron, the mill manufactured only the latter metal.[16] Bollman experimented with making tin plate (sheet iron with a tin coating), but calculated that the mill would make more money by selling hoops and nail rods than any other product.[17] Unfortunately, during Bollman's tenure as manager, the mill was unprofitable because the Schuylkill engine ran poorly. The boiler was very leaky, requiring far more coal than

---

9. 2 February 1799, 1 August 1799, Minutes of the Select Council, Phil. Archives; Watering Comm., *Report* (1799), p. 33; *Proposals for Establishing a Company*, p. 1.
10. See quote from Farey, *Steam Engine*, pp. 48–49 above.
11. 27 October 1800, Cope Diaries, Haverford College; Hamlin, *Latrobe*, p. 169.
12. BHL to Peters, 14 February 1807, Letterbooks; 8 July 1800, Cope Diaries, Haverford College.
13. See pp. 167–72 below.

14. *Proposals for Establishing a Company*, pp. 1–4. It seems likely that this prospectus was BHL's production, but no authorship is indicated.
15. Fritz Redlich, "The Philadelphia Water Works in Relation to the Industrial Revolution in the United States," *Pennsylvania Magazine of History and Biography* 69 (1945): 243–46, 247, 249; Fritz Redlich, *Essays in American Economic History* (Ann Arbor, Mich.: Edwards Brothers, Inc., 1944), pp. 19, 22–24.
16. BHL to Ingersoll, 9 December 1811, Letterbooks.
17. Memoranda of 29 October–3 November 1802, Eric Bollman Letter and Memorandum Book, George Bollman Collection, Hist. Soc. Pa. Although Bollman did not specifically state that he was experimenting with tin plate manufacture, the process he described is the same. See Rees, *Cyclopaedia*, s.v. Rolling-Mill.

expected, while providing too little steam pressure for efficient operation. Added to this were a defective steam cylinder and gearing which did not provide sufficient speed to the rolling apparatus. According to Bollman's account, "the Mill therefore must Stop nearly every 20 Minutes to give the Steam time to accumulate[;] of Course, the Men stand half of their Time idle whilst the Wages run on."[18]

Bollman never saw a change in these conditions while he was manager because in the summer of 1803 the firm of Eric and Lewis Bollman officially went bankrupt and their share in the rolling mill was sold along with their other assets. William Cramond, a Philadelphia merchant, and Samuel Mifflin, of a prominent Pennsylvania family, became Roosevelt and Latrobe's new partners. Mifflin replaced Bollman as manager.[19]

The two original partners remained active in the concern. Roosevelt provided a new cylinder and two new cast-iron boilers that rectified the engine problems to some degree, while Latrobe provided some capital to keep the mill going.[20] Latrobe, moreover, took the opportunity of his work for the federal government to promote the mill's product. His plans for the Naval Drydock, the President's House (White House), the Capitol, and other buildings included sheet iron roofing to be purchased from the Schuylkill mill.[21] He also secured a personal order from President Jefferson for sheet iron to roof Monticello.[22] Although he clearly had a personal interest in promoting the sale of sheet iron when the Schuylkill mill was operating, he advocated its use for years afterward as the cheapest and most easily installed roofing material.[23]

The rolling mill's major product was sheet iron, but it produced hoops and nail and spike rods as well.[24] The mill also had a foundry which cast two heating stoves for the Senate chamber of the Capitol.[25] At some time in 1803 or 1804 the partners installed a boring mill in the Schuylkill Engine House and, with James Smallman as engineer, quoted prices for steam engines and pumps.[26]

In spite of new equipment and steady sales to the federal government in 1803 and 1804, the rolling mill continued to lose money.[27] It is likely that the conditions which existed during Bollman's tenure continued to plague the mill since the Schuylkill engine had frequent equipment failures.[28] Another problem was that Roosevelt's lease on the engine house and lots was never formally commenced because he never satisfied the stipulations of his contract for the waterworks steam engines. The operation of the rolling mill was technically illegal, although the city never brought suit against Roosevelt. He repaired the engines and pumped water without any compensation from the city, and at the same time paid no rent.[29] Roosevelt and his partners were the losers by the arrangement because the city's payments would have exceeded the rent.[30]

Nonetheless, the situation was also unsatisfactory for the city, and it was decided that for the security of the city's water supply, city councils should purchase whatever right Roosevelt had to the Schuylkill Engine House. By June 1804 Stephen Girard represented the city in negotiations, and Samuel Mifflin acted as the representative of the rolling mill partners.[31] The negotiations dragged on for months because Roosevelt held out for a large sum, and the city had no reason to accede to his terms. Latrobe advised Roosevelt to threaten to stop pumping water, but in light of the anomalous legal position, that apparently had little effect other than angering the city.[32]

Finally, in mid-September 1805, a newspaper story aroused the city by

---

18. Redlich, "The Philadelphia Water Works," p. 251.
19. Ibid., pp. 249, 256; Redlich, *Essays*, pp. 79–89; BHL to Peters, 14 February 1807, Letterbooks.
20. BHL to Bollman, 11 May 1816, Letterbooks; *Report of the Watering Committee to the Select & Common Councils, November 1, 1803*.
21. [Jefferson], *Message from the President*, pp. 10, 17; Redlich, "The Philadelphia Water Works," p. 255.
22. BHL to Mifflin, 2 October 1803, Letterbooks. This is the first of many letters to Mifflin about "the President's iron."
23. BHL, "Observations on the foregoing communications [on the construction of buildings in India]," *Transactions of the American Philosophical Society* 6 (1809): 390; BHL to President and Directors of the Bank of Washington, 24 July 1810 and 31 July 1810, Letterbooks; BHL to Mills, 14 September 1812, Letterbooks. For the early years of sheet iron roofing in the United States, see Charles E. Peterson, "Iron in Early American Roofs," *Smithsonian Journal of History* 3 (1968–69): 42–43.
24. 20 November 1804, Journal, Comm. of Works, Salem Co. Hist. Soc.; *Report of the Watering Committee to the Select and Common Councils, October 7th, 1805*, accounts; receipt, Samuel Mifflin & Co. to the City of Philadelphia, 18 September 1805, Watering Committee Bills Received, Phil. Archives.
25. BHL to Mifflin, 20 October 1803, 27 October 1803, and 7 November 1803, Letterbooks.
26. BHL to Bollman, 6 February 1804, Letterbooks; BHL to Gobert, 16 March 1804, Letterbooks; BHL to Paine, 13 September 1804, Letterbooks.
27. BHL to Bollman, 13 August 1804, Letterbooks.
28. Redlich, "The Philadelphia Water Works," pp. 250–51.
29. 10 March 1805, Cope Diaries, Haverford College; Watering Comm., *Report* (1801), pp. 9–10; *Report of the Watering Committee to the Select & Common Councils, November 1, 1803*, p. 5 ff; BHL to Roosevelt, 13 August 1804, Letterbooks.
30. *Proposals for Establishing a Company*, p. 1.
31. [Girard] to Mifflin, 25 June 1804, series 2, reel 31, Stephen Girard Papers, APS.
32. BHL to Roosevelt, 13 August 1804 and 2 November 1804, Letterbooks; BHL to Roosevelt, extract, 23 July 1805, Letterbooks; 10 March 1805, Cope Diaries, Haverford College.

relating that during a fire the Schuylkill engine was so unready to pump that the Centre Square reservoirs were not replenished in time to prevent serious property loss. A few days later city councils placed an attachment on the engine house, and the county sheriff put it under the protection of the Watering Committee.[33]

In January 1806 the city agreed to purchase Roosevelt's rights to the Schuylkill engine's extra power, the boring mill, and outbuildings for $16,000.[34] The rolling equipment remained the property of the partners, and went up for sale.[35]

The Schuylkill rolling mill was a business failure largely because of the faulty steam engine, but there is no evidence that the rolling works were a technical failure. Latrobe's drawings show rollers, slitters, and shears very similar to those used in English and American mills, so it is unlikely that the equipment had innovative features which had to be tested.[36] One or two of the mill's employees had experience in English ironworks and were regarded by Latrobe as exceptionally competent workmen.[37] There is no record of complaints about the mill's products.[38]

As the first steam-powered rolling mill in the United States, the Schuylkill mill probably had a substantial impact on the transfer of steam rolling mill technology to the United States. Two of the Philadelphia workmen were instrumental in establishing the first steam rolling mill in Pittsburgh in 1812. When Latrobe saw it two years later he called it an "exact copy" of the Schuylkill mill. Other steam-powered rolling mills may have been established by others who worked at or closely observed the ill-fated enterprise with which Latrobe was associated.[39]

33. *Aurora, General Advertiser* (Philadephia), 24 September 1805; *United States Gazette* (Philadelphia), 1 October 1805.

34. BHL to Roosevelt, 8 January 1806, Letterbooks. The city rented the boring mill to James Smallman, Oliver Evans, and others for the next several years.

35. BHL to Lea, 28 January 1806, Letterbooks. Some equipment may have been sold earlier: see account in BHL to Peters, 14 February 1807, Letterbooks.

36. See, for example, Rees, *Cyclopaedia*, s.v. Iron Manufacture, plates IV–V; William E. Harrison, "The First Rolling-Mill in America," *Transactions of the American Society of Mechanical Engineers* 2 (1881): facing p. 104.

37. BHL to Hazlehurst, 30 March 1805, Letterbooks; BHL to Bollman, 21 June 1809, Letterbooks.

38. Of course, virtually the only source regarding the mill is BHL's papers, hardly an unbiased authority. A decade after the mill's closing Thomas Cooper reported that American sheet iron was defective "in consequence of the scales they acquire in the process of heating for the purpose of being rolled." Thomas Cooper, "Sheet Iron, Ordnance. Iron Cables," *Emporium of the Arts and Sciences*, new series, 3 (1814): 321.

39. James M. Swank, *Introduction to a History of Ironmaking and Coal Mining in Pennsylvania* (Philadelphia, 1878), p. 55; BHL to Fulton, 8 February 1814 and 1 March 1814, Letterbooks.

Washington Navy Yard Steam Engine

Latrobe's most important activity as engineer to the Navy Department, an appointment he held from 1804, was the installation of a steam-driven forge, sawmill, and block mill in the Washington Navy Yard. In his earliest planning for the Navy Yard, Latrobe recognized the importance of such machines for a viable naval facility.[40] In 1808 or 1809 Latrobe began negotiations with James Smallman, who lived in Philadelphia, concerning the manufacture of an engine for the yard, and in August 1809 Latrobe received authorization to go ahead with the project.[41]

Smallman built the engine in Philadelphia during the latter part of 1809 and the early months of 1810.[42] In January and March 1810 he prepared several drawings of the engine which he forwarded to Latrobe, who needed them to plan the dimensions of the engine house.[43] Late in June the completed engine arrived in Washington, accompanied by Smallman and a team of millwrights and smiths. They put the engine in operable condition by mid-September, but it was another two months before the forge was attached and set to work. In the meantime the building was also completed, and Latrobe found a reliable supply of water to cool the condenser.[44]

The Navy Yard engine was a low-pressure Boulton and Watt condensing engine with a 21-inch steam cylinder. It originally had one iron boiler $11\frac{1}{2}$ feet long and 6 feet wide, providing steam at 12 pounds of pressure. Later a second iron boiler 12 feet long and $7\frac{1}{2}$ feet wide was added. The air pump was a double American type similar to that used for the Philadelphia Waterworks engines.[45]

Two of the men whom Smallman brought from Philadelphia to erect the

40. BHL to Tingey, 18 May 1805, Letterbooks.

41. When negotiations began is unclear. A memorandum to Henry Latrobe of 31 May 1809 (Letterbooks) mentions "the Engine of the Navy Yard," and in 1811 BHL claimed that he "labored for 5 Years to erect a steam engine" at the Navy Yard (BHL to Moore, 20 January 1811, Letterbooks). The Smallman drawings of 1808 are evidence that discussions had begun. See pp. 214–22 below.

42. BHL noted in February (BHL to Hamilton, 19 February 1810, Letterbooks) that the cylinder had been bored. It seems likely that it was one of the seven cylinders which Smallman had bored at the Schuylkill Engine House during 1809. *Report of the Watering Committee to the Select & Common Councils, November 2d, 1809* (Philadelphia, 1809).

43. Smallman to Hamilton, 22 August 1810, Miscellaneous Letters Received (M-124), RG 45, NA; see pp. 223–33 below for the drawings.

44. BHL to Hamilton, 1 July 1810, 12 September 1810, 20 October 1810, and 19 November 1810, Letterbooks.

45. BHL to Hazlehurst, 5 February 1815, Letterbooks; BHL to Johnson, 4 January 1813, Letterbooks; BHL to Fulton, 2 February 1813, Letterbooks.

engine remained to operate it and keep it in repair. Mahlon Cooper, a master millwright, competently directed the erection of all the equipment attached to the engine, and later assumed the position of sawyer.[46] Jonathan Criddle was Cooper's chief assistant and remained at the Navy Yard as a pattern-maker.[47] Three other men who assisted in the erection of the engine apparently returned to Philadelphia afterward.

In the fall of 1810 Samuel Ellis, an Englishman who was formerly an engine keeper at the Philadelphia Waterworks, came to Washington to supervise the operation of the engine. After considerable negotiation concerning his salary, Ellis became a regular employee of the Navy Yard and remained there for many years. Latrobe thought enough of Ellis's capabilities that in 1816 he tried to induce him to move to New Orleans to direct the waterworks.[48]

Of the work of Cooper, Criddle, and Ellis, Latrobe stated: "[The engine] was built by Smallman, in a cheap way, but having plenty of good Mechanics at command, I never rested untill I had every thing in the most perfect order, so that we could use all the power which the framing and the strength of the gearing would bear. By the time I had her in perfect order she stood us in about $8500...."[49]

The Navy Yard engine worked an impressive array of equipment. Near the forge was a heating furnace with two bellows driven by the engine. The forge had a 1600-pound stamper which turned heated scrap iron into homogeneous balls, and a 3700-pound tilt hammer to work the balls into useful products. The hammer made 96 to 100 strokes per minute. Both the hammer and the anvil which it struck were cross-faced, giving more working edge so that they were "worth in effect two common forge hammers." In the first four days of operation the forge turned 5600 pounds of scrap into usable iron, which according to Latrobe cost $148.50 less to make than it would have cost to purchase it. At the end of the first full year of operation Latrobe reported to Congress that the forge had saved $6,994.[50]

The loss at sea of equipment sent from Philadelphia impeded the setting up of the sawmill, but in June 1811 it was finally started. It operated at about 150 strokes a minute with a 2-foot stroke, and once sawed 10,000 feet of timber in a 13-hour day. The engine also powered a rope which dragged logs from the dock to the mill, and a fan (installed by Henry Latrobe) which blew the sawdust out of the mill.[51]

Although Latrobe early considered attaching machinery for making wooden pulley blocks, not until March 1812 did he learn of Louis Rose, the first man whom he thought competent to undertake it. Rose had worked in the naval yards of France for nineteen years, and currently had a business in Philadelphia. With Rose's assistance Latrobe calculated that building a block mill would take about six months and cost about $10,000; he thought that in two or three years the money spent would be recovered.[52] Latrobe noted that the particular advantage of a mechanical block mill would be

> that every block of a particular size & use required by the Navy, will take in exactly every sheave intended for that particular kind of block, tho' there were many thousands of them delivered promiscuously, and that also every pin may be indiscriminately put into any block of the kind for which it is intended and fit to a hairbreath and that every sheave will run true and square with the block exactly filling up the mortice as it ought, tho' shifted 1,000 times.[53]

Rose's machinery may have had elements similar to parts of the block milling machines which were installed at Portsmouth, England, in 1807. They are often cited as the first machines to produce interchangeable parts.[54] Rose's block mill was completed in 1813, but was destroyed the following year when the British burned Washington.[55] The sawmill, the forge, and much of the steam engine were also destroyed.[56] (See figure 21, p. 54.)

Having resigned his appointment as engineer to the Navy Department in

---

46. BHL to Hamilton, 1 July 1810, Letterbooks; BHL to Jones, 2 February 1813, Letterbooks. Cooper is listed in the Philadelphia directory of 1809 as a "nailer," and in that of 1810 as a "millwright."

47. BHL to Hamilton, 1 July 1810, Letterbooks; BHL to Foxall, 1 July 1813, Letterbooks. Criddle later went to Pittsburgh to work for BHL.

48. BHL to Smallman, 21 October 1810, Letterbooks; BHL to Tingey, 5 January 1811, Letterbooks; BHL to Ellis, 3 May 1816, Letterbooks; David Bailie Warden, *A Chorographical and Statistical Description of the District of Columbia* (Paris, 1816), p. 64.

49. BHL to Hazlehurst, 5 February 1815, Letterbooks.

50. Ibid.; BHL to Johnson, 4 January 1813, Letterbooks; BHL to Hamilton, 3 December 1810, Letterbooks; BHL to Ogden, 3 March 1814, Letterbooks.

51. BHL to Armstead, 25 February 1811, Letterbooks; BHL to Hamilton, 29 June 1811, Letterbooks; BHL to Johnson, 4 February 1813, Letterbooks; BHL to Hazlehurst, 5 February 1815, Letterbooks. Hamlin's summary of a letter from BHL to Martha Sellon of 15 November 1817 mentions the sawdust fan installed by Henry: Hamlin, *Latrobe*, p. 602. (The Letterbook copy is missing.)

52. BHL to Hamilton, 11 September 1810, Letterbooks; BHL to Tingey, 3 March 1812, Letterbooks; BHL to Hamilton, 23 March 1812, Letterbooks.

53. BHL to Tingey, 3 March 1812, Letterbooks.

54. Rees, *Cyclopaedia*, s.v. Machinery for manufacturing Ships' Blocks; Cardwell, *Turning Points in Western Technology*, p. 118.

55. BHL to Jones, 2 February 1813, Letterbooks.

56. Jones to Anderson, 11 November 1814, in *ASP, Naval Affairs*, 1:361.

The Engineering Practice of Benjamin Henry Latrobe

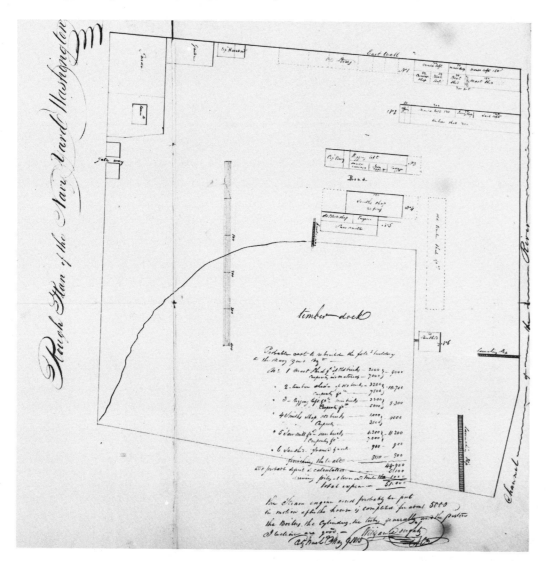

Fig. 21. Contemporary map of the Washington Navy Yard, 1815. Note the complex of the steam engine, smith shop, block shop, and saw mill, slightly above and to the right of center. (RG 45, NA.)

equipment as well. After another decade of dependable work the engine was replaced by a new engine of greater horsepower.[58]

The Washington Navy Yard steam engine operated one of the more unique and complex industrial establishments of the era, and its history demonstrates Latrobe's intense interest in the application of steam power to industry.

Blydensburg Loom

The episode of the Blydensburg loom was one of the worst of Latrobe's financial ventures. It began when Samuel Blydensburg, a New Englander, came to Washington late in 1809 to promote a cotton loom which he claimed to have invented.[59] It was a power loom: that is, one suitable for operation by water or steam power, and capable of automatic weaving. Power looms at the time were still in the developmental stage, producing only a small fraction of the cotton cloth manufactured in England and America.[60] They had the potential of vastly increasing the rate and quantity of cotton manufactures if they could be made to work reliably and produce acceptable cloth.

A congressman introduced Blydensburg to Latrobe and asked if Blydensburg's loom might be set up and exhibited in a vacant part of the Capitol. When Latrobe obliged, Blydensburg put on a public demonstration, attended by Latrobe, during which the loom actually produced cloth when operated by a crank.[61] Blydensburg interested several congressmen in his loom, but he was unable to demonstrate to them that he could put it into sound working order.[62]

In August 1810 Latrobe became seriously interested in the device, and made an agreement with Blydensburg to purchase forty looms at $100 each. Since Blydensburg had made a patent application, Latrobe also committed himself to paying a one-time fee of $50 per loom as a royalty. At first, Washingtonians Jeremiah Mason and Thomas Ewell were partners in this venture, and later the Baltimore merchant William Lorman joined in, but all

1813, Latrobe had nothing to do with the reconstruction of the Navy Yard, which did not take place until 1816.[57] Samuel Ellis then rebuilt the steam engine, and other English mechanics also established themselves at the yard. By 1820 the engine again drove the forge and sawmill, and possibly other

---

57. BHL to Criddle, 26 May 1816, Letterbooks; Warden, *Description of the District of Columbia*, pp. 62–63.

58. Patterson to Chauncey, 3 November 1838, in Thomas C. Cochran, ed., *The New American State Papers. Science and Technology*, 14 vols. (Wilmington, Del.: Scholarly Resources, 1972), 7:355–56; W. Faux, *Memorable Days in America* (London, 1823), reprinted in Reuben Gold Thwaites, ed., *Early Western Travels*, 32 vols. (Cleveland: A. H. Clark Co., 1904–7), 11:119; *ASP. Naval Affairs*, 1:715; Warden, *Description of the District of Columbia*, pp. 62–64.

59. BHL to Hazlehurst, 1 September 1810, Letterbooks. The *National Intelligencer* (Washington), 15 August 1810, said Blydensburg was "a native of Connecticut."

60. David J. Jeremy, "Innovation in American Textile Technology during the Early 19th Century," *Technology and Culture* 14 (1973): 41, 45; T. S. Ashton, *The Industrial Revolution, 1760–1830*, 2d ed. (London: Oxford University Press, 1968), p. 53.

61. BHL to Hazlehurst, 1 September 1810, Letterbooks.

62. *National Intelligencer* (Washington), 15 August 1810.

shortly withdrew. Only Latrobe actually committed money to the venture.[63]

Latrobe designed a mill to house the forty looms, and foresaw a work force of a superintendent, a master weaver, and sixteen women and girls. Within a few weeks of making his agreement with Blydensburg, he rented a mill seat on the Potomac Company's canal about three miles above Georgetown, and began erection of a two-story mill 40 by 18 feet. A former millwright at the Washington Navy Yard supervised construction, which was completed by December.[64]

Blydensburg had meanwhile accepted a $1,000 cash advance from Latrobe as payment for the first ten of the forty looms.[65] Early in September he left Washington for New York, presumably to set up a shop for manufacturing his machines. Thereafter he communicated with Latrobe sporadically, usually to ask for more money, to offer excuses for not delivering looms, or to report that he had some looms nearly ready and could send them soon. But Blydensburg never delivered a single loom to Latrobe.

Although Latrobe fancied himself knowledgeable in textile manufacture,[66] he apparently knew very little about recent trends and was unable to evaluate Blydensburg's loom accurately. For example, Blydensburg led Latrobe to believe that one of his loom's central features, the grooved cylinders which functioned as cams triggering the shuttle, was his innovation and patentable.[67] Yet shortly after Blydensburg's departure Latrobe learned that the new Washington Manufacturing Company had a loom with a similar device.[68] In November he wrote frankly to his associate of his growing knowledge of power loom technology: "You must also be aware, that the cylindrical grooved Cam is used in the old Loom in the patent office for throwing the Shuttle, and I believe in the English Loom."[69] Yet Latrobe believed that other features of the Blydensburg loom were patentable, and for a time remained Blydensburg's advocate at the patent office. He even attempted to apply for French and English patents.[70]

By the summer of 1811 Latrobe became disillusioned by the failure of Blydensburg to supply any looms. In July he recapitulated his investment in the scheme and found that he had sunk about $2,500 into payments for the looms, expenses related to the mill, and incidentals. Not a cent was ever returned. His last correspondence with Blydensburg was in the summer of 1812 when it appeared that two looms might finally arrive, but that last hope vanished like the rest. At Latrobe's bankruptcy in December 1817 he listed a debt of $2,731.32 from Samuel Blydensburg as "uncollectible."[71]

The entire experience must have been a bitter one for Latrobe, especially when he knew from an early point that he might be "a sort of speculator who has got bit."[72] Nonetheless, his ebullient nature prevailed, and he remained interested and active in textile manufacture. In 1812 and 1813 he wanted to supply a steam engine to the new Hamilton cotton manufactory in Baltimore, although he was unable to obtain the order.[73] And in the next year he became the engineer of the Steubenville Woolen Mill.

Of the subsequent activities of Samuel Blydensburg little is known. In 1815 he actually received a patent for a power loom of unknown specifications, and fourteen years later he also patented a throstle spinner. In the 1840s he was the editor and author of publications on silk production.[74] Blydensburg's continuing interest in textiles, however, gave him no more historical fame than his early and abortive attempt to manufacture a workable power loom.

Steubenville Woolen Mill

The construction of a woolen mill in Steubenville, Ohio, provided Latrobe with his only opportunity to participate in the development of the growing American textile industry. The mill was largely the project of Bezaleel Wells, a founder of Steubenville and a proprietor of its largest bank. Wells's partners in the woolen enterprise were Samuel Patterson of Steubenville and James Ross and Henry Baldwin of Pittsburgh, the latter being important Pennsylvania politicians. It was probably through his friendship with Baldwin that Latrobe

---

63. BHL to Blydensburg, 14 August 1810, Letterbooks; BHL to Hazlehurst, 1 September 1810, Letterbooks; BHL to Ewell, 15 August 1810, Letterbooks; BHL to Blydensburg, 26 September 1810 and 8 December 1810, Letterbooks; Hamlin, *Latrobe*, p. 368.
64. BHL to Blydensburg, 18 August 1810, Letterbooks; BHL to Glasgow, 12 December 1810, Letterbooks; BHL to Blydensburg, 18 January 1811 and 7 September 1812, Letterbooks.
65. BHL to Blydensburg, 14 August 1810 and 4 July 1811, Letterbooks.
66. BHL to Blydensburg, 1 September 1810, Letterbooks.
67. BHL to Blydensburg, 25 October 1810, Letterbooks.
68. BHL to Blydensburg, 11 September 1810, Letterbooks.
69. BHL to Blydensburg, 19 November 1810, Letterbooks.
70. Ibid.; BHL to Blydensburg, 8 December 1810 and 7 March 1811, Letterbooks.

71. BHL to Blydensburg, 4 July 1811, 16 June 1812, 4 July 1812, 13 August 1812, and 20 August 1812, Letterbooks; BHL, bankruptcy petition, 19 December 1817, RG 21, NA.
72. BHL to Blydensburg, 7 March 1811, Letterbooks.
73. BHL to McKim, 20 March 1813, Letterbooks; BHL to Graf, 20 August 1812, Letterbooks; BHL to Gwynn, 24 March 1813, Letterbooks; BHL to McKim, 5 April 1813, Letterbooks; Richard W. Griffin, "An Origin of the Industrial Revolution in Maryland: The Textile Industry, 1789–1826," *Maryland Historical Magazine* 61 (1966): 32.
74. Henry L. Ellsworth, *A Digest of Patents Issued by the United States, from 1790 to January 1, 1839* (Washington, 1840), pp. 109, 125; Hamlin, *Latrobe*, pp. 367n–368n. Blydensburg also shared a patent for "clocks" in 1833: Ellsworth, *A Digest of Patents*, p. 232.

received a commission in 1814 to oversee the equipping of the mill. At the time, the building (110 by 28 feet and 3 stories) was under construction and a decision had been made to power the machinery by steam.[75]

Latrobe designed a 16-horsepower engine of the type patented by Francis Ogden in 1813. It had two 12-inch steam cylinders at right angles to one another, and worked on the expansive force of steam after admitted to the cylinders, rather than the original pressure of the steam. It did, however, use a condenser to evacuate the cylinders. The boiler was also somewhat unusual, having a teakettle form rather than the rectangular shape of the standard wagon boiler. Like other industrial engines with which Latrobe was associated, this engine had a cast-iron flywheel, in this instance 14 feet in diameter. The working beam was also cast iron. To insure rapid completion of this engine Latrobe distributed contracts for its parts among the three Pittsburgh foundries of Thomas Copeland, George Evans, and Anthony Beelen.[76]

Fig. 22. BHL's plan of the Steubenville engine, 1814. (BHL to Wells, 7 November 1814, Letterbooks.)

In August 1814 Latrobe visited Steubenville for the first time, examining the space allocated for the engine so that he could plan its installation. His major interest was setting up the wooden frame to which the engine would be anchored.[77] By late fall almost all parts of the engine were ready, and Latrobe took steps to have it put up. First he sent Thomas Copeland to Steubenville to erect the boiler, and then he sent most of the parts. In the second week of December he dispatched a group of six men to put the engine together. One was Jonathan Criddle, who had helped assemble the Washington Navy Yard engine in 1810, and who had since come to Pittsburgh to work on the Ohio steamboat. Also included was Isaac Carey, another "steam engineer." Latrobe visited Steubenville later in December to observe their work.[78]

Problems in boring the steam cylinders caused some delay in completing the engine. They were cast and probably bored by George Evans at the Pittsburgh Steam Engine Company's shop. The first cylinder was acceptable, but two others were cast, bored, and found faulty before the fourth came out clean. In late January 1815 the two good cylinders were sent to Steubenville. During the rest of the winter and spring the workmen were busy erecting and adjusting the engine.[79]

After a third visit to the mill in March, Latrobe took the responsibility for placing an order with a "Mr. Price" (probably William Price, an English immigrant) in Pittsburgh for a wool carding machine. He also discussed with him his idea for a mechanical quilling machine which could wind thread onto quills for the weaving shuttles. Although both Price and Latrobe thought the idea ingenious, it is not clear that the device was tried. If it was, it antedated by more than a year the patenting of a similar device by two Massachusetts mechanics.[80]

Latrobe anxiously awaited news of the engine's trial in early April, especially when he knew that he would soon be returning to Washington as architect of the U.S. Capitol. He was distressed to learn that at first the engine did not operate satisfactorily. Apparently that fact occasioned his fourth visit to Steubenville late in May, when he finally put things in order.[81] Thereafter the engine worked well, and Latrobe heard in 1816 that it was "the best in the Western Country."[82]

---

75. W. H. Hunter, "The Pathfinders of Jefferson County," *Ohio Archaeological and Historical Publications* 6 (1898): 210, 234–35; *Biographical Directory of the American Congress, 1774–1961* (Washington: Government Printing Office, 1961), pp. 508, 1543; BHL to Baldwin, 16 March 1814, Letterbooks.

76. Pursell, *Early Stationary Steam Engines in America*, pp. 115–16; BHL to Wells, 22 April 1814, 21 November 1814, and 29 December 1814, Letterbooks; BHL to Cooper, 10 July 1814, Letterbooks; BHL to Commissioners of the Navy, 20 December 1817, RG 45, NA; *ASP. Finance*, 4:198; Hunter, "Pathfinders," p. 235. The unusual shape of the boiler noted by Hunter seems confirmed by the outline BHL gives it in his sketches. See fig. 22 above.

77. BHL to Cowan, 5 August 1814, Letterbooks; BHL to Wells, 6 August 1814 and 15 October 1814, Letterbooks.

78. BHL to Wells, 7 November 1814, 1 December 1814 [misdated 31 November 1814], and 8 December 1814, Letterbooks; *The Pittsburgh Directory for 1815* (Pittsburgh, 1815; reprint ed. [Pittsburgh]: The Colonial Trust Company, 1905), p. 17; BHL to Roosevelt, 26 December 1814, Letterbooks.

79. BHL to Wells, 29 December 1814, Letterbooks; BHL to Criddle, 5 January 1815, Letterbooks; BHL to Ogden, 23 January 1815, Letterbooks.

80. BHL to Roosevelt, 22 March 1815, Letterbooks; BHL to Orth, 27 March 1815, 2 April 1815, and 5 April 1815, Letterbooks; Jeremy, "Innovation in American Textile Technology," p. 54. For references to William Price, see Joseph D. Weeks, *Report on the Manufacture of Glass* (New York, 1883), p. 85; Edward Thornton Heald, *Bezaleel Wells: Founder of Canton and Steubenville, Ohio* (Canton, O.: Stark County Historical Society, 1948), p. 167.

81. BHL to Ogden, 3 April 1815, Letterbooks; BHL to Mary E. Latrobe, 21 April 1815, Mrs. Gamble Latrobe Collection, MdHS; "Letters. June 1815," Letterbooks.

82. BHL to Criddle, 26 May 1816, Letterbooks.

Industrial Works

Although it changed ownership, the mill which Latrobe helped to get under way operated until 1867 when it was destroyed by fire.[87]

Ohio Steamboat Company

At one time Latrobe believed that any attempt to power a boat with a steam engine would be an economic failure,[88] but he changed his mind after the success of Robert Fulton's *North River* (later known as the *Clermont*). Latrobe had become acquainted with Fulton within a year of Fulton's return from Europe in December 1806, probably through their common friendship with Joel Barlow. Fulton lived at the Barlow residence, *Kalorama*, while he was in Washington conducting his patent affairs.[89]

Steam engines and steamboats soon became a major topic of conversation between the two engineers. Latrobe was called upon to negotiate between Fulton and Nicholas Roosevelt when Roosevelt claimed that he had a prior agreement on steamboat manufacture with Fulton's partner, Robert R. Livingston. Eventually Roosevelt (Latrobe's son-in-law) became the agent for Fulton's Mississippi Company and in 1811 built and operated the *New Orleans*, the first steamboat to navigate the Mississippi River.[90]

In 1812 Fulton and Livingston made Latrobe their agent for two new steamboat companies in the Washington area. One was to operate between Georgetown and Alexandria, the other between Washington and the mouth of the Potomac Creek above Fredericksburg, Virginia. In return for half of the royalty payments on their steamboat patent, Latrobe agreed to raise the capital for the companies.[91] He succeeded in obtaining sufficient stock subscriptions to organize the Washington-Fredericksburg line (the Potomac Steam Boat Company), but the British blockage of the Chesapeake made it impossible to bring one of Fulton's boats down from New York and the enterprise languished until after the war. When it was reorganized and put into

Fig. 23. Steubenville Woolen Mill, 1815. Watercolor by BHL dated 9 March 1815. (Latrobe Sketchbook XI.)

The Steubenville Woolen Mill was one of the earliest steam-powered woolen mills in the United States, as well as one of the most prominent of the era.[83] In 1820 the factory contained "5 carding machines, 1 picker, 572 spindles, 8 shearing machines, working 16 pairs of hand shears by steam power; 3 shearing machines with spiral blades; 4 fulling stocks, 12 broad looms, 6 narrow looms, 1 bobbin winder, 2 warping mills, [and] 2 machines for nap raising."[84] *Niles' Weekly Register*, a newspaper of national circulation, referred to the mill in 1825 as "the celebrated establishment of Messrs. B. Wells and company, [which sends] to the Atlantic states many thousand dollars' worth of superior superfine cloths every year and a large amount of other woolen goods."[85] Some of the mill's success was due to its skilled work force. Christopher Orth, a German mechanic who patented a cloth-shearing machine in 1814, was the first manager of the mill, and many of the original employees were European workmen.[86]

---

83. Arthur Harrison Cole, *The American Wool Manufacture*, 2 vols. (Cambridge, Mass.: Harvard University Press, 1926), 1:195–96, 251n.
84. *ASP, Finance*, 4:198.
85. *Niles' Weekly Register* (Baltimore), 28 (1825): 82n.
86. J. A. Caldwell, *History of Belmont and Jefferson Counties* (Wheeling, W.Va., 1880), p. 493; Ellsworth, *A Digest of Patents*, p. 116.

87. Joseph B. Doyle, *Twentieth Century History of Steubenville and Jefferson County, Ohio* (n.p.: Richmond-Arnold Publishing Co., 1910), p. 282.
88. BHL, "Steam Engines in America," pp. 90–91.
89. H. W. Dickinson, *Robert Fulton: Engineer and Artist, His Life and Works* (London: John Lane, 1913), pp. 206, 289; BHL to Fulton, 28 December 1807, Letterbooks; Hamlin, *Latrobe*, pp. 313–14.
90. BHL to Fulton, 7 February 1809, Letterbooks; BHL to Roosevelt, 7 February 1809, Letterbooks; J. H. B. Latrobe, *The First Steamboat Voyage on the Western Waters*, Maryland Historical Society Fund Publication, No. 6 (Baltimore, 1871). Roosevelt married Lydia Latrobe on 15 November 1808.
91. BHL, Robert Fulton, and Robert Livingston, indenture, 10 October 1812, LeBoeuf Collection, N-Y Hist. Soc. (not in microfiche ed.); BHL to Fulton, 3 January 1813, Letterbooks; BHL to DeLacy, 29 December 1814, Letterbooks.

operation in 1815, Latrobe was no longer Fulton's and Livingston's agent.[92] In the meantime, however, Latrobe embarked on a fateful steamboat adventure in Pittsburgh.

He began the adventure indirectly. In 1810 Latrobe had obtained the franchise for the New Orleans Waterworks, and had begun to make the steam engine and other equipment in Washington. In 1812 he shipped some articles to New Orleans, but the blockade at the beginning of the War of 1812 made further shipments hazardous. Committed to the New Orleans project, and with his work in Washington reduced because federal appropriations were diverted to military expenditure, Latrobe decided that he should move to Pittsburgh to complete the equipping of the waterworks.[93] Pittsburgh was already famous as an industrial city and was certain to have adequate facilities for the project; more importantly, it was at the head of an almost uninterrupted internal navigation to New Orleans.

Almost immediately Latrobe recognized that his plan coincided with Robert Fulton's great design to establish a system of steamboats on western rivers. Roosevelt's famous trip to New Orleans in 1811 had demonstrated that it was possible to build a steamboat in Pittsburgh which could operate commercially, and Fulton began gathering capital for more boats. He possessed charters for two companies, the Ohio Steamboat Company for the Ohio River and the Mississippi Steamboat Company for the Mississippi, which together could establish a chain of regularly scheduled boats connecting Pittsburgh and New Orleans.[94]

The Ohio Steamboat Company was incorporated in 1810 by an act of the Indiana legislature to encourage the development of steam navigation on the Ohio River. The incorporators included Fulton, Livingston, and Nicholas Roosevelt.[95] This company was to build and operate steamboats between Pittsburgh and the dangerous falls at Louisville, Kentucky, where transfer of passengers and goods was necessary. Fulton's plan called for the Ohio Company to build a large boat with a powerful engine and passenger accommodations, and for a smaller freight boat which would be towed by the engine boat.[96] In the summer of 1812 Fulton was taking steps to implement this great scheme, and in July Latrobe wrote to offer his services.[97]

Over the fall, winter, and spring of 1812/13, Latrobe and Fulton corresponded about steamboat topics, slowly coming to an agreement about what Latrobe's role might be in promoting Fulton's western operations. At first Latrobe suggested that they become partners in a foundry and engine shop which would build boat engines, the waterworks engine, and sugar mill engines for Louisiana. But Fulton did not like involving anyone else so directly in the affair, and finally decided to assign one-third of his patent royalties of the Ohio Steamboat Company to Latrobe and to give him an annual salary of $2,000.[98] In return Latrobe was to act as Fulton's and Livingston's agent in Pittsburgh, helping to raise funds and to supervise construction. Fulton remained in New York and was responsible for raising most of the capital.

By the summer of 1813 Fulton's Mississippi Steamboat Company had established shops in Pittsburgh and had its first boat, the *Vesuvius*, on the stocks. Fulton was anxious for the Ohio Company to begin work, but Latrobe found that his personal finances were so confused that he could not leave Washington without being threatened with lawsuits. Early in July he wrote to Fulton listing his assets and debts, recording a theoretically favorable balance, but commenting: "This side [of] the grave however there is not a poorer man."[99] Shortly thereafter he went to New York to thoroughly discuss the future arrangements in Pittsburgh. Fulton gave him drawings of the steamboat and officially assigned one-third of the patent rights for the Ohio Steamboat Company to Latrobe.[100]

August and September Latrobe spent clearing up personal affairs and getting ready to move. He resigned his government appointment, sent some equipment and men to Pittsburgh, organized his family, and tried to realize his assets to cover his debts. He was unable to fully satisfy his creditors, and finally borrowed $1,500 from Fulton so that he could leave Washington.

---

92. Account of meeting of the subscribers to the Potomac Steam Boat Company, 2 January 1813, and notice of the suspension of subscriptions to the Potomac Steam Boat Company, [12 September 1813], Papers of Joseph Gales and William W. Seaton, LC; BHL to DeLacy, 29 December 1814, Letterbooks; *National Intelligencer* (Washington), 30 December 1815.
93. The first mention of the move is in BHL to Henry Latrobe, 10 June 1812, Letterbooks.
94. Louis C. Hunter, *Steamboats on the Western Rivers: An Economic and Technological History* (Cambridge, Mass.: Harvard University Press, 1949), p. 309.
95. Hunter, *Steamboats*, p. 309n.
96. The reasoning behind the plan for both an engine boat and a freight boat on the Ohio is not clear, although during Fulton's experiments in Paris in 1803 he suggested using "an engine boat intended for towing one or several freight boats each of which shall be so close to the preceding that water does not flow between to create resistance." Dickinson, *Fulton*, p. 152.
97. BHL to Fulton, 5 July 1812 and 13 July 1812, Letterbooks.
98. BHL to Fulton, 31 July 1812, 21 March 1813, 15 March 1814, and 17 May 1814, Letterbooks.
99. BHL to Fulton, 6 July 1813, Letterbooks.
100. BHL to Baldwin, 10 October 1814, Letterbooks; BHL to Fulton, power of attorney, 17 May 1814, Letterbooks.

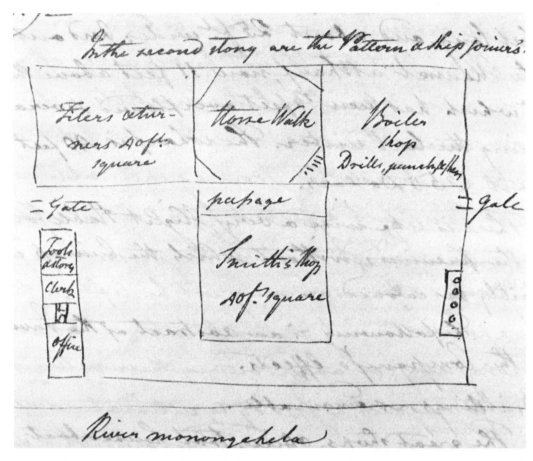

Fig. 24. Plan of BHL's machine shop at Pittsburgh, 1814. (BHL to Fulton, 21 May 1814, Letterbooks.)

After brief family visits the Latrobes were on their way over the Appalachians. They arrived in Pittsburgh on the last day of October.[101]

Latrobe found work for the Ohio boat well under way. On hand were the steam cylinder, air pump, and assorted brass castings for the engine. Latrobe immediately ordered additional ironwork, including boiler plates, and began to think about putting the engine together. Fulton had assured him that the Mississippi Company's shops would have room enough for him, but Latrobe found it crammed with the work of the two boats the company then had under way. He decided it was impossible to begin operations in the Mississippi Company's shops.[102]

At first Latrobe rented the idle engine manufactory of Thomas Copeland,[103] where he "got many of our principal tools and some of the Machinery underway."[104] But Copeland terminated the arrangement within a few weeks, and Latrobe decided to build his own shops on the land along the Monongahela which he had originally rented for a shipyard. By early January 1814 he had raised a simple timber structure containing a boiler shop, smith shop, and a room for lathes and filing. In the center was a horse gin with gearing to drive lathes and drills.[105] On adjacent land was the foundry of Anthony Beelen, one of Latrobe's principal suppliers.[106]

By the time his shop was ready Latrobe had hired a competent group of skilled men to make the engine and boat. He intended to bring several of the men from his Washington engine shop to Pittsburgh, but none came. He did send on a "filer and finisher" from the Washington area to join his staff, and near the end of the work Jonathan Criddle, formerly a millwright at the Washington Navy Yard, arrived. Latrobe also employed several Pittsburgh smiths and engine makers.[107]

Like most people engaged in shipbuilding on western waters, Latrobe brought from the East an experienced shipwright. William Hurley, who for several years was the second-ranking shipwright at the Washington Navy Yard, directed the procurement of timber and the construction of both boats. There was at least one other eastern boatbuilder working under Hurley, but such talent was scarce enough that in January 1814 Latrobe requested the Army to assign to him a British prisoner of war whom he believed to be a "very capital shipwright."[108]

Latrobe also looked ahead to the final outfitting and operation of the Ohio boat, and brought Nicholas D. Baker to Pittsburgh for these purposes. Baker had been Roosevelt's engineer on the *New Orleans*, and Latrobe had a high opinion of his abilities. In October 1812 Latrobe had inquired after Baker in

---

101. BHL to Fulton, 15 August 1813 and 5 September 1813, Letterbooks; BHL to Roosevelt, 2 November 1813, Letterbooks; Fulton to Cooke, 24 October 1814, Gilbert H. Montague Collection, New York Public Library, New York, N.Y.

102. BHL to Fulton, 2 November 1813, 11 November 1813, and 15 March 1814, Letterbooks.

103. Zadok Cramer, *The Navigator: Containing Directions for Navigating the Monongahela, Allegheny, Ohio and Mississippi Rivers* (Pittsburgh, 1814), p. 60.

104. BHL to Fulton, 15 March 1814, Letterbooks.

105. Ibid.; BHL to Baker, 27 January 1814, Letterbooks; BHL to Stackpole, 24 May 1814, Letterbooks; BHL to Baldwin, 10 October 1814, Letterbooks.

106. BHL to Fulton, 11 November 1813, Letterbooks; Cramer, *Navigator*, p. 57.

107. BHL to Fulton, 15 August 1813, Letterbooks; BHL to Baker, 5 November 1813 and 27 January 1814, Letterbooks; BHL to Criddle, 30 April 1814, Letterbooks.

108. "Pay Roll for Boat-Builders, employed in the Navy Yard, Washington, in the month of July, 1811," entry 258, RG 45, NA; BHL to Sinclair, 1 January 1815, Letterbooks; BHL to Tannehill, 8 February 1814, Letterbooks; BHL to Mason, 27 January 1814, RG 45, NA; Hunter, *Steamboats*, pp. 67, 69.

connection with the Potomac Steam Boat Company, but as the Pittsburgh venture became more certain he made it clear that he wanted Baker for the Ohio boat. Not needing him until the final stages of work, Latrobe first employed him to erect a steam engine in Baltimore, but finally brought him west in May 1814.[109]

It was this team of artisans that Latrobe supervised at the shop and shipyard on the Monongahela. The boat they were to build was described in the drawings Fulton had given Latrobe as 116 feet long with a 25-foot beam. The engine was a Boulton and Watt type with a 34-inch steam cylinder. Latrobe modified Fulton's plans only slightly, making the hull 4 feet longer, and designing a new boiler for the engine.[110]

The engine boat was framed in January, planked and caulked in April, and launched in May. Latrobe was pleased to find that on launching she drew only 2 feet 3 inches of water, and he thought her "a most capital vessel." Named the *Buffalo*, she was rated at various times from 250 to 300 tons, and probably had side paddle wheels.[111] The freight boat progressed more slowly than the *Buffalo*, not having her keel laid until late in May, nor being planked until July. She had a length of 100 feet with a beam of 20 feet, and was rated at 136 to 154 tons. Although Fulton did not design her to have an engine, Latrobe made some alterations so that she could carry one with a 24-inch steam cylinder. She was named the *Harriet*, after Fulton's wife.[112]

Unfortunately, Latrobe never saw either boat enter commercial use because he and Fulton had a dispute which left the Ohio Steamboat Company without working capital. At the root of the problem was both parties' lack of knowledge of the situation in Pittsburgh. Before Latrobe left for the west he had agreed with Fulton that the Ohio boats could be built as an adjunct to the Mississippi Company's operations for a total cost of about $25,000, the amount of the Ohio Company's capitalization. But Latrobe expended $4,000 to build independent shops, and something more to purchase tools. In addition, the rapid inflation of the last of the war years drove prices and wages to heights never dreamed of during the planning for the Ohio boats. Latrobe was forced to call upon Fulton for frequent advances, and the cost quickly rose above the original ceiling.[113]

As early as March 1814 Latrobe had a letter from Fulton complaining about the financial situation, and two months later Fulton refused to accept any further drafts from Latrobe. Ultimately he accused Latrobe of extravagance, especially in building separate shops and in renting land for a shipyard. It also rankled him that Latrobe had never repaid his $1,500 personal loan. In September he discharged Latrobe from the agency for the Ohio Steamboat Company.[114]

Latrobe responded to Fulton's initial displeasure by asserting that ". . . I did not believe that altho' the company's capital at which a dividend is to be made is only 25.000$ that that was to be the limit of the cost of the boat to be built by me. I took it always for granted that the object was to be accomplished *at all events*. . . ."[115] Latrobe then consulted with John Livingston, agent for the Mississippi Steamboat Company, and learned that each of the engine boats which that company had under way would ultimately cost about $45,000. He could not imagine that Fulton could expect him to do much better than that, especially when Fulton had prescribed the size of the boat and engine.[116]

Latrobe conceded that expenses had exceeded original estimates, and attempted to ease the strains on Fulton's capital. With his own earnings from architectural endeavors in Pittsburgh he paid for some tools and wages. He made clear that he would purchase the shop from the Ohio Company so that its cost would actually be recovered. He gave Fulton a power of attorney to mortgage his share of the patent right to raise more money, and he convened the Pittsburgh stockholders in an unsuccessful attempt to get a loan from them.[117]

But during the summer of 1814 work on the Ohio boats ground slowly to a halt as Fulton refused to make further advances and Latrobe's personal

109. J. H. B. Latrobe, *First Steamboat Voyage*, p. 14; BHL to Roosevelt, 27 October 1812 and 8 May 1814, Letterbooks; BHL to Fulton, 26 May 1813, Letterbooks; BHL to Graf, 12 December 1814, Letterbooks.

110. BHL to Fulton, 20 January 1814, 3 June 1814, 3 July 1814, and 21 August 1814, Letterbooks.

111. BHL to Roosevelt, 20 January 1814, Letterbooks; BHL to Fulton, 1 May 1814, 14 May 1814, and 21 August 1814, Letterbooks; BHL to Graf, 25 May 1814, Letterbooks; David Whittet Thomson, "The Great Steamboat Monopolies, Part I: The Mississippi," *The American Neptune* 16 (1956):34n.

112. BHL to Fulton, 29 May 1814, 9 August 1814, and 21 August 1814, Letterbooks; BHL to McNeven, 24 July 1814, Letterbooks; Jean Baptiste Marestier, *Memoir on Steamboats of the United States of America*, Sidney Worthington, trans. (Mystic, Conn.: The Marine Historical Association, Inc., 1957), p. 60. Hunter, *Steamboats*, p. 309, rates the *Harriet* at 54 tons, which must be an error.

113. BHL to Baldwin, 10 October 1814, Letterbooks.

114. BHL to Fulton, 15 March 1814 and 17 May 1814, Letterbooks; BHL to Baldwin, 10 October 1814, Letterbooks; Fulton to BHL, 26 October 1814, LeBoeuf Collection, N-Y Hist. Soc. (not in microfiche ed.).

115. BHL to Fulton, 15 March 1814, Letterbooks.

116. Ibid.

117. BHL to Fulton, 1 May 1814, 17 May 1814, and 27 May 1814, Letterbooks; BHL to McNeven, 6 September 1814, Letterbooks.

resources ran dry. On being ordered by Fulton to turn over all of the Ohio Company's assets to the officers of the Mississippi Company, Latrobe put a lien on the property as a creditor. A friendly sheriff deputized Hurley, the shipwright, to take charge of the boats, and Baker, the engineer, to take possession of the shop and engine.[118]

This action made Latrobe's break with Fulton irreconcilable, and the matters of the Ohio Steamboat Company were thereafter out of his hands. Fulton's own quarrel with the New York stockholders gave Latrobe some personal satisfaction, but in December 1814 a Pittsburgh court ruled that Latrobe had no right under Pennsylvania law to put a lien on the property and he lost any chance for a good bargaining position. In fact, he never did recover the money advanced to the Ohio Company, and his personal bankruptcy in 1817 resulted partly from his unsettled claim of over $4,000.[119] For eight months before his departure from Pittsburgh in April 1815 his boats and shop on the Monongahela lay idle, along with his hopes for an important role in the development of western steamboating. His boats eventually went into service after the Ohio Steamboat Company's assets were sold at a sheriff's sale.[120]

It is significant that after a few months in Pittsburgh Latrobe realized that the design of the boats he was building for Fulton was unsuited for Ohio River navigation. The *Buffalo* was essentially the same sort of vessel that the first western steamboat, the *New Orleans*, had been: broad-beamed, deep-hulled, generally on the lines of an oceangoing ship. After descending the Mississippi, the *New Orleans* remained to run between New Orleans and Natchez, partly because its deep hull made it difficult to operate in the shallower waters of the upper Mississippi basin.[121]

Latrobe was struck by the rapid changes in the stages of the rivers at Pittsburgh and by the long summer and fall season of low water which made it impossible for boats the size of his *Buffalo* even to go downstream. During his first summer in Pittsburgh he reached the conclusion that the *Harriet*, with a hull depth of $6\frac{1}{2}$ feet compared to the *Buffalo*'s $9\frac{1}{2}$ feet, should be converted to a steamboat and used on the Ohio in the summer. He went on to conclude generally that Fulton's boats were far too large and had excessive drafts; further, he promised to design smaller boats with flat bottoms which would be suitable for the Ohio River traffic.[122] It is probable that Latrobe's designs were influenced by the Pittsburgh mechanicians and entrepreneurs who were ushering in the steamboat age on western rivers. Yet it was two decades before such designs became dominant and the essential elements of the classic western steamboat were settled.[123] Had Latrobe's Pittsburgh enterprise not failed, he might have remained in Pittsburgh and changed the history of steamboats.

6. CONCLUSION

Engineering as practiced by Benjamin Henry Latrobe was a complex profession. It was scientific in its general concepts of topography, geology, and hydrology; it was mechanical, involving pumps, pile drivers, industrial machinery, and steam engines; it was precise, requiring surveying, detailed cartography, and fine drawing; and it was managerial, including the organization and supervision of those who carried out Latrobe's designs. Examination of these aspects of Latrobe's engineering reveals the state of the art in his time.

Surveying

Surveying was a basic engineering skill and had a sound mathematical basis by the end of the eighteenth century. Transportation projects, the most numerous and remunerative employment for English engineers, always required accurate instrument surveys to lay out level and direct routes.[1] Latrobe's two maps of the proposed Maldon branch of the Chelmer–Blackwater Navigation are fine expressions of the state of his surveying skill when he left England.[2]

When he came to the United States, Latrobe sent some of his instruments by a vessel which was captured by the French, but brought others with him on

---

118. BHL to McNeven, 30 September 1814, Letterbooks; BHL to Cooke, 1 October 1814, Letterbooks.

119. BHL to Fulton, 13 November 1814 and 22 November 1814, Letterbooks; BHL to Cooke, 27 December 1814, Letterbooks; Hamlin, *Latrobe*, p. 480n.

120. Hunter, *Steamboats*, pp. 16, 19, 19n, 309; Marestier, *Memoir on Steamboats*, pp. 59–60; H[enry] McMurtrie, *Sketches of Louisville and its Environs* (Louisville, 1819), p. 201. McMurtrie speaks of the smaller boat as the *James Monroe* rather than the *Harriet*.

121. Hunter, *Steamboats*, p. 70.

122. BHL to Fulton, 29 May 1814, 5 June 1814, and 21 August 1814, Letterbooks; BHL to Livingston, 27 November 1814, Letterbooks. See also BHL's later plans for a steamboat, BHL to Farish, 26 October 1816, Letterbooks.

123. Hunter, *Steamboats*, p. 69.

1. A. W. Richeson, *English Land Measuring to 1800: Instruments and Practices* (Cambridge, Mass.: Society for the History of Technology and M.I.T. Press, 1966), pp. 187–88; Rees, *Cyclopaedia*, s.v. Canals, the British, and s.v. Road, in Rural Economy.

2. See pp. 114, 115 below.

the *Eliza*. These included a theodolite made by Jesse Ramsden, the foremost instrument maker in England, and a sextant. During his American career he also used other theodolites (including another English one purchased in Philadelphia) and a circumferentor. Latrobe's other surveying equipment included a surveyor's compass, measuring rods, chains, and tapes.[3] When he discussed the possible survey of the future route of the Erie Canal with the New York state commissioners he listed the equipment required for a survey. "It will also be necessary to provide instruments. Those which I possess, I should be unwilling to risk in so rough an operation: they are too expensive. Two telescopic spirit levels [theodolites], with a compass would be wanted, value about 100 to 120$. One I should be glad to have made here on Rittenhouses plan. 4 distance staves 20$ and 2 four pole chains—7$."[4] With similar equipment, Latrobe performed several topographical surveys in the United States, including those for the Susquehanna navigation, the Chesapeake and Delaware Canal, and the Washington Canal.

When he accepted a commission for a survey he consulted the best available map to determine the most likely route of the improvement. In the case of the Chesapeake and Delaware Canal, he used the maps of the area made in 1769 by a party sponsored by the American Philosophical Society.[5] Latrobe began the survey by fixing the altitudes of important points. On one occasion he actually determined elevations after starting the survey from a base point at the tide line.[6] Then he proceeded to lay out possible routes of the improvement using instruments to determine changes in elevation and horizontal angles. On the Susquehanna survey the route of improvement was predetermined, and Latrobe had the time to obtain significant accuracy in measurement of distances by triangulation. He kept the results of his surveys in carefully laid out field books.[7]

Not all preliminary data came from instrument surveys. In several instances Latrobe used a weight and line to sound the depth of rivers.[8] Twice he calculated the volume of water carried by a stream.[9] He also tested river bottoms by probing with long poles, and sampled soil quality by digging pits.[10] In at least one instance, the Susquehanna River survey, Latrobe also collected visual information. His sketches of panoramas, bird's-eye views, and details contributed significantly to the creation of the map.

For all surveying and fieldwork Latrobe regarded qualified assistants as indispensable. He thought it best to hire surveyors who were "Good plain sensible Men, [who are] accustomed to run lines with the Compass and chain, and to keep fair fieldbooks."[11] Such men, who could usually be employed at reasonable rates, enabled Latrobe to survey much more than he could have done by himself. Latrobe found a number of American surveyors whom he regarded as excellent, including John Thomson and Daniel Blaney for the Chesapeake and Delaware Canal; Eugene Leitensdorfer for the Washington Canal and Columbia Turnpikes; and Joseph Jeffers for the Jones' Falls improvement survey.[12]

Drawings

From survey records Latrobe drew maps. Those maps which survive are carefully prepared, highly detailed, and in the best contemporary tradition of European cartography. Representation of natural and man-made features is rendered through different colors and symbols. Routes of the planned canals are often given in section on a portion of the map so that the viewer has a clearer idea of the changes in altitude involved. Some maps have heavy written annotation. Looking at his maps, one feels that Latrobe thought topographically, and wanted his maps to show landscapes rather than just survey data.

Latrobe also made drawings of buildings, machinery, and other objects to make his engineering proposals understandable. (Presumably he made rough

---

3. BHL to Chequiere, 12 November 1804, Letterbooks; BHL to Gilpin, 10 October 1803 and 4 November 1803, Letterbooks; Washington Canal Co. to BHL, account, 25 September 1810, Letterbooks; BHL to Briggs, 30 May 1810, Letterbooks; BHL, *Journals*, 1:28; BHL, bankruptcy petition, 19 December 1817, RG 21, NA; Richeson, *English Land Measuring*, pp. 169–72. Theodolites and sextants measured vertical or horizontal angles between two objects perceived by an observer, while circumferentors measured only horizontal angles.

4. BHL to Fulton & Eddy, 23 May 1811, Letterbooks.

5. 16 December 1803, Minutes of the American Philosophical Society, APS.

6. 2 June 1804, C & D Engineer's Reports, Hist. Soc. Del.

7. The only field books extant are those for the Susquehanna survey, August-November 1801, PBHL, MdHS, and the New Castle (Del.), 1804–05, Robert Mills Papers, Tulane.

8. BHL to Gallatin, 16 March 1808, in Cochran, *New ASP. Transportation*, 1:266; BHL to Tatnall, 23 April 1804, Letterbooks; BHL to Harper, 9 February 1818, Mrs. Gamble Latrobe Collection, MdHS.

9. 26 January 1804, C & D Engineer's Reports, Hist. Soc. Del.; BHL, "Report on the Improvement of Jones' Falls, Baltimore," 31 May 1818, PBHL, MdHS.

10. BHL to Poulson, 24 October 1801, in *Poulson's American Daily Advertiser* (Philadelphia), 27 October 1801; BHL to Harper, 9 February 1818, Mrs. Gamble Latrobe Collection, MdHS; Washington Canal Co. to BHL, account, 25 September 1810, Letterbooks; BHL to Cochran, 20 March 1810, Letterbooks; BHL to Henry Latrobe, 16 August 1812, Letterbooks.

11. BHL to Gales, 16 April 1818, Archibald D. Murphey Papers, North Carolina State Archives, Raleigh, N.C.

12. BHL to Gilpin, 19 October 1803, Letterbooks; BHL to Bradley, 30 January 1811, Letterbooks; BHL, "Report on the Improvement of Jones' Falls, Baltimore," 31 May 1818, PBHL, MdHS.

preliminary sketches to aid in his development of proposals, but only a few from his letterbooks are known to survive.) Having arrived at a mature design, Latrobe created measured technical drawings which depicted it. He took sheets of paper from one to two feet square and fastened them in layers of two or more to drawing boards. Then he mathematically reduced dimensions to scale and pencilled his design on the paper. He pierced his drawing with tiny holes at certain key points in order to transfer the points to the paper beneath, where this pricking would serve as the basis for a duplicate drawing. When Latrobe was satisfied with the final form, he inked over the pencil lines and watercolored appropriate areas. Generally he used a standard code of colors, such as light red for brick, blue for stone, and yellow for wood, but these were seldom flat colors. With delicate shadings, the illusion of a light source and shadows, and often by adding background or foreground elements, Latrobe could create dramatic effects. Although many drawings stand incomplete, or were not intended for public viewing, those drawings on which Latrobe focused his artistic powers are works of compelling beauty.

Latrobe used several types of views in rendering his engineering designs. He often put plans, sections, and elevations, the basic views of technical drafting, on separate sheets, but sometimes he included two or all of them on a single sheet. Additionally, Latrobe employed split sections in which a symmetrical object was divided on the midline, exhibiting a deeper section on one side than the other. To enhance his range of visual communication, Latrobe's drawings included details (enlargements of certain parts), exploded views (parts separated, but in proper relationship for assembly), and broken elements (parts shortened for drawing; indicated by zig-zag break).[13]

Latrobe apparently depended upon a large number of drawings to ensure correct execution of his designs. Preliminary renderings were followed by revisions and details as the work progressed. Some of his drawings were produced partly or completely by pupils and assistants under his supervision, which not only freed him from the drawing board but also taught them a necessary skill.

Latrobe's insistence on scaled drawings (reduced from actual size) was exceptional for his era. Most mechanical engineers, for example, worked from full-size drawings chalked directly on boards, since the pattern-makers and other craftsmen were not used to working from scale.[14] Oliver Evans, in fact, told the young George Escol Sellers that Latrobe "introduced a higher standard of mechanical drawing [to America] that... stimulate[d] our native mechanics."[15]

Personnel

If Latrobe's plans for any given undertaking were acceptable and he was certain of his appointment as engineer, he then organized a group of men to whom he could entrust the execution of the project. If Latrobe could not be on the site of operations continuously he usually requested that a clerk of the works (a position common on English engineering projects) be appointed and paid by the funding group. The clerk of the works was responsible for daily oversight of all work, including certification that a contractor had fulfilled his agreement, keeping all accounts relating to wages, and purchasing tools and materials. To be competent enough to hold such a position a man had to have some technical experience himself. Serving as Latrobe's clerks of engineering works were Frederick Graff, who had training as a builder, John Davis, with engineering and architectural experience in England, Robert Brooke, who had worked under an English engineer, and Henry Latrobe, who was trained by his father.[16]

For the execution of some projects Latrobe also relied upon a permanent staff of skilled craftsmen. The most formal arrangement was on the Chesapeake and Delaware Canal where, apparently following standard English practice for canal construction, Latrobe hired a master carpenter, a master mason, and a master blacksmith. They were expected to live near the works and oversee all jobs requiring their skills. In other instances the arrangement was less precise, but Latrobe still relied on master workmen like William Hurley, the shipwright for the Ohio Steamboat Company.

In a different relationship to a project were contractors, who engaged to perform certain services for a set fee rather than wage rates. The Philadelphia Waterworks and Chesapeake and Delaware Canal were undertaken by men

---

13. See Charles E. Brownell's excellent and more extensive analysis of BHL's artistic skill and rendering technique in BHL, *Journals*, 2:457–66; and Charles E. Brownell, ed., *The Architectural Drawings of Benjamin Henry Latrobe*, in this series. See examples of BHL's use of special views in the catalogue of engineering drawings, below: split sections, nos. 13, 17, 57, 97, 98; details, nos. 5, 11, 16, 37, 56, 79; exploded view, no. 17; broken elements, nos. 17, 56.

14. For this information I have utilized notes on the subject by Eugene S. Ferguson. See also Peter Jeffrey Booker, *A History of Engineering Drawing* (London: Chatto & Windus, 1963), p. 128.

15. Ferguson, *Early Engineering Reminiscences*, p. 38.

16. Graff, Bank of Pennsylvania; Davis, Philadelphia Waterworks; Brooke, Chesapeake and Delaware Canal; Henry Latrobe, Washington Canal. On clerks of the works, see BHL to Jefferson, 29 February 1804, Letterbooks.

who agreed to a rate for each cubic yard of earth excavated, foot of masonry laid, and so forth. James Smallman agreed to accept a stipulated sum of money for the manufacture, erection, and initial operation of the Washington Navy Yard steam engine. Latrobe himself was a contractor several times, such as when he made a boiler for the Navy Yard engine, and when he made an engine for William Hartshorne's flour mill in Baltimore. The contractors had to gather and pay the men who actually did the labor.

Latrobe's pupils usually did not have official status in the organization of work, even though they assisted him in all phases of engineering. Officially Latrobe charged his pupils a substantial fee for instruction, but he usually forgave it in exchange for their services. Occasionally he provided them with room and board. Robert Mills, William Strickland, and possibly Lewis DeMun lived with the Latrobe family during the construction of the Chesapeake and Delaware Canal feeder. Latrobe found William Tell Poussin a room in a vacant part of the Capitol for two years and paid him a small salary. Sometimes his pupils received direct compensation, such as the few weeks when Lewis DeMun served as acting engineer for the Chesapeake and Delaware Canal.[17]

Latrobe maintained close relationships with his pupils, but he was never very comfortable with the company directors or government officials who were his superiors. Generally he regarded them as unable to understand the engineering details which he, as a professional, could comprehend. He was always surprised by their concern with immediate, short-run costs, and, as he saw it, lack of appreciation for the long-run savings produced by his techniques. The internal politics of board and committee meetings seemed to make his work more difficult, and he took notice of it only when absolutely necessary.[18] This attitude hindered Latrobe's career in early nineteenth-century America, where politics was everybody's business and litigation was a national pastime.

Materials and Power

After Latrobe had organized a suitable staff to undertake a project, one of his major concerns was determining what materials could be used for construction. Stone was nearly always his preferred material, since he regarded it as strong, visually attractive, and "in the end the cheapest."[19] Given the great disadvantage of the weight of stone, Latrobe was always interested in locating quarries which were near his projects and on routes of water transportation. In Philadelphia that was relatively easy, since there were operating quarries on the Schuylkill and Delaware rivers and wharves in the city.[20] When he worked on the Chesapeake and Delaware Canal he found suitable stone near the beginning of the feeder and used it to supply stone for the culverts, bridges, and aqueducts.[21] Before he became surveyor of the public buildings the government had opened quarries at Aquia Creek in Virginia, but Latrobe also found other sources. In 1815 he located a beautiful puddingstone along the upper Potomac, and had it brought to Washington to make the new columns for the House of Representatives.[22]

Brick was probably Latrobe's second choice as a building material, although he used it less in engineering than in architecture. The conduit between the Schuylkill and Centre Square engine houses was brick, and is one of the few Latrobe engineering structures still extant.[23] In Pittsburgh he erected a temporary brick kiln on his property, probably to supply his flourishing business as an architect-contractor.[24] The absence of significant comment on brick suggests that Latrobe had no objection to it, but regarded it as distinctly different from, and in some ways inferior to, stone.[25]

Being committed to masonry construction, Latrobe took great interest in cement, particularly hydraulic cement, which could endure constant immersion in water. In the latter part of the eighteenth century John Smeaton conducted the first published tests of hydraulic cement, finding that a substance of volcanic origin known as *pozzuolana* (or *tarras*, or *trass*) mixed with

---

17. BHL to Ormond, 20 November 1808, Letterbooks; 5 December 1804–24 December 1804, 4 May 1805–20 May 1805, Journal, Comm. of Works, Salem Co. Hist. Soc.

18. These attitudes are particularly evident in BHL's canal experiences. BHL to Hazlehurst, 4 February 1804 and 10 June 1804, Letterbooks; BHL to Tatnall, 21 February [1805], Letterbooks; BHL to Fox, 11 May 1805, Letterbooks; BHL to Caldwell, 17 January 1810 [misdated 1809], Letterbooks; BHL to Law, 7 July 1813, Letterbooks.

19. BHL, "Report on the Improvement of Jones' Falls, Baltimore," 31 May 1818, PBHL, MdHS. In this case, he was referring to the material for a bridge across the falls.

20. Ashmead, *Delaware County*, p. 753; Henry Darwin Rogers, *The Geology of Pennsylvania*, 2 vols. (Edinburgh, 1853), 1:215; Mease, *Philadelphia*, p. 321; 21 November 1803, Minutes, Comm. of Survey, Hist. Soc. Del.

21. BHL to Vickers, 10 February 1804, Letterbooks; 10 May 1804, Minutes, Comm. of Survey, Hist. Soc. Del.; BHL to Vickers, 20 May 1804, Letterbooks; 20 November 1804, 30 May 1805, C & D Engineer's Reports, Hist. Soc. Del. The Maryland Geological Survey of 1902 designated the stone at the feeder quarry as "granite-gneiss." Currently there is commercial quarrying on the site.

22. Hamlin, *Latrobe*, pp. 443–44; *National Intelligencer* (Washington), 24 January 1817.

23. See fig. 14, p. 32 above.

24. BHL to Henry Latrobe, 8 May 1814, Letterbooks.

25. Note that BHL originally thought of using bricks for the Chesapeake and Delaware Canal feeder if suitable stone could not be located. BHL to Vickers, 10 February 1804, Letterbooks.

lime and water created a satisfactory hydraulic cement. He first used it for the Eddystone Lighthouse and certain canal locks. Pozzuolana was imported from Italy, but similar substances were found in the Rhine valley and at St. Eustatius in the West Indies.[26]

Latrobe may have learned of Smeaton's work while still associated with him, although his only surviving notebook from his English practice contains excerpts and condensations from Smeaton's later published account of his cement experiments.[27] He also mentions tarras mortar in one of his Chelmer-Blackwater Navigation reports.[28] Knowledge of the substance arrived in the United States about when he did, the first documented use of it being on the Middlesex Canal in Massachusetts in 1797.[29]

Throughout his American career Latrobe promoted the use of hydraulic cement.[30] At his insistence the Watering Committee hired Thomas Vickers to build the brick conduit of the Philadelphia Waterworks because while working under William Weston he had gotten experience in hydraulic cements. Latrobe then made Vickers the chief mason on the Chesapeake and Delaware Canal feeder, and under his direction a team of masons used tarras mortar for the construction of seven culverts and two bridges.[31] Latrobe urged its use at the Washington Navy Yard, and probably applied it to the new stone lock and wharves built for the Washington Canal in 1816 and 1817.[32] He also purchased a small quantity of an imported waterproof "Roman Cement" in 1817 and used it with good results for the masonry of the Capitol.[33] Clearly, Latrobe appreciated the value of hydraulic cement and did not hesitate to employ it.

Latrobe believed iron was a good building material, though only for certain purposes. After he became involved with the rolling mill at the Schuylkill Engine House he never ceased to advocate sheet iron for roofing and used it on several of his buildings. He regarded it as valuable because it was lightweight compared to slate, fireproof, and durable if painted regularly.[34] Yet he was skeptical of iron as a major construction element because of the high incidence of imperfectly made castings and forgings. In the bridge design which he hoped to patent he gave a minor role to iron elements. He was aware of the need to have effective tests for the chains of suspension bridges, and he thought cast-iron bridges had "great defects."[35]

In the mechanical realm Latrobe knew firsthand the basic dependence of steam engines and other equipment on high quality, well-cast and -forged iron. He showed preferences for special types of iron for certain purposes ("English" iron for boiler plate, for example), and took care to reject imperfectly made items. Although he did not urge that iron replace wood in machinery wherever possible, he seemed to recognize the benefits of iron's durability.[36] Latrobe's constant involvement with steam engines as a consultant, purchaser, and manufacturer made him intimately acquainted with the state of iron manufacture and fabrication in America.

Wood was the material which Latrobe liked least to use for most purposes. In many instances he associated it with cheapness, impermanence, and false economy. He agreed to build the locks, bridges, and other structures of the Washington Canal of wood, but never forgave the directors of the company for forcing him to do so. In contrast, he was proud of the masonry construction of the Chesapeake and Delaware Canal feeder.[37] There is some evidence that he even regarded the use of wood as a major construction element to be beneath the talents of a true civil engineer. Twice he told prospective clients that if they built of wood their engineering works required "only a New England bridge builder."[38]

---

26. Banks, *Reports of Smeaton*, 3:414–16; Smeaton, *Edystone* [sic] *Lighthouse*, pp. 102, 109, 111; McKee, *Masonry*, pp. 66–67.

27. BHL, notebook, c. 1793–94, Osmun Latrobe Collection, LC.

28. BHL, "Report upon the Practicability and advantage of making the River Blackwater navigable . . .," 30 December 1793, Essex Record Office.

29. Roberts, *Middlesex Canal*, pp. 96–98.

30. BHL was not the first to promote the use of hydraulic cement in the United States. Certainly William Weston, who came to the United States in 1793, was familiar with it and probably employed it on many of his projects. Tarras mortar was used on the Schuylkill Permanent Bridge (1801–06) and the Trenton Bridge (1804), and probably other major masonry works in water at this time. The idea that Canvass White introduced hydraulic cement into the United States in 1818 is erroneous.

31. Watering Comm., *Report* (1801), p. 14; pp. 14–18 above.

32. BHL to Tingey, 28 May 1807, Letterbooks; BHL to Mark, 7 December 1815, Letterbooks; BHL to Dawson, 28 April 1817, Letterbooks.

33. BHL to Dawson, 21 July 1817, Letterbooks.

34. For example, BHL to President and Directors of the Bank of Washington, 24 July 1810, Letterbooks.

35. BHL, "Specifications of a new method of building Bridges of Stone or Brick," [21 February 1806], Letterbooks; BHL to Greenleaf, 4 January 1813, Letterbooks; BHL, "Report on the Improvement of Jones' Falls, Baltimore," 31 May 1818, PBHL, MdHS.

36. BHL to Graf, 9 December 1812, Letterbooks; BHL to Henry Latrobe, 11 May 1816, Letterbooks; BHL to Wells, 1 December 1814 [misdated 31 November 1814], 29 December 1814, and 5 January 1815, Letterbooks; BHL to Criddle, 5 January 1815, Letterbooks.

37. BHL to Gallatin, 16 March 1808, in Cochran, *New ASP. Transportation*, 1:268–69; BHL to President and Directors of the Washington Canal Company, 6 February 1810, Letterbooks; BHL to Fulton, 31 July 1811, Letterbooks; BHL to May, 10 February 1814, Letterbooks; BHL to Randall, 19 August 1815, Letterbooks.

38. BHL to Fitzsimmons, 6 June 1807, Letterbooks; BHL to Caldwell, 17 January 1810 [misdated 1809], Letterbooks.

Yet wood was a time-honored material, valued for its lightness and workability; and it was cheap and abundant in the United States. Like any engineer of the era Latrobe routinely used it for many purposes, including pipes for waterworks, centering for the erection of masonry arches, temporary buildings, framing for steam engines, and boats. Often Latrobe preferred a certain type of wood for special purposes. For example, when he wanted Henry Latrobe to make the working beam for the New Orleans Waterworks engine, he told him to find timber "about the strength and elasticity of yellow pine."[39] If Latrobe liked masonry because it produced structures whose strength, durability, and beauty agreed with his engineering principles, wood was nonetheless integral to his techniques and he appreciated its qualities in many instances.

Power was also a central concern of Latrobe's when he undertook any project. In most circumstances human muscle provided the power to excavate earth, move and place stone, fit timbers together, and forge iron. Latrobe took that for granted and always attempted to gather good and reliable workmen. In directing them, quality of workmanship, rather than productivity, seems usually to have occupied Latrobe's attention. Such an approach was not conducive to keeping labor costs to a minimum.[40]

However, Latrobe knew ways of reducing the muscle-power required in some situations. Following English precedent he completed the easiest parts of canals first, partially filled them with water, and used those sections as waterways.[41] He proudly reported his achievement on the Chesapeake and Delaware Canal feeder to Joshua Gilpin:

> The greatest part of the stone at the [aqueduct] embankment has been boated to the place from the rocky hill, to which place the Canal has now been for some time navigated. One wooden legged man at 75 Cents pr day (with two men to assist in loading and unloading) does here the business of 3 two horse carts, at [$]2.25 each which require at least two additional hands to load. The saving thus effected is 6$.75 cts—[minus] 75 cts., or 6 Dollars pr day.[42]

On the feeder Latrobe also used railways with four-wheeled carts (probably pushed by hand) to aid in removing earth from deep cuts.[43]

Latrobe's practical engineering knowledge included the use of muscle-powered machinery. In Philadelphia he used a special crank for raising stone, and in Washington he designed a derrick for raising the columns for the House of Representatives from boats onto a dock.[44] In two instances he employed horse gins to operate equipment requiring continuous motion.[45]

Men and animals were the major sources of power for an engineer of the time, but steam interested Latrobe far more. His promotion of steam engines is a major theme of his career, and he was fascinated with the "incredible" power of which they were capable.[46] Coming to the United States where steam engines were virtually unknown, he saw immense potential just in using them to pump water and drive industrial machinery, their two most common uses in Britain. His promotion of steam engines centered on those fields, except for the construction of the steamboat *Buffalo*.

When he left England, Latrobe was familiar with the superior engineering of the standard Boulton and Watt engine, and he never lost his admiration for it. He recognized that years of experimentation had gone into the engine and that its basic form would be unchanged for years to come. Moreover, the alternative in the United States was the high-pressure Oliver Evans engine, which Latrobe regarded as nearly useless. Clearly the Boulton and Watt type of steam engine had superior dependability and was considerably safer to operate, but Latrobe never understood that the lightness and great power variability of the Evans engine made it better for many purposes.[47]

Latrobe's enthusiasm for steam engines prevented him from developing an interest in water power. When he made arrangements to purchase cotton looms from Samuel Blydensburg he purchased a mill seat on the Potomac Canal above Georgetown to power them, but was almost apologetic when writing of it. "My Mill is quite up. It is an humble thing, and I consider it merely as an experiment. It has an admirable water Wheel. . . . If the looms succeed tolerably, I shall put up a steam engine in this city [Washington]."[48]

---

39. BHL to Henry Latrobe, 31 May 1812, Letterbooks.
40. BHL to Hamilton, 3 July 1812, Letterbooks. See the criticism of BHL's relationship with workmen in Leslie to Jefferson, 10 January 1803, Jefferson Papers, LC.
41. Burton, *Canal Builders*, p. 191. For the Washington Canal, see BHL to President and Directors of the Washington Canal Company, 2 June 1810, Letterbooks.
42. BHL to Gilpin, 26 August 1804, Letterbooks.

43. See p. 17 above.
44. See p. 242 below.
45. Archimedean screw pump at the Philadephia Waterworks basin and the machine shop in Pittsburgh.
46. For example, BHL to Stewart, 29 October 1799, in *Federal Gazette* (Baltimore), 21 February 1801.
47. My assessment does not differ from that in Hamlin, *Latrobe*, p. 551.
48. BHL to Blydensburg, 18 January 1811, Letterbooks. BHL's papers indicate at least three other

Latrobe's diffidence about using water power was in contrast to the realities of American industry, which during his career was completely dominated by water power.[49]

Latrobe had some familiarity with a third source of inanimate power—the explosive, black powder. He supervised its use for the breaking up of small formations of rock on three projects: the Susquehanna navigation, the Philadelphia Waterworks, and the Chesapeake and Delaware Canal.[50] However, blasting rock formations with black powder was an expensive and slow process reserved only for those circumstances where it could not be avoided.

Theory and Principles

Latrobe possessed not only a substantial practical knowledge of materials and power, but also considerable theoretical knowledge. He had read or owned copies of standard engineering publications in English, French, and German, among them John Smeaton's *Edystone* [sic] *Lighthouse* and Bernard Belidor's *Architecture Hydraulique*. Few engineering books of the time were theoretical in the modern sense of being the formulation of general principles based on experimentation or mathematical logic. Rather, they were compilations of the best and most successful projects in a field, often those executed by the author. The theory which Latrobe knew was therefore the best European practice of the time, and to him it represented a goal to which Americans might strive. When he introduced theory into his discussion of the effect of Thomas Moore's wing dams on the Potomac, he commented:

> In this opinion, I am supported by the testimony of the ablest engineers who have treated on the subject, especially by Fabre, whose work on the theory of rivers, streams, and torrents, is invaluable, and in which the system laid down, is not less ably proved by mathematical demonstration, than by the facts which he exhibits from his own experience, and that of others, in works actually erected for the last 2,000 years.[51]

It is also noteworthy that the one time Latrobe borrowed the contents of an engineering treatise almost entirely—in his essay on repairing breaches in levees—he promoted the adoption of a very mundane and completely practical technique. Certainly Latrobe's wide acquaintance with European engineers' publications made him appreciative of the historical lessons of engineering.

Using his practical and theoretical knowledge, Latrobe brought a large body of information to bear on any problem to which he devoted his attention. He visualized the problems and solutions in a more systematic and comprehensive pattern than most European and American engineers of the era. His analyses of the problems of coastal North American rivers emphasized constant and rapid sedimentation as a long-term effect of agricultural development of the hinterland.[52] Similarly, his plan for the Philadelphia Waterworks took into account the sources of pollution of the city's water, and the future water requirements of a growing population.[53] His essay on turnpikes considered the factors of road surface composition, soil erosion, and grades in reaching a conclusion about how the ideal turnpike should be built.[54] The relatively infant state of civil engineering allowed Latrobe to make wide-ranging recommendations, since he did not make the modern distinction between mechanical and civil engineering, fields which were combined in both his waterworks.

Latrobe's engineering training made a substantial contribution to his comprehensive approach, particularly because he learned in England that permanence was one of the most important goals of engineering. He could only attempt a permanent solution of an engineering problem by paying close attention to all factors, and understanding past lessons. He revealed his approach in a letter to the directors of the Washington Canal Company.

> The law requires your Canal to be 20 feet wide and 3 feet deep. The slope of the Canal was formerly stated by me [in his 1804 report on the Washington Canal] to be as 4 to 5. Since then I have had so much

---

instances when he may have had something to do with a water mill: BHL, *Journals*, 1:23; BHL to Gilpin, 2 December 1805, Letterbooks; BHL to Brent, 29 March 1813, Letterbooks.

49. It is generally believed that total industrial steam power in America exceeded total industrial water power only after 1860. George Rogers Taylor, *The Transportation Revolution, 1815–1860* (New York: Harper & Row, 1968), p. 224; Nathan Rosenberg, *Technology and American Economic Growth* (New York: Harper & Row, 1972), pp. 66, 158.

50. BHL to Smith, 21 March 1804, Letterbooks; Powder Sales Book, 9 July 1804–17 October 1804, item 1500, series Y, part II, Accession 500, Eleutherian Mills; 10 May 1804, 1 June 1804, Minutes, Comm. of Survey, Hist. Soc. Del.

51. BHL, *Opinion*, p. 21.

52. Ibid., p. 21; BHL to Fitzsimmons, 25 May 1807, Letterbooks; BHL to Bate, 3 February 1810, Letterbooks; BHL, "Report on the Improvement of Jones' Falls, Baltimore," 31 May 1818, PBHL, MdHS; BHL to Thompson, 24 March 1806, Letterbooks.

53. "Designs of Buildings . . . in Philadelphia," [1799], Hist. Soc. Pa. The Centre Square Engine House had a design providing for a pipe to supply the western part of the city, and the Schuylkill engine had "extra power" to fulfill the future demand for water.

54. BHL, "To the Editor of the Emporium," *Emporium of the Arts and Sciences*, new series, 3 (1814): 284–97.

experience of the effect of frost and wash upon the banks of the feeder of the Chesapeake and Delaware Canal that I cannot advise you to a less slope than as 2 to 3, which is the slope of that work,—more especially as your canal is to be so narrow.[55]

This can be taken as evidence of his extension of the English principle of permanence to the American environment.

It is clear, however, that Latrobe seldom attempted to adapt his engineering values to American society. Although accurate surveying and sound knowledge of materials and processes were equally important on both sides of the Atlantic, for example, the ideal of permanence was not nearly as useful and necessary here. The reality and the expectation of rapid territorial expansion and economic growth made temporary improvements much more acceptable in America. Who knew where the next great city of America would rise, and whether its rise would make a formerly growing city merely a satellite to the new metropolis? This provisional view of the future often made Americans prefer cheaper wood to more expensive stone as a construction material, even though they knew that wood might deteriorate rapidly. In the same spirit, ornament and aesthetic considerations were ultimately secondary to the value of economic expediency. Thus Latrobe's beautiful Centre Square Engine House, a building which today would be an object of admiration, was torn down less than thirty years after it was erected. A typically American improvement, by contrast, was the Erie Canal, which used wood in many places where the English would have preferred masonry, and which was rebuilt within twenty years of its opening.[56] Latrobe never accepted this American viewpoint, and his career and reputation suffered as a result. As he admitted to DeWitt Clinton, "The fault which the public have found with my professional character, is that my ideas and projects are too extended and magnificent to be practicable for some centuries to come."[57]

Transfer of Engineering Concepts and Skills

However poorly he adapted his engineering principles to American society, Latrobe made a substantial contribution to the United States in the transfer of engineering concepts and skills. His promotion of new concepts was his most important transfer to America, such as when he convinced the people of Philadelphia to solve their water supply difficulties with a steam-powered waterworks. Latrobe's knowledge of English engineering made him capable of proposing solutions to some American problems which native engineers could not, and his experience and ability permitted him to persuasively promote his proposals. In the last decade of his American career, however, Latrobe was too far removed from his English experience to have the same conceptual impact as he had at first, and in those years he turned more to the promotion of American innovations, such as Blydensburg's loom and Fulton's steamboat.

Latrobe insured his lasting effect on American engineering by training several pupils and associates. During the first decade of his American career he accepted several pupils into his office, including Frederick Graff, Adam Traquair, William Strickland, Robert Mills, and Lewis DeMun. Later he trained his son Henry and had more fleeting arrangements with several other young men. These pupils he took in as apprentices, taught them such basic skills as drafting and surveying (if they did not know them already), and gave them valuable experience in supervising his engineering projects.[58] Only one of his pupils, Frederick Graff, became famous for a career strictly in engineer-

55. BHL to President and Directors of the Washington Canal Company, 6 February 1810, Letterbooks.
56. Ronald E. Shaw, *Erie Water West* (Lexington, Ky.: University of Kentucky Press, 1966), pp. 96–97, 306–7, 312, 324.
57. BHL to Clinton, 20 January 1812, Letterbooks.
58. The table below indicates the direct line of BHL's influence through the apprenticeship system of education. Most of these men are noticed in the *DAB*. For Henry Latrobe, see the relevant parts of Hamlin, *Latrobe*. For Ellwood Morris there is no biographical sketch, but for confirmation of his work under Strickland, see Thomas U. Walter et al., "Obituary of John C. Trautwine," *Journal of the Franklin Institute*, 3d series, 116 (1883):391. For John C. Trautwine, Jr., see *Who Was Who in America*, 6 vols. (Chicago: A. N. Marquis Company, 1942–76), 1:1251. This group of Latrobe's professional descendants includes some of the major figures in American engineering of the nineteenth century, among them a president of the American Society of Civil Engineers (Frederic Graff) and the author of a standard manual of engineering data (John C. Trautwine).

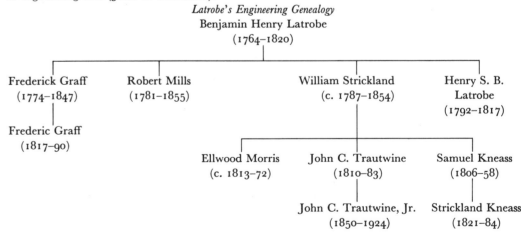

Latrobe's Engineering Genealogy

ing, but, like their mentor, William Strickland and Robert Mills successfully combined engineering with architecture. Graff, Strickland, and Mills had considerable influence on the development of American civil engineering, the former two training another generation of prominent engineers. In these succeeding generations of men whose engineering talents were derived from Latrobe it is possible to see the same love of fine drawing, reliance on sound knowledge of materials, and a desire to build works of lasting value and beauty.

Throughout his American career Latrobe influenced technical thinking by his creation of outstanding engineering drawings and by teaching others his methods. If, as Eugene Ferguson has suggested, technical creativity often involves nonverbal thinking (i.e., thought not reproducible in speech or writing), and drawings allow one technician to communicate some of his nonverbal thinking to another technician, Latrobe's engineering drawings played a significant role in the diffusion of his knowledge. Certainly they must have had a major impact during his own lifetime when much American technical rendering was, by comparison, awkward and untutored. Moreover, the survival of Latrobe's drawings in his students' and associates' collections (sometimes with later annotation) suggests that his drawings were reservoirs of nonverbal knowledge, as well as models of technical rendering, for later engineers.[59]

Of Latrobe's transfer of skills to those many other Americans who worked under him in various capacities, little is known. He valued several of his workmen and contractors so highly that he found them employment in more than one of his projects. To the extent that such men became able to read high quality engineering drawings and execute work to Latrobe's standards, other engineers who employed them probably found their work easier.

Latrobe also promoted the transfer of engineering skills to America by preferring to give employment to skilled Englishmen and Americans trained by Englishmen. For the Philadelphia Waterworks he insisted on the right to select a man paid by the city to supervise daily activities, and he chose John Davis, who had architectural and engineering experience in England. Robert Brooke, who worked under Latrobe on the Chesapeake and Delaware Canal, was a Philadelphia surveyor who had been an assistant to the English engineer William Weston. Latrobe exhibited the same interest in the English talents of the carpenters, masons, and contractors who actually performed the labor. At the rolling mill he hired John Parkins, an ironworker from Sheffield; at the Philadelphia Waterworks he employed Thomas Vickers, a mason who had worked under Weston; and on the Chesapeake and Delaware Canal he made a contract with Charles Randall, whom he described as "a practical road- and Canal-maker from England."[60] Latrobe gave many such skilled individuals their first employment on a major American engineering work, and several went on to distinguished careers in the United States.

Promotion of Technical Projects

Another aspect of Latrobe's career was his enthusiastic promotion of technical projects. In the reports on the Maldon branch of the Chelmer and Blackwater Navigation there are promotional elements, and in America he soon demonstrated his ability to write enthusiastic propositions and optimistic reports of works-in-progress.[61] While these were usually associated with projects that he directed, Latrobe also produced general essays on technical development, such as those on internal improvements which he wrote for Albert Gallatin.[62]

Latrobe invested time and money in promotion. He was for years a lobbyist for the Chesapeake and Delaware Canal, attempting to obtain a government investment to revive the company. To obtain his New Orleans Waterworks charter he had to conduct negotiations with the territorial legislature of Louisiana, the New Orleans City Council, and Congress. In Pittsburgh he raised money for the Ohio Steamboat Company, and got subscriptions for a steamboat line on the Cumberland River. He personally invested in many projects, including the Schuylkill rolling mill, the Blydensburg loom, and the New Orleans Waterworks. He owned so much stock in the Washington Canal that he was elected a director of the company for two years. Latrobe's promotional activities as an engineer showed that he easily acquired (if he did

---

59. Eugene S. Ferguson, "The Mind's Eye: Nonverbal Thought in Technology," *Science* 197 (August 26, 1977): 827–36. The collections I refer to are the Trautwine and Graff collections at the Franklin Institute (Philadelphia, Pa.), the Stoudinger Collection at the New Jersey Historical Society (Newark, N.J.), and the John Davis Collection, which is now part of the Latrobe Papers at the Maryland Historical Society (Baltimore, Md.). See, for example, drawing no. 40, below.

60. Watering Comm., *Report* (1801), p. 14; BHL to Hazlehurst, 30 March 1805, Letterbooks; BHL to Gilpin, 2 September 1804, Letterbooks; BHL to Madison, 8 April 1816, Letterbooks.

61. BHL, "Report upon the Practicability and advantage of making the River Blackwater navigable . . .," 30 December 1793, Essex Record Office; BHL, "Report upon the intended Improvement of the rivers Blackwater and Chelmer . . .," 28 October 1794, Essex Record Office; BHL, *American Copper Mines*.

62. Cochran, *New ASP. Transportation*, 1:261–70; BHL, "To the Editor of the Emporium," pp. 284–97.

not already possess) the American traits of unbounded enthusiasm in the future and the expectation of financial gain from economic growth.

Yet he did not seem to understand that many American investors were speculative (anticipating short-term gains) and unsympathetic to his engineering principle of permanence, which entailed heavy initial investment and long amortization. As Daniel Calhoun has pointed out, Americans expected an engineer to make an investment in his project as a sign that he had the same goals as they. Latrobe often did invest, but still found his engineering questioned because his long-range planning conflicted with the immediate financial goals of American investors.[63]

The American Technical Community

To assess Latrobe's role as an engineer completely, his membership in the technical community in America must be considered. His election to the American Philosophical Society in 1799 was a formal recognition of his acceptance as a major figure in American technology. Moreover, Latrobe knew, corresponded with, or worked with a sizeable number of American technicians, including most of his famous contemporaries.

The preserved correspondence between Latrobe and such men as Eric Bollman, Nicholas Roosevelt, Thomas Jefferson, and Robert Fulton suggests that there was a fertile interchange of ideas and information in the American technical community. Of course letters were an imperfect means of communication and the really important discussions must have taken place when the men were together; but of such meetings there is virtually no record. It is probable that in informal gatherings Latrobe heard not only the news from Europe, but also about the ideas and plans of his friends and about recent publications.

His elevated rank in the American technical community did not prevent Latrobe from having critics and antagonists, however. Thomas Pym Cope, a Quaker merchant and politician of Philadelphia, served on the Watering Committee while Latrobe directed construction of the Philadelphia Waterworks, and Cope developed a strong dislike for him. Used to strict adherence to contractual arrangements, Cope was angered when Latrobe was unable to construct the waterworks within the time and expense limits which the engineer had initially set. Neither could he find any justification in his system of moral standards for Latrobe accepting a partnership in the Schuylkill rolling mill with Nicholas Roosevelt, the engine contractor, while he gave the committee advice on whether Roosevelt's engines were adequate.[64] Cope came to the conclusion that "Latrobe is a cunning, witful, dissimulating fellow—possessing more ingenuity than honesty," and that "he undertook more than he understood and has been making experiments at the expense of the City."[65]

Oliver Evans, the designer of an automatic flour mill and the inventor of the high-pressure Columbian steam engine, also became a bitter enemy of Latrobe. Although the two men barely knew one another and never had any business relationship, controversy between the two lasted more than ten years. It began because Latrobe's 1803 report to the American Philosophical Society on steam engines in America took scant notice of Evans's achievements and, without mentioning him by name, took pains to show how ridiculous were the claims of those who believed in the possibility of profitable land and water carriage by steam. That pronouncement hurt Evans deeply, and he never forgot it.[66] As late as 1814 he published an article on his steam engines which recalled:

> I also communicated my plans [for steamboats and steam wagons] to B. H. Latrobe, esq...; who publicly pronounced them chimeral, and attempted to demonstrate the absurdity of my principles, in his report to the *Philosophical society of Pennsylvania* on steam engines;... The liberality of the members of the society caused them to reject that part of his report which he designed as demonstrative of the absurdity of my principles;... and ordered it to be stricken out.[67]

Latrobe denied the latter circumstance, and to him the falsehood was only another in a series perpetrated by Oliver Evans.[68] He had for many years discredited Evans's engines because he thought they were based on principles tested and discarded in England, and because they were impractical and expensive in actual service.[69] He personally petitioned Congress asking that Evans's engine patent be voided.[70]

---

63. Calhoun, *The American Civil Engineer*, p. 16; Condit, *American Building*, p. 25.

64. 2 March 1799, Minutes of the Common Council, Phil. Archives; 8 July 1800, 10 October 1800, 13 October 1800, 15 August 1801, Cope Diaries, Haverford College.

65. 28 January 1801, 5 March 1801, Cope Diaries, Haverford College.

66. Ferguson, *Early Engineering Reminiscences*, p. 37.

67. Oliver Evans, "On the Origin of Steam Boats and Steam Wagons," *Emporium of the Arts and Sciences*, new series, 2 (1814): 210.

68. BHL to Cooper, 19 June 1814, Letterbooks; BHL to Vaughan, 19 June 1814, Letterbooks.

69. BHL to Patterson, 28 March 1804, Letterbooks; BHL to Prentis, 24 March 1813, Letterbooks; BHL to Ingersoll, 17 January 1814, Letterbooks; BHL to Roosevelt, 20 January 1814, Letterbooks.

70. BHL to U.S. House of Representatives, 1 March 1814, Letterbooks.

Cope and Evans, Robert Fulton, William Thornton, Robert Leslie, and others found ample and numerous reasons to strongly criticize Latrobe. Their charges ranged from simple incompetence to gross financial mismanagement. The virulence of the controversies suggests that Latrobe's manner and personality were tinged with pride and not a little arrogance, making compromise or reconciliation difficult.

If Latrobe sometimes had a divisive effect on the American technical community, he more often unified and strengthened it. Carroll Pursell has noted that Latrobe "served as an intermediary between those whose inclinations or livelihoods led them to an interest in steam," receiving inquiries on engines and directing them to the proper persons who might answer them.[71] Latrobe also received many inquiries concerning canals, and in several cases gave advice about engineers or contractors who might be employed on them.[72] And, as Talbot Hamlin has pointed out, he showed almost constant interest in a variety of mechanical inventions and improvements, sometimes going to great lengths to promote them.[73] In an era without significant technical institutions such as professional societies, universities, or large corporations, the continued growth and integration of American technology depended upon men like Latrobe who were focal points in the network of activity.

Summary

For a number of reasons, then, Benjamin Henry Latrobe must be accorded a fundamental role in the early development of American engineering. He was the first engineer with European training to become an American and to bring his skills to bear fully on American problems. He designed and superintended the construction of many seminal projects which provided employment and training for men who went on to other important engineering works. He played a major part in the diffusion of new skills and ideas by participating in a network of American technicians who consulted and disputed with one another. Significantly, he had a knack for gathering around him men whose skills were necessary to the exercise of his talents, and Latrobe's papers illuminate the intimate creative relationships which existed between him and them. Latrobe's career provides deep and significant insights into the development of early American technology.

---

71. Pursell, *Early Stationary Steam Engines in America*, pp. 36–37.
72. BHL to Smith, 21 March 1804, Letterbooks; BHL to President and Managers of the Salem Creek Canal, 27 May 1805, Letterbooks; BHL to Randall, 19 August 1815, Letterbooks; BHL to Madison, 8 April 1816, Letterbooks; BHL to Smith, 25 April 1816, Letterbooks; BHL to Seaton, 3 September 1817, Letterbooks.
73. Hamlin, *Latrobe*, p. 554. BHL's promotion of Fulton's patents is probably the best example of this interest.

The Susquehanna River Survey Map

THE SUSQUEHANNA RIVER bisects Pennsylvania, and with its tributaries drains half of the state's surface. Recognized as one of the major river systems of the Atlantic coast from the earliest European settlement, the Susquehanna had little importance until the settlement of the interior of Pennsylvania in the latter eighteenth century. Then pioneers used the valleys of the Susquehanna, and its largest tributary, the Juniata, to penetrate the Appalachian ridges which cross the state. Once settled, they sent simple wooden rafts, keelboats, and arks loaded with coal, grain, whiskey, and iron downriver to market.[1]

But navigation of the Susquehanna River was not easy. Much of the river is shallow except during spring freshes, and where it cuts through the ridges there are ribs of rock across the river which presented constant danger to the boatmen. Prior to the construction of modern reservoirs, the lower portion of the river had a series of transverse ledges of rock over which water descended rapidly to the tide, making it a particularly hazardous area for rivercraft.[2]

There were also political obstacles to the Susquehanna navigation. Though it tapped the rich hinterland of Pennsylvania, its mouth was in Maryland at the head of the Chesapeake Bay, far from any Pennsylvania port. Philadelphia merchants were particularly concerned that the developing central portion of the commonwealth might find its natural commercial ally to be Baltimore, thus depriving them of trade which they thought rightfully theirs. In the early 1790s the Philadelphia merchants set in motion several internal improvement companies meant to channel trade in their direction. Two connecting canals were laid out from Philadelphia to the Susquehanna, but construction on them was suspended within three years after it had begun.

Meanwhile Maryland had incorporated the Susquehanna Canal Company to build a canal from near the state line to below the last rapids. Although it was slow in completion, it brought to everyone's attention the alternate route from the interior to Philadelphia by way of that canal and the proposed Chesapeake and Delaware Canal. They would bring traffic down the Susquehanna to its mouth, across the waters of the upper Chesapeake Bay, over the Delmarva Peninsula by canal to the Delaware Bay, and thence north to Philadelphia. Though roundabout, it offered the possibility of Pennsylvania,

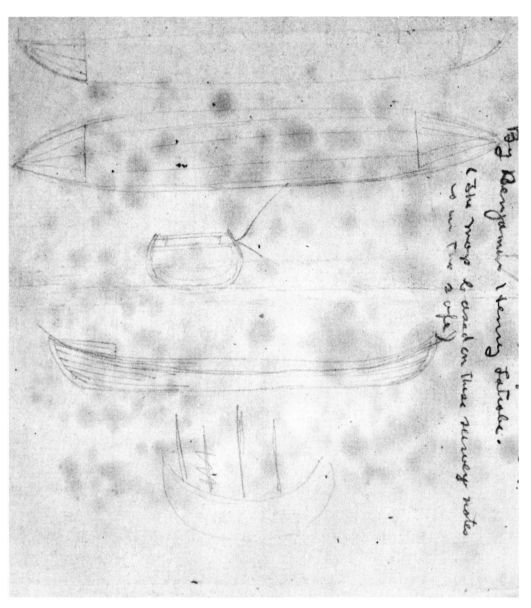

Fig. 25. Susquehanna River keelboat: plan, sections, and elevation. Unknown vessel at bottom. September 1801. Pencil. (Field Book 1.)

1. George W. Lightner, *Susquehanna Register of Arks, Rafts, &c. &c. Arriving at Port Deposit [Maryland], in the Year 1822* (Baltimore, 1823), is a later list of such craft and their cargoes.
2. Sylvester K. Stevens, *Pennsylvania: Birthplace of a Nation* (New York: Random House, 1964), pp. 11–12; James Weston Livingood, *The Philadelphia-Baltimore Trade Rivalry, 1780–1860* (Harrisburg, Pa.: The Pennsylvania Historical and Museum Commission, 1947), pp. 1–12, 28–29.

Susquehanna River Survey Map

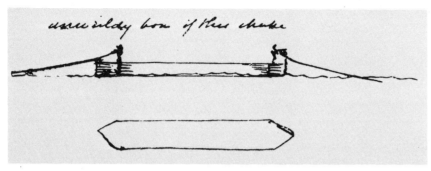

Fig. 26. Ark on the Susquehanna River: plan and elevation, 1809. (Joseph E. Walker, ed., *Pleasure and Business in Western Pennsylvania: The Journal of Joshua Gilpin, 1809* [Harrisburg, Pa.: The Pennsylvania Historical and Museum Commission, 1975], p. 12.)

Maryland, and Delaware pooling capital for construction, as well as having fewer engineering problems than the other route.[3]

In 1799 the Pennsylvania Assembly offered to clear the lower Susquehanna for navigation if Maryland and Delaware would incorporate the Chesapeake and Delaware Canal Company. After nearly two years this condition was complied with, and the assembly authorized Governor Thomas McKean to allocate up to $10,000 for navigational improvement of the river below Columbia. In July 1801 McKean made a contract with Colonel Frederick Antes to carry out the assembly's wishes. Antes was Benjamin Henry Latrobe's maternal uncle, and he appointed Latrobe his engineer and surveyor. Simultaneously the Susquehanna Canal Company borrowed $5,000 from the state of Maryland to improve the river in Maryland and chose Sebastian Shade, an experienced riverman, and Christian Hauducoeur, the company's engineer, to direct the work. The Pennsylvania and Maryland groups cooperated with each other.[4]

Colonel Antes, a Revolutionary War hero and former commonwealth treasurer and assemblyman, got his contract through political influence as well as through his experience with the improvement of internal navigation. While treasurer he was appointed as a delegate to a convention in Maryland to take measures for the improvement of the Susquehanna River, and he had served as a commissioner for the exploration and survey of the headwaters of the Delaware, Lehigh, Schuylkill, and North Branch of the Susquehanna rivers. He had also been a contractor for the Susquehanna Canal Company in 1798, opening a channel near the east shore of the Susquehanna in the area of Wright's Falls (river mile 39.2/kilometer 63.1) and Turkey Hill (river mile 34.5/kilometer 55.5).[5]

Work on his second Susquehanna improvement formally commenced on 1 August 1801, when Colonel Antes in the company of Sebastian Shade made an initial reconnaissance of the river from Columbia to the Maryland state line, a distance of about 25.9 river miles (41.6 kilometers). At this time decisions were made about priorities in channel clearance and the removal of obstructions for improvement of navigation courses. It appears that arrangements for work were also begun. Christian Wisler was engaged as a supervisor in the Turkey Hill area, and black powder and brimstone were purchased on 6, 15, and 28 August. Money was advanced on 15 August to Joseph Turner for erecting a smith's shop for repairing tools and a contract was made for work below Peach Bottom (river mile 16.6/kilometer 26.8) on 26 August.[6]

Benjamin Henry Latrobe left Philadelphia on 5 September 1801 to assume his new position. He arrived at Lancaster, Pennsylvania, two days later, and learned that his uncle lay ill at Slaymaker's Tavern in the city. During the fourteen days of Antes's illness prior to his death on 20 September, Latrobe assisted him and helped to plan the work on the river. Latrobe also made detailed observations on Antes's condition which, in light of modern diagnosis, suggest that he died of a combination of typhoid fever, locally endemic at the time,[7] and kidney stones.[8]

While bedridden, Antes gave Latrobe particulars of his subcontracts for channel clearance, describing "the characters of George Stoner of Burkholder's ferry, and of Christian Wisler of Turkey hill, to whom he had com-

---

3. Livingood, *The Philadelphia-Baltimore Trade Rivalry*, pp. 10–11, 32–36, 161.

4. Ibid., p. 33; Ralph D. Gray, "The Early History of the Chesapeake and Delaware Canal—Part I," *Delaware History* 8 (1959): 231–39; *Pennsylvania Archives*, ed. Gertrude MacKinney, 9th series, 10 vols. ([Harrisburg, Pa.]: Department of Property and Supplies, 1931), 3:1755–56; *The Oracle of Dauphin, and Harrisburgh* (Pa.) *Advertiser*, 15 March 1802.

While it seems likely that Hauducoeur had European training, there is no evidence to support Talbot Hamlin's assertion that he was a Frenchman. Hamlin, *Latrobe*, p. 547. Hauducoeur's training may be inferred from an advertisement for his Susquehanna River map of 1799 which stated: "This performance is in a style entirely new in this country, and which is in regard to engraving one of the best executed in America." *Porcupine's Gazette* (Philadelphia), 16 July 1799.

5. Edwin MacMinn, *On the Frontier with Colonel Antes* (Camden, N.J.: S. Chew & Sons, 1900); annotation on map sections 1 and 2, pp. 91, 93. River mile/kilometer ordinates are on the map borders.

6. BHL, Field Book 1, August–September 1801, PBHL, MdHS.

7. Charles Trcziyulny, *Report of Charles Trcziyulny, appointed to explore the River Susquehanna, in Pursuance of an Act of the General Assembly, with a View to its Improvement, from the New York to the Maryland Line* (Harrisburg, Pa., 1827), p. 10; Benjamin Rush, "An Enquiry into the Cause of the Increase of Bilious and Intermitting Fevers in Pennsylvania, with Hints for Preventing Them," *Transactions of the American Philosophical Society* 2 (1786): 206–12.

8. This diagnosis was provided by Linda R. Sessums, M.D., University of Maryland Medical School, Baltimore, Md.

mitted the work in those places;—and also gave his opinion of the talents and character of Sebastian Shade, with his reasons for the degree of confidence he placed in him."[9] On 13 September Latrobe was joined by Antes's son, Henry, and together they settled accounts of work done in the course of Antes's illness. Latrobe assumed Colonel Antes's contract with the commonwealth.

On 21 September Colonel Antes was buried, and on the following day Latrobe and Henry Antes proceeded from Lancaster to Pequea (river mile 27.6/kilometer 44.4), where they met Sebastian Shade. They commenced a general study of the project area, inspecting the extent and progress of work previously contracted, settling the accounts of Colonel Antes, and letting out new contracts.[10] Returning to Pequea on 30 September, Latrobe examined the work completed by George Stoner on 12 September, and made additional contracts in that area. From there he moved downstream, arriving on 2 October at Octoraro Creek (river mile 8.0/kilometer 12.8). There Latrobe probably visited the office of the Susquehanna Canal Company and arranged for Christian Hauducoeur to survey the river from the Maryland state line north to McCall's Ferry, a distance of 9.5 miles (15.2 kilometers). Hauducoeur was the canal company's engineer and in 1799 had published a map of the Maryland section of the Susquehanna River.

Arriving at Havre de Grace, at the mouth of the Susquehanna, on 4 or 5 October (executing sketch sec. 10, fig. 1) Latrobe proceeded to Baltimore, perhaps by coach. There he met with the governor and the directors of the Susquehanna Canal Company, and other interested persons to discuss how to raise additional money for the improvement of the Susquehanna River.[11]

By 17 October Latrobe had returned to Philadelphia, where he spent much of his time defending himself against attacks on his fiscal responsibility while he had been engineer of the Philadelphia Waterworks. Late in October Latrobe returned to the Susquehanna and he remained there until all his work was completed.

Channel Clearance

The contracted project was administered and conducted as two separate and sequential operations: channel clearance and river survey. Channel clearance began in August 1801 and appears to have been near completion by late October 1801. Work areas stretched from Columbia (river mile 40.1/kilometer 64.5) to the Maryland state line (river mile 14.2/kilometer 22.8). In this phase of the project Latrobe functioned as the general contractor and coordinator, with individual improvements executed under the immediate supervision of Sebastian Shade, George Stoner, and Christian Wisler.[12]

Latrobe described the conception of the project as follows:

> ... all my exertions were bent to force through all obstructions, a channel clear of rocks, of 40 feet [12.2 meters] wide close to the Eastern shore, never leaving any rock upon which a vessel could be wrecked between the channel and the shore,—so that in the most violent freshes a boat should always be safe, by keeping close in shore. Rocks of immense magnitude were therefore blown away, in preference to the following a crooked channel more cheaply made, but more difficult and dangerous, and varying in safety and practicability, according to the degree of rise on the river.[13]

To this end he expended the full $10,000 appropriated by the legislature. His accounts contain a partial record of this expenditure, concentrating on contracts, receipts, and unspecified financial transactions.

In the course of the work, sections were examined to determine the nature and extent of the obstructions, a plan for clearance was developed, and a contract for the work was let out. Contracts were normally made with an individual or a small group of local residents for a fee which was fixed in writing. Often cash advances were made to cover initial expenses and materials. In the few cases where the contractors incurred excessive expenses due to underestimation of work difficulty, fee adjustments were made.

Channel clearance frequently involved blasting with black powder, followed by removal of debris and further excavation. Labor was supplied predominantly by local farmers seeking wages to supplement their income. Tools included augers, crowbars, sledgehammers, picks, wedges, needles (long, slender, pointed pry rods), scrapers, hammers, and rammers. The tools appear to have been provided by both the Commonwealth of Pennsylvania and those contracted.[14]

---

9. BHL, Field Book 1, 12 September 1801, PBHL, MdHS.
10. Henry Antes accompanied BHL until about 26 September 1801, and Sebastian Shade accompanied him until about 30 September.
11. Susquehanna Canal Company, *Report of the Governor and Directors to the Proprietors of the Susquehanna Canal* (Baltimore, 1802), p. 17.

12. BHL, Field Book 1; Field Book 2, October–November 1801, PBHL, MdHS; BHL to McKean, 27 January 1802, Susquehanna River Improvement File, Pa. Hist. Mus. Comm.
13. BHL to Gallatin, 16 March 1808, in Thomas C. Cochran, ed., *The New American State Papers. Transportation*, 7 vols. (Wilmington, Del.: Scholarly Resources, 1972), 1:266.
14. BHL, Field Book 2, PBHL, MdHS.

Susquehanna River Survey Map

In spite of the extensive channel clearance in Pennsylvania and Maryland, carried out at an expense approaching $20,000, the Susquehanna River remained a treacherous waterway. Only the most dangerous rocks and obstructions were removed from a narrow channel, and even that was expected to be safely navigable only during the spring high water. Latrobe estimated that Pennsylvania could spend another $100,000 to make the river "fit for the common purpose of convenient intercourse," a sum which was never provided. The Susquehanna and Tidewater Canal from Wrightsville, Pennsylvania (opposite Columbia), to Havre de Grace, Maryland, finally provided a safe water navigation for the lower Susquehanna valley when it opened in 1840. The Susquehanna itself has never been tamed for navigation, although three dams have made the river much quieter.[15]

River Survey

The river survey phase of the project took place in October and November 1801, benefitting from an unusually low water level and the reduced understory vegetation of autumn. Two parties conducted the survey: one, under the direction of Latrobe, worked downstream from Columbia (river mile 40.1/kilometer 64.5) to McCall's Ferry (river mile 23.7/kilometer 38.2); the other, under the direction of Christian Hauducoeur,[16] worked upstream from the Maryland state line (river mile 14.2/kilometer 22.8) to McCall's Ferry. The work of the Latrobe and Hauducoeur survey parties was within the highest standards of contemporary practice, and was far more sophisticated than the work of the common American land surveyor of the period.[17]

For the survey Latrobe employed his brother-in-law Robert Hazlehurst (1774–1804) as assistant surveyor. Acting as an aide was William Paul Crillon Barton (1786–1856), the son of Latrobe's friend Judge William Barton, a prominent Pennsylvania lawyer. They were supported in the field by chainbearers, axemen, and canoemen.[18] Hauducoeur presumably employed a similar survey crew as he included in his account the following item: "By expenses of Assistants and Canoe Men, . . . $51.50."[19]

Latrobe conducted his portion of the survey over a period of fourteen days, from 28 October to 12 November. The party did not work on Sundays. They took their food and lodging at riverside inns and farmers' homes. They progressed downstream, sometimes backtracking to improve observation points and to strengthen the triangulation network (sec. 7, fig. 1).

Latrobe left a personal description of the working conditions in a letter to his wife of 10 November 1801:

> Yesterday we made a great exertion to forward our work, and returned at night so entirely fatigued, that none of us could write or you would have had a more circumstantial letter now. But we must depart immediately, or do no business today. I expect to meet Mr. Hauducoeur a few miles below Mr. Reeds' [river mile 25.1/kilometer 41.4]. As soon as we meet my labours are at an end.

> If I could bring myself to write to any one but you, I could furnish a letter entertaining enough. The little incidents of our journey have been often extremely laughable, and almost always curious. The very reception we have met with has been so various, that I could fill a letter with descriptions of character, and manner that would often make you laugh. And as to the natural Scenery in which we have been engaged, it is so Savage in many instances, and so beautiful in others, that I could not fail to find in that alone matter enough for twenty letters.[20]

Survey Method

Due to the rugged terrain and the numerous obstacles to a clear line of vision, Latrobe employed the triangulation method of surveying. Triangulation was widely used in the eighteenth century (as it is today) for topographical surveys. It is a system which creates interlocking triangles to accurately determine distances between points which are not observable from one another. Only one baseline and two angles need be found for each triangle, so that measure-

---

15. Susquehanna Canal Company, *Report of the Governor and Directors*, pp. 18–19. The total expenditure by the state of Pennsylvania on the Susquehanna improvement was $12,126, including the original authorization of $10,000 on 11 April 1799, and a supplemental $2,126 authorized 6 April 1802. See "General Statement of the Contracts for opening and improving roads and rivers," 1801–2 (a printed form with handwritten entries), Department of Community Affairs, Bureau of Land Records, Harrisburg, Pa.

16. Hamlin incorrectly states that Hauducoeur was "assistant engineer and surveyor." Hamlin, *Latrobe*, p. 183.

17. John Barry Love, "The Colonial Surveyor in Pennsylvania" (Ph.D. diss., University of Pennsylvania, 1970), p. 197.

18. BHL, Field Book 2, PBHL, MdHS; BHL to Barton, 14 November 1801, Mrs. Gamble Latrobe Collection, MdHS; *DAB*, s.v. Barton, William Paul Crillon; BHL to Boileau, 20 August 1817, Letterbooks.

19. BHL, Field Book 2, PBHL, MdHS.

20. BHL to Mary E. Latrobe, 10 November 1801, Mrs. Gamble Latrobe Collection, MdHS.

ments across difficult terrain or rivers may be reduced or eliminated.[21] In his field application of this system Latrobe made numerous though not highly accurate determinations of base lines, and produced a map which varies somewhat from modern measurements.

During his career Latrobe owned a circumferentor, a "Ramsden" transit, theodolites, a sextant, a surveyor's compass, and rods and chains.[22] On the Susquehanna survey he might have used any of the first four instruments to take horizontal bearings in degrees and minutes such as those recorded in the field notes. Yet the absence of any measurements of vertical angles in the field notes suggests that Latrobe utilized a circumferentor, which could measure only horizontal angles, as opposed to a theodolite, transit, or sextant, which could measure both horizontal and vertical angles. A circumferentor is basically a mounted compass with a sight, occasionally with a telescopic optic. The compass on Latrobe's instrument was marked in compass quadrants rather than in an azimuthal fashion. Latrobe had at least two angle-determining instruments with him on the survey because on 12 November his group broke into two parties, each working a different portion of the survey (sec. 7, fig. 1).

Latrobe's team measured baselines with the surveyor's, or Gunter's, chain. A surveyor's chain is 66 feet (20.12m) long and composed of 100 links joined to one another by three rings. It is normally used with offset staffs and pins. The offset staff, a piece of wood ten links in length (7.92 feet/2.41m), measures distances from the chain to various adjoining points, trees, fences, houses, etc. The pins are spikes of metal used to mark points on the chain and at its ends; for increased visibility they have loops at the top where small flags are attached.[23]

There are numerous possibilities for instrumental error (in addition to user error) when these instruments are employed in the course of survey. With the transit a major problem is compass variation, that is, the tendency of the compass needle to stray from a true northern orientation. In Latrobe's time magnetic variation could be determined by astronomical methods, and allowance made in calculation. Because of the dependence of the compass upon magnetism, deviations are also caused by local iron ore deposits (many occur along the lower Susquehanna) or by small pieces of iron in the instrument itself. "Variation in the magnetic needle," stated the early American surveyor Andrew Ellicott, "is the most important source of [surveying] inaccuracy."[24] But other errors occurred. The length of the chain could vary with temperature, rings could become flattened, the points of contact could be altered by mud, ice, and vegetation, the chain could be kinked or worn. Ellicott also warned surveyors, "It is difficult to find two [chains] of the same length."[25] The offset staff could warp, shrink, or stretch because of variations in humidity.

Latrobe recorded his survey data (fig. 27) in a format including compass bearings, distances, and field notes. Such a pattern is advocated in Love's *Geodaesia*, and a similar format is used in the 1737 field book entries of Lewis Morris, an English surveyor.[26]

Field Sketches

Latrobe's practice of drawing field sketches to supplement and clarify field observation and measurements went beyond conventional bounds of professional skill and entered the realm of artistic presentation. Latrobe employed a variety of illustrative views: panoramas, bird's-eye views (*vues d'oiseau*), general views, and detailed sketches. In his notes he includes drawings of both physical and cultural features.[27]

Depth Soundings

In addition to conducting the surface survey, Latrobe took a series of depth soundings in which he located the submerged "deeps" of the Susquehanna.[28] He provided these accounts of the phenomena:

---

21. Louis DeVorsey, "A Background to Surveying and Mapping at the Time of the American Revolution: An Essay on the State of the Art," in W. Bart Greenwood, comp., *The American Revolution, 1775–1783: An Atlas of 18th Century Maps and Charts. Theatres of Operations* (Washington: Government Printing Office, 1972), p. 12; Raymond E. Davis and Joe W. Kelly, *Elementary Plane Surveying*, 4th ed. (New York: McGraw-Hill Book Company, 1967), p. 216.

22. BHL to Gilpin, 10 October 1803 and 4 November 1803, Letterbooks; BHL to Voight, 10 March 1804, Letterbooks; BHL to Caldwell, 14 March 1810, Letterbooks; BHL to Briggs, 30 May 1810, Letterbooks; BHL, bankruptcy petition, 19 December 1817, RG 21, NA.

23. Henri Michel, *Scientific Instruments in Art and History*, trans. R. E. W. Maddison and Francis R. Maddison (London: Barrie and Rockcliff, 1966); Paul Ballard, "Early Surveying Instruments in Relation to Australian Cartography," *Cartography* 8 (1974): 193–98. Both authors provide excellent illustrated reviews of surveying instruments.

24. Andrew Ellicott, *Several Methods by Which Meridian Lines may Be Found with Ease and Accuracy* (Philadelphia, 1796), p. 5.

25. Ibid.

26. Silvio A. Bedini, *Early American Scientific Instruments and Their Makers* (Washington: Smithsonian Institution, 1964), p. 10; A. H. W. Robinson, *Marine Cartography in Britain* (Leicester: Leicester University Press, 1962), p. 78, fig. 17.

27. See figs. 25, 27, 28, 29, 30 for examples.

28. Edward B. Matthews, "Submerged 'deeps' in the Susquehanna River," *Geological Society of*

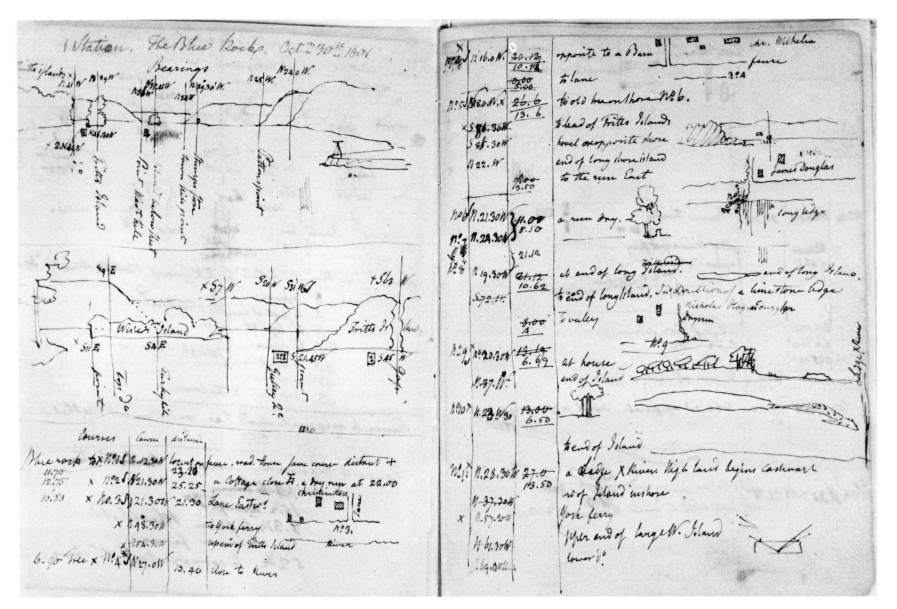

Fig. 27. BHL's observations made from the Blue Rock (river mile 36.4/kilometer 58.6), 30 October 1801. On the top portion of the left-hand page are two sketches made from the survey point: the upper to the north, the lower to the south. Note the survey instrument on the right side of the upper sketch. The remainder of the two pages contains the survey record: reading number or station (first column), angle of reading (second column), distance in chains to the object sighted (third column), and field notes (fourth column). (Field Book 2.)

At the Neck [river mile 24.2/kilometer 39] the water is so deep that in one place no bottom was found with a line 180 feet [54.9 meters] long.

... and at a place called McCall's ferry [river mile 23.7/kilometer 38.2], it narrows to the width of 16 perches [264 feet/80.5 meters]. Here I attempted to find bottom with a line of 180 feet [54.9 meters], but failed, notwithstanding every precaution taken to procure a perpendicular descent of the weight attached to it.[29]

America Bulletin 28 (1917): 335–46. In 1766 Charles Mason, one of the surveyors of the Mason and Dixon Line, recorded the existence of the deeps in his journal. See Hubertis M. Cummings, *The Mason and Dixon Line* (Harrisburg, Pa.: Commonwealth of Pennsylvania, 1962), p. 53.

29. BHL to Gallatin, 16 March 1808, in Cochran, *New ASP. Transportation*, 1:266; map annotation, section 4.

Fig. 28. "View at the Bear Island at the head of Niels falls," 13 November 1801. Pen and ink. The canoe in the foreground is probably that of BHL's survey party. River mile 21.7/kilometer 35.1. (Field Book 2.)

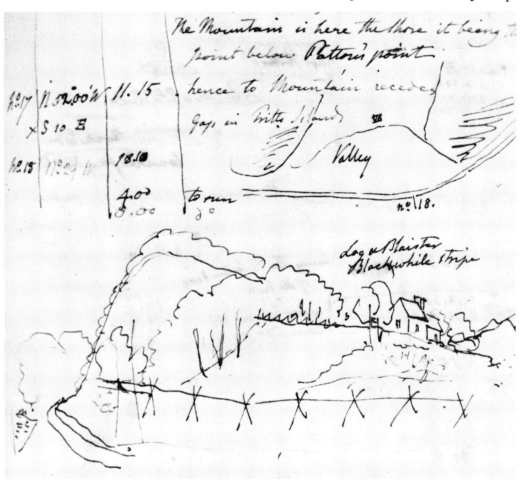

Fig. 29. "Log and Plaister; Black & white stripe," 30 October 1801. Pen and ink. BHL's notation refers to the house in the right background which had a strong pattern of dark logs and white plaster between the logs. Note the rail fence in the foreground. The sketch is at the bottom of a page of field notes. River mile 38.0/kilometer 61.1. (Field Book 2.)

For these measurements he probably employed a lead weight secured by a measured knotted line.[30]

Expenditure of Time

By liberal estimate Latrobe devoted a total of thirty-four days to administering the Susquehanna channel clearance and survey; he spent an undetermined amount of time in Philadelphia preparing his report and maps. His time was employed as follows:

1.	Preliminary Survey	3 days	8, 9, 10 September
2.	Settling Col. Antes's Accounts and Inspecting Works	15 days	21 September to 4 October
3.	Survey	14 days	28 October to 12 November
4.	Meeting with Hauducoeur	2 days	13, 14 November (uncertain)

Latrobe paid $120 to Christian P. Hauducoeur for "... 30 days surveying from the Maryland line to McCalls' on both sides of the river, and through the Islands and traveling to Phila. and back at $4 a day."[31] Most of Haudu-

---

30. DeVorsey, "A Background to Surveying and Mapping," p. 13.

31. BHL, Field Book 2, PBHL, MdHS.

Susquehanna River Survey Map

Fig. 30. "View Among the Bear Islands, Head of Niels Falls," undated. BHL probably derived this pencil sketch from fig. 28. River mile 21.7/kilometer 35.1. (Latrobe Sketchbook VIII.)

coeur's time must have been allotted to his trip to Philadelphia, since Latrobe spent a total of fourteen days surveying an area nearly twice the size of that covered by Hauducoeur and immensely more difficult due to treacherous currents and great topographic relief. In addition, Latrobe paid Hauducoeur for six days in Philadelphia during which they outlined the lower portion of the river. Undoubtedly the accuracy of the map was improved by the data from Hauducoeur's recent survey, as well as from his former surveys for the Susquehanna Canal Company.

Map Description

The product of the survey was a large-scale triangulated, strip-type map approximately seventeen feet long and two feet wide. The horizontal features were based on field measurements, while the vertical relationships were displayed by artistic means with relative values being developed from sketches in the field notes. The map was rendered in pencil, pen and ink, and watercolor on white paper. A series of subtle naturalistic watercolor hues were employed in a technique adapted from architectural illustration. Fine details include the painting and shadowing of individual rocks, trees, and row patterns in fields.

Since the map was prepared for navigational purposes, shorelines were clearly delimited, as were tributary streams, falls, and rapids. Navigation courses are precise, with notation supplied as to the nature of the waters. The geology of the adjacent area is depicted, with the type and extent of prominent formations also noted (sec. 8, fig. 1). Trees and vegetation were recorded in field books.

In format and style the Susquehanna survey map is a high-quality example of the eighteenth-century approach to the problem of topographical mapping, best exemplified by military maps.[32] At this time in continental Europe, especially in France, military schools served as the principal training ground for those entering the engineering profession.[33] Latrobe himself at one time aspired to be a military engineer, and as his first commission to the Federal government, in 1798, he prepared maps for the reconstruction of fortifications at Norfolk, Virginia.[34]

The Susquehanna map was compiled primarily from field survey notes, field sketches, and a series of large-scale rough drawings made during the Latrobe survey of October and November 1801. In developing the map Latrobe also made extensive use of the Hauducoeur map of 1799 (sec. 10, fig. 2) and the Hauducoeur survey of October and November 1801. Cartographic drafting instruments probably employed in its rendering closely resemble those of today, including proportional dividers, mapmeters, parallel rulers, pantographs, sectographs, protractors, ruling pens, bow pens, and drop bow compasses.[35] The map itself may be favorably compared with the subsequent maps of the same area by Poppleton (1824), Scott (1824), Trcziyulny and Haines (1827), and Kirk and Fairland (1830), especially considering Latrobe's success in illustrating the environment.[36]

---

32. A topographic map is a systematic representation of a small part of the land surface, having physical features (e.g., relief and hydrography) and cultural features (e.g., roads and political boundaries). Such maps present both vertical and horizontal features.

33. Frederick B. Artz, *The Development of Technical Education in France, 1500–1850* (Cambridge, Mass.: Society for the History of Technology and M.I.T. Press, 1966).

34. BHL to Jefferson, 4 July 1807, Letterbooks; see p. 247 below.

35. Paul Ballard, "Cartographic Drawing Instruments: The Eighteenth and Nineteenth Centuries," *Cartography* 8 (1974): 140–46.

36. T. H. Poppleton, "Plan and Section of the Susquehanna River between Columbia Bridge and the Maryland Canal...," 1824, Commonwealth of Pennsylvania, Department of Community Affairs, Bureau of Land Records, Harrisburg, Pa.; Joshua Scott, *Map of Lancaster County, Pennsylvania* (n.p., 1824), copy in Lancaster County Historical Society, Lancaster, Pa.; Charles Trcziyulny and George

A unique aspect of the Susquehanna survey is the survival of the two original field survey notebooks,[37] supplemented by a sketchbook of pencil, pen and ink, charcoal, and watercolor views.[38] It appears that the only original cartographic documents lacking are what Latrobe described as "rough drawings being upon loose sheets and not so easily put together and understood."[39]

The map was constructed with regard only to northward orientation. No attempt was made to tie the map into any plane or spherical coordinate system (longitude and latitude, for example), which would cause for correction in angular relationships due to the earth's curvature. In constructing the map Latrobe employed a series of five "Magnetic Meridians," parallel lines of unspecified longitude. These lines are oriented nearly parallel to true north since the value of local magnetic declination in 1801 was 1° East ($\pm 1°$). The failure to fix true geographic coordinates resulted in the most serious error on the map, the misdesignation of the value of the Mason and Dixon Line: Latrobe providing the value "39°41' nearly" as opposed to a modern value of 39°43'26.3" North.[40]

The map follows the conventions of the period in its depiction of relief and portrayal of physical and cultural features. In his rendering of the map section from the Maryland state line (river mile 14.2/kilometer 22.8) downstream to Havre de Grace (river mile 0.0/kilometer 0.0) Latrobe dispensed with the extensive use of watercolor so significant in the Pennsylvania portion and executed the area in pencil and pen and ink only. However, Latrobe continued to carefully delimit the channel and the old and new navigation courses, providing local annotations. He rendered relief with rudimentary hachuring rather than with the shading technique he used for the upstream portion.[41]

It is interesting that the Maryland section is a generalized form of Hauducoeur's 1799 map with some updating, such as the progress of the canal, alteration of navigation channels, and changes in the form and location of some islands. It should be noted that no mention is made by Latrobe of Hauducoeur's excellent engraved map of the same section, although it is of great interest to those concerned with the lower Susquehanna.

Color

Latrobe complied with standardized coloring conventions of his period when preparing the map.[42] For example, he showed houses in red and main roads in ochre. In the depiction of trees and rocks he applied the formulas provided by Smith (1788): "... if there be any woods, dab every tree with the point of a very fine pencil dipped in grass-green, made of copper-green, tempered with the gumboge [bright yellow]; ... Rocks must be done with tincture of myrrh [yellow brown to reddish brown], and the trees, some with copper-green, some with dark grass-green, and some with thin umber."[43] Latrobe also conformed to convention when portraying the general landscape: "If you are to color any representations of land, do the lightest parts over with very thin yellow, that represents a straw color, shading it with orpiment [lemon yellow]; and in other parts, do it with light green and shade it with a deeper green."[44]

Place Names

Geographical place name changes since 1801 have been few, the majority being alterations in spelling or simplifications. A distinct exception has been the innumerable changes in the names of alluvial islands,[45] which can be attributed to their geomorphic pattern of continual modification, destruction,

---

Haines, "Charts & Profile of the Susquehanna River in Its Passage Through the State of Pennsylvania ...," 1827, Commonwealth of Pennsylvania, Department of Community Affairs, Bureau of Land Records, Harrisburg, Pa.; C. Kirk and Jonas P. Fairland, *Columbia, Wrightsville, and Marietta, Pennsylvania* (n.p., 1830), copy in Hist. Soc. Pa.

37. BHL, Field Books 1 and 2, PBHL, MdHS.
38. Latrobe Sketchbook VIII, PBHL, MdHS.
39. BHL to Caton, 25 May 1817, Letterbooks.
40. The Mason and Dixon Line, surveyed between 1763 and 1768 by the English surveyors Charles Mason and Jeremiah Dixon, served to demarcate the disputed boundary between Maryland and Pennsylvania. It appears that Mason and Dixon did not ascertain the true latitude of the boundary, and by BHL's time it had still not been established.

The value of local magnetic declination in 1801 was provided by E. B. Fabiano of the United States Geological Survey.

41. Yolande Jones, "Aspects of Relief Portrayal on 19th Century British Military Maps," *The Cartographic Journal* 11 (1974): 19–33. Hachuring is a system of lines drawn in the direction of slope,

---

producing a shadow effect. Shading is the use of a continuous graded tone which simulates the appearance of shade and shadow.

BHL used hachuring on other maps, among them his "Section of the Northern Course of the Canal from the Tide in the Elk River at Frenchtown to the forked [oak] in Mr. Rudulph's Swamp," 3–10 August 1803, and his maps of the proposed Potomac Canal extension in December 1802, but he usually employed shading.

42. E.g., John Hammond, *The Practical Surveyor*, 2d ed. (London, 1731), pp. 139–40; John Muller, *A Treatise Containing the Elementary Part of Fortification, Regular and Irregular* (London, 1756), pp. 15–18.

43. John Smith, *The Art of Painting in Oil to Which is Added the Whole Art and Mystery of Colouring Maps*, 9th ed. (London, 1788), quoted in Arthur L. Humphreys, *Old Decorative Maps and Charts* (London: Halton & Truscott Smith, 1926), pp. 45, 47.

44. Ibid., pp. 46–47.

45. M. Luther Heisey, "The Susquehanna Islands in Lancaster County," *Papers of the Lancaster County Historical Society* 59 (1955): 25–55.

and rebuilding under the influence of floods and ice jams.[46] Most complete name transformations stem from localities that bore personal names, the most significant being the name alteration of the large bedrock island at the head of the Chesapeake Bay from Palmer's Island to Garrett Island. The correspondence of features with place names as plotted by Latrobe is correct without exception. However, Latrobe occasionally neglected to label a locality which he depicted in his field books and sketchbooks. For example, he drew and redrew a log cabin which he did not identify (see sec. 6, figs. 3 and 4) but which may be Quiggler's house (river mile 33.5/kilometer 54.0; see sec. 6, fig. 1).

History of the Map

The map was drafted in Philadelphia and delivered to Governor McKean on 1 November 1802; the report to the governor had been submitted to his office almost a year earlier.[47] Latrobe retained a copy of the map along with his preliminary drawings and field books. In making the maps Latrobe was briefly assisted by Christian Hauducoeur, and probably had the help of Adam Traquair, his clerk.[48]

The Susquehanna map was employed as a legislative and technological document at the corporate, state, and national levels. The fate of his copy and that delivered to the Commonwealth of Pennsylvania was explained by Latrobe as follows:

> ... I have sent my own Copy of the Map of the Susquehannah to Mr. Caton [of the Library Company of Baltimore]. It is a document of the utmost importance,—that is, *it would cost much money to create it* if it did not exist. An exact copy, was sent to Governor McKean in 1802, with a long report on the whole subject of the improvement of the Navigation of the Susquehannah. This report must be among the public papers at Harrisburg, but at the time when the subject of Canal improvement was before Congress, the map found its way to the Committee [in Washington] and I have seen it before the Committee of the Senate or of the House of Representatives, in the scandalous condition of Fly dirt and Smoke into which it was put by hanging up for a Year or two in the [Pennsylvania] House at Lancaster, where it occupied 17 feet of Wall next to the ceiling. [This map "remained probably among the records of the House of Rep. U.S. untill it was destroyed in their general destruction by the British in 1814."] My own copy, now in the hands of Mr. Caton is in excellent preservation and I earnestly request that it be taken good care of while in the hands of the Commissioners.[49]

Latrobe had a high opinion of the cartographic data employed in making the map. "Few people, who are likely to be employed will take the trouble I did and many will not possess the necessary knowledge and few the disinterested spirit of enterprise, to encounter the labor I underwent."[50]

Latrobe's personal copy was stored in a trunk for several years and, upon request, transferred to Richard Caton for use in the development of plans for internal improvements in Maryland. This copy was later returned to Latrobe and he presented it to the Library Company of Baltimore on 17 June 1817. At this point a signed and dated annotation was added by Latrobe (map sections 2 and 3), the misinterpretation of which gave rise to Talbot Hamlin's incorrect statement that "later (in 1817), when Latrobe was living in Baltimore, he made a replica of it [the Susquehanna survey map] from his notes and from memory and presented it to the Library Company of Baltimore...."[51] Eventually the map came into the possession of Charles H. Latrobe, grandson of Benjamin Henry Latrobe, who on 12 October 1885 presented the map to the Maryland Historical Society, its current owner.[52]

Natural History

Latrobe acquired his knowledge of natural history (botany, geology, mineralogy, soils, and zoology) from his Moravian education, his knowledge of current general literature, field observations, and his involvement in scientific circles, including the American Philosophical Society.[53] He recognized the engineering applications of information derived from natural history, partic-

---

46. Stephen F. Lintner, "The Historical Physical Behavior of the Lower Susquehanna River, 1801 to 1976" (Ph.D. diss. in progress, Johns Hopkins University).

47. Written on the back of the map at the upstream end is the notation, "The Original delivered to Govr. McKean 1st Nov. 1802."

48. Adam Traquair was BHL's clerk from about 1800 to 1802.

49. BHL to Morgan, 7 June 1817, Letterbooks; map annotation, sections 4 and 5.

50. BHL to Caton, 25 May 1817, Letterbooks.

51. Hamlin, *Latrobe*, p. 183.

52. *Magazine of American History Illustrated* 14 (1885): 624–25.

53. See Gordon L. Davies, *The Earth in Decay: A History of British Geomorphology, 1578–1878* (London: Macdonald & Co., 1969), p. 126; V. A. Eyles, "The Extent of Geological Knowledge in the Eighteenth Century, and the Methods by Which it Was Diffused," in Cecil J. Schneer, ed., *Toward a History of Geology* (Cambridge, Mass.: M.I.T. Press, 1969), pp. 159–83.

In his writings BHL referred to such works as Erasmus Darwin's *Zoonomia* (1794–96); Thomas Brown's *Observations on the Zoonomia of Erasmus Darwin* (1798); John Clayton's *Flora Virginica* (1739–43); André

ularly with regard to hydraulic engineering, and his observations included the economic and utilitarian use of the environment. He noted at one point on the Susquehanna map: "The Slate of this ledge is of very good quality but [it is] thick, the quarry has scarcely been opened, and it is highly probable, that on getting deeper into the Rock, the Slates will prove an immense acquisition to the Cities on the Chesapeake."[54]

Latrobe's geological thinking was greatly influenced by Constantin F. C. de Volney and William Maclure. Latrobe met both men while they were in Richmond, Virginia, and he developed a deep admiration for their abilities. In 1803 Volney published his *Tableau du climat et du sol des États-unis d'Amerique*, which has graphic presentations of stratigraphic sections and river terraces paralleling those of contemporary practice. He was especially knowledgeable of what is now termed *regional geology* and *soil-plant relations*. Maclure wrote the first extensive account in English of American geology and published the first geological map of the United States in 1809.[55]

Latrobe embraced the Neptunian theory of the preeminent German geologist Abraham Werner, who held that rock formations were uniformly distributed over the globe and that all rocks had been precipitated from a universal ocean. Latrobe's field observations led him to believe that Werner's theory could explain the geology of the Atlantic coastal United States, and he employed the Wernerian system for classification of rock formations. But his support for this theory was not uncritical, and his "Memoir on the Sand Hills of Cape Henry in Virginia" stated that contemporary coastal processes were creating a geological formation that could be erroneously interpreted by some future natural philosopher using Wernerian concepts.[56]

Latrobe also believed that past geologic forces had greater energy than present ones, and in this he was allied with the geological catastrophists. The catastrophists thought that "the causes of geological changes in the past *differ not in kind*, though they may sometimes *differ in energy*, from those now in operation."[57] In describing the formation of the puddingstone being quarried for the columns of the United States House of Representatives, he conjectured:

> Imagine these pebbles rounded and mingled by attrition for ages, and then to have been left, and cemented by some matter filling all interstices, ... so as to become a solid mass. Suppose then that the valley become the bed of a mighty torrent running from S.W. to N.E. over this cemented mass, wearing it down in the direction of its current unequally, according to the velocity of its veins; and employing, (as in all our rivers) the agency of loose stones, to whirl deep basins into the solid mass, and thus giving to the rocks, now separated into distinct masses, that specific character, which the rocks of all our rapids acquire by the action of water, *and which character cannot possibly be mistaken, or derived from any other known agency*. Imagine then that this torrent cease, leaving its bed dry, and the rocks bare, but covered in its lower parts with alluvial soil.[58]

Latrobe recognized the processes of contemporary river systems in the development and modification of landscapes, but he saw them as factors acting to finish and refine features rather than as primary agents. It is probable that like many contemporary observers he thought that the age of the earth was insufficient for currently observable processes to have modified more than a minute portion of the surface. Nonetheless, he frequently discussed the impact of fluvial processes on the landscape. For example, in agreement with modern geological theory, he believed the Susquehanna River had cut its channel through the mountains.[59]

> Another narrow ridge of granite hills crosses the river immediately below Columbia, over which the river falls rapidly, and then enters the wider limestone valley known by the name of Jochara valley. The river spreads here to the width of three miles [4.8 kilometers], its stream is gentle though rapid, and it abounds with beautiful and fertile islands.

Michaux's *Flora Boreali-Americana* (1803); Thomas Nuttall's *Genera of North American Plants* (1818); Claude C. Robin's and Constantine S. Rafinesque's *Florula Ludoviciana* (1817); Lazzaro Spallanzani, *Viaggi alle Due Sicilie* (1792–97); and Alexander Wilson's *American Ornithology* (1808–14).

54. See annotation on map section 6.

55. Hamlin, *Latrobe*, pp. 79–80; Constantin François Chasseboeuf, Comte de Volney, *Tableau du climat et du sol des États-Unis d' Amerique*, 2 vols. (Paris, 1803); Charles C. Gillespie, ed., *Dictionary of Scientific Biography*, 14 vols. (New York: Charles Scribner's Sons, 1970–76), s.v. William Maclure; BHL, "Observations on the foregoing communications [on the construction of buildings in India]," *Transactions of the American Philosophical Society* 6 (1809):283.

For extensive examinations of BHL's geological thinking, see BHL, *Journals*, 2:384–90, and Stephen F. Lintner and Darwin H. Stapleton, "Geological Theory and Practice in the Career of Benjamin Henry Latrobe," in C. J. Schneer, ed., *Two Hundred Years of Geology in America* (Hanover, N.H.: University Press of New England, 1979), pp. 107–119.

56. Davies, *Earth in Decay*, passim; BHL, "Memoir on the Sand Hills of Cape Henry in Virginia," *Transactions of the American Philosophical Society* 4 (1799): 439–44; Alexander G. Ospovat, "Reflections on A. G. Werner's 'Kurze Klassifikation,'" in Schneer, ed., *Toward a History of Geology*, pp. 242–56.

57. Reijer Hooykaas, "Catastrophism in Geology, Its Scientific Character in Relation to Actualism and Uniformitarianism," *Mededelingen der Koninklijke Nederlandse Akademie van Wetenschappen* 33 (1970): 273.

58. BHL to Gales & Seaton, 18 January 1817, in *National Intelligencer* (Washington), 24 January 1817.

59. Luna B. Leopold, M. Gordon Wolman, and John P. Miller, *Fluvial Processes in Geomorphology* (San Francisco: W. H. Freeman and Co., 1964).

It then suddenly contracts *and is received into* a narrow ravine which it has sawed down into the granite hill called Turkey hill. . . .[60]

He also attributed the development of minor valleys on the Delmarva Peninsula to gradual erosion,[61] and explained in a strikingly modern fashion the causes and impact of culturally accelerated erosion and sedimentation in the rivers of the middle Atlantic states. In Latrobe's words:

> Since the clearing of the country, the heads of the navigation have gradually become choaked with the alluvion of the upper country deposited at the place where the velocity of the upper water is checked by the opposition of the tide, and by the expansion of the stream in the wide water below the rocky ridge of the falls.
>
> Thus by degrees, and without exception, I believe, have the ports immediately below the falls of all our rivers, gradually lost their shipping business.[62]

Latrobe had an awareness of general rock types and successfully constructed an accurate map of the basic geology of the Susquehanna valley.[63] His classification of rocks proceeded from an understanding and recognition of their mineralogy. He described many forms by their crystal structure, but further refined his terminology by the use of modifying terms such as *schistous*.[64]

He recognized the variable resistance within and between rock types to weathering, which he attributed principally to the mechanical forces of water and ice. Latrobe probably underestimated the magnitude and extent of chemical weathering although no doubt he was familiar with the varying environmental stability of common building stones such as granite, limestone, marble, slate, and sandstone.[65] Continuing his earlier observations on the weathering of granite rocks on the piedmont plateau of Virginia,[66] he characterized the relief relationships of rock and water along the lower Susquehanna by observing: "Wherever the river crosses a valley of limestone or slate [softer rock], the rocks are worn into a smoother and wider bed: but when it has to cross a ridge of granite [harder rock], its course is immediately broken by irregular masses of large rocks; its bed is narrow and enclosed by precipices, and its torrent furious and winding."[67]

After geology, Latrobe's comments on the natural history of the lower Susquehanna focused on botany and plant ecology. He knew that trees are involved in the retention of soil, stabilization of river banks, and obstruction of water flow, so he ordered them removed from certain areas to help open the Susquehanna channel. Latrobe noted the pruned treetops and branches and the inclined posture of many trees on the floodplain, an effect of ice jams and floods. He also observed the common phenomenon of multitrunked trees, caused either by stump sprouting or the coalescence of dense saplings.[68]

In the field books of the Susquehanna survey, Latrobe referred by common names to the trees which served as surveying stations or observation points, and in so doing provided a basic list of the composition of the Susquehanna valley forests.[69] Most frequently noted was the buttonwood, or sycamore—a large tree whose distinctive mottled brown bark flakes off in large pieces to expose a yellow-white underbark. It is the most distinctive species of the floodplain. Latrobe made no comment on the understory vegetation other than to describe it as "bushy."

Hydrology was another area in which Latrobe exhibited some special knowledge and interest. Early informal study under a German engineer, his work under Smeaton, and his reading of technical treatises had taught him

---

60. BHL to Gallatin, 16 March 1808, in Cochran, *New ASP. Transportation*, 1:266. Maclure and Volney believed that the major gaps in the Appalachians were the result of a vast transmontane lake breaking through the Appalachians in an earlier geologic period.
61. 21 October 1803, C & D Engineer's Reports, Hist. Soc. Del. See also BHL, *Journals*, 2:348–50.
62. BHL to Thompson, 24 March 1806, Letterbooks. Cf. Stanley W. Trimble, *Man-Induced Soil Erosion on the Southern Piedmont, 1700–1970* (Ames, Iowa: Soil Conservation Society of America, 1974).
63. See fig. 1, p. 104 below.
64. Susquehanna map annotations; Latrobe Sketchbook VIII, PBHL, MdHS.
65. BHL to Gales & Seaton, 18 January 1817, in *National Intelligencer* (Washington), 24 January 1817.
66. BHL, *Journals*, 2:386.
67. BHL to Gallatin, 16 March 1808, in Cochran, *New ASP. Transportation*, 1:265.
68. See fig. 31. R. S. Sigafoos, *Botanical Evidence of Floods and Flood-Plain Deposition*, USGS Professional Paper 485-A (Arlington, Va.: United States Geological Survey, 1964); Lintner, "The Historical Physical Behavior of the Lower Susquehanna River."

69. | Latrobe's Common Name | Genus and Species |
|---|---|
| Aldar [Alder] | *Alnus* ssp. (L.) |
| Beech | *Fagus grandifolia* (L.) |
| Birch [probably River Birch] | *Betula nigra* (L.) |
| Black Cedar | [possibly *Chamaecyparis thoides* (L.)] |
| Buttonwood [Sycamore] | *Plantanus occidentalis* (L.) |
| Chestnut | *Castanea dentata* (Marsh) |
| Chestnut Oak | *Quercus prinus* (L.) |
| Hiccory [Hickory] | *Carya* ssp. (K. Koch) |
| Linden [Basswood] | *Tilia americana* (L.) |
| Locust | *Robinia pseudoacacia* (L.) |
| Pine | *Pinus* ssp. (L.) |
| Span[ish] Oak | *Quercus* ssp. (L.) |
| Walnut | *Juglans nigra* (L.) |

(From Field Books 1 and 2, PBHL, MdHS.)

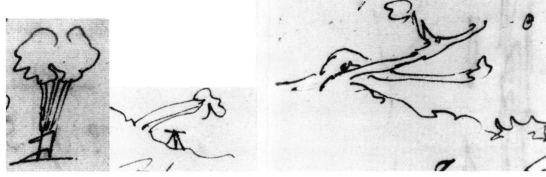

Fig. 31. Composite of three of BHL's sketches of trees in his surveying notes. At left is a multitrunked tree and at right is a tree growing nearly horizontally, both of which are common phenomena on the Susquehanna floodplain. In the center a surveying instrument (possibly a circumferentor) stands beneath a recumbent tree, an example of how field sites were identified by prominent trees. (Field Book 2.)

something of hydraulic engineering. In several writings he cited such major works on hydraulic engineering as those by Bernard Belidor and Jean Antoine Fabre.[70]

Latrobe had a keen awareness of the dynamic relationship between hydrologic factors and channel morphology, and sought to employ natural forces rather than engineering structures to maintain or modify river channels. In the improvement of the Susquehanna he employed this principle by attempting to create a channel which would clear itself of silt and debris.[71] He also had some notion of groundwater hydrology, and his observations on the subsurface conditions of Philadelphia illustrated his understanding of how subterraneous water and surficial geology are interrelated with water supply, sewage disposal, and contamination.[72]

70. Belidor, *Architecture Hydraulique, ou l'Art de Conduire, d'Elever et de Ménager les Eaux pour les Différens Besoins de la Vie...*, 2 parts in 4 vols. (Paris, 1737–53); Fabre, *Essai sur la Théorie des Torrens et des Rivières* (Paris, 1794); BHL to Mayor and Council of New Orleans, 10 June 1816, Mrs. Gamble Latrobe Collection, MdHS; BHL to Moore, 20 January 1811, Letterbooks; Asit K. Biswas, *History of Hydrology* (New York: American Elsevier Publishing Co., 1970), pp. 261, 262, 279; BHL, *Opinion*, pp. 17, 21, 24–25; BHL, *Remarks on the Address of the Committee of the Delaware and Schuylkill Canal Company* (Philadelphia, 1799), p. 10.

71. Leopold, Wolman, and Miller, *Fluvial Processes*, passim; BHL to Fitzsimmons, 25 May 1807, Letterbooks; BHL to Bate, 21 November 1809, Letterbooks.

72. BHL, *Journals*, 2:379–81; BHL, "Designs of Buildings... in Philadelphia," [1799], Hist. Soc. Pa.

Susquehanna River Survey Map

Landscape Change 1801–1978

When considering Latrobe's Susquehanna map in relation to the modern landscape one must bear in mind that significant changes have occurred in land use patterns in the survey area and upstream; tilled fields have returned to woodlands, alluvial islands have been continuously modified by flood and ice, and roaring gorges have given way to power dams. The sediment in the channel, formerly a golden sand, is now a mixture of sand, slag, and fine coal dust derived from upstream mining wastes. The channel itself is reduced in width, while the adjacent floodplain level has risen rapidly in elevation due to the massive influx of coal-laden sediment.[73]

Approximately 60 percent of the river channel mapped by Latrobe is submerged by the reservoirs of three twentieth-century dams.[74] They control the river's flow so that the radical seasonal fluctuations in water level which once characterized it have been eliminated. The massive summer pools of stagnant water which provided breeding grounds for mosquitoes and the development of unsanitary conditions are also gone. Control also may have reduced the frequency and extent of floods and ice jams, which caused property damage and loss of life.

Conclusion

The Susquehanna River map by Benjamin Henry Latrobe is an outstanding engineering, scientific, and artistic document. It may be appreciated as a historical document of exceptional detail and beauty, as well as a rare profile of hydrology, geology, and biology in a part of America just beginning to feel the impact of extensive agricultural and commercial development. Although his work on the Susquehanna was one of Latrobe's shorter engineering engagements, it produced the most beautiful and fascinating of his engineering drawings.

73. Lintner, "The Historical Physical Behavior of the Lower Susquehanna River."

74. Holtwood Dam (1910) at river mile 23.7/kilometer 38.1; Conowingo Dam (1928) at river mile 8.5/kilometer 13.6; Safe Harbor Dam (1931) at river mile 30.5/kilometer 49.

THE SUSQUEHANNA RIVER SURVEY MAP BY BENJAMIN HENRY LATROBE

MAP DATA

Title: "The Susquehanna from Columbia to the Pennsylvania line & thence to Havre de Grace"

Date: 1801–2

Cartographer: Benjamin Henry Latrobe, with the assistance of Christian P. Hauducoeur and probably Adam Traquair.

Sources: Survey by Benjamin Henry Latrobe, October–November 1801, and surveys by Christian P. Hauducoeur, fall of 1798, and October–November 1801.

Type: Manuscript strip map

Dimensions: 22 inches by 206 inches (55.9 centimeters by 523.2 centimeters).

Scale: 1:12,672

Construction: Composed of 6.33 sheets of white paper, each full sheet measuring 22 inches by 32.75 inches (55.9 centimeters by 83.2 centimeters). The sheets are glued to a continuous sheet of linen, and rolled as a scroll with handles.

Technique: Pencil, pen and ink, and watercolor.

Area: Lower Susquehanna River from Columbia, Pa., to Havre de Grace, Md., a distance of 40.7 river miles (65.5 kilometers).

Purpose: "This survey was made by direction of his Excellency Thos. McKean, Govr. of Pennsa. in concert with the Susquehanna Company of Maryland, in order to ascertain the best means of rendering the River safely navigable for Rafts and Arks downwards during Spring freshes, and also with a view to the future improvement of the Navigation so as to enable boats to ascend the River by means of towing paths." (Susquehanna Map, sections 4 and 5.)

Repository: Maryland Historical Society, Baltimore, Md.

Designation of modern impoundment areas, map sections 1–10.

——————— Primary area of reservoir ponding

– – – – – – Secondary area of reservoir ponding

COLOR VALUES OF FEATURES

Cultural	Color	Munsell Notation*
fields	light gray	10 YR 7/2
structures	light red	2.5 YR 6/4
roads	strong brown	7.5 YR 5/6
Physical		
trees	olive	5 Y 5/2.5
green understory	pale yellow or very pale green	5 Y 7/3
northeast bank	olive gray	5 Y 5/2
river	light blue	10 B 7/3
granite	reddish brown	5 YR 5/3.5
limestone	gray	10 YR 6/0.5
slate	gray	2.5 Y 5/0

*Munsell Color Co., Inc., Baltimore, Maryland.

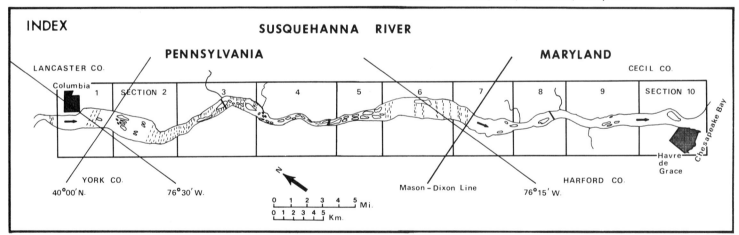

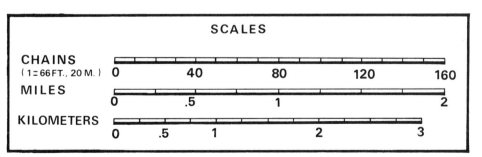

The scale on this page measures distances between any two points in the map sections which follow. On the horizontal borders of each section are the indications "river miles" and "river kilometers": these measure distances along the center line of the river from the modern Conrail bridge at Havre de Grace, and thus vary considerably from the straight-line distances measured by this scale.

Susquehanna River Survey Map: Section 1

Sec. 1, fig. 1 Panorama, looking north from the middle of the Susquehanna River between Columbia and Wright's Ferry, and survey notes for the station. 2 November 1801. (Field Book 2.)

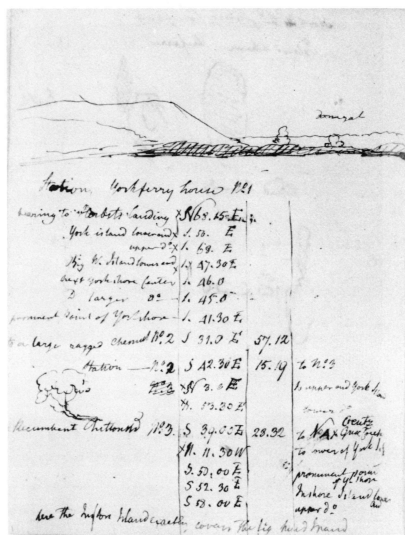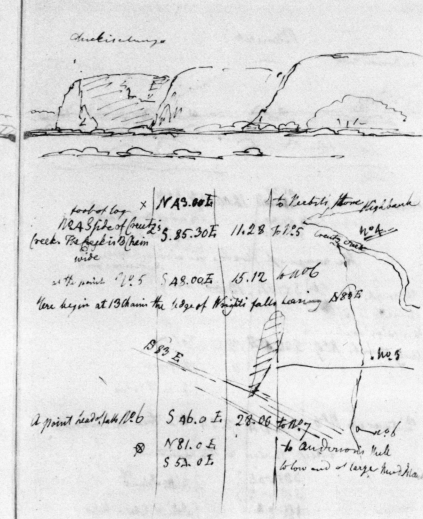

At the north (left) end of the map on the east shore is Columbia, the end of navigation for many of the Susquehanna rivercraft. Here the backcountry navigators sold their goods to merchants who took them principally to eastern markets. Many goods went overland to Lancaster, a few miles distant, and then on to Philadelphia by the Lancaster Turnpike. Joshua Gilpin, a Philadelphia merchant, took special notice of the Susquehanna emporium during a journey in 1809, when conditions were much the same as when Latrobe was there in 1801.

Columbia is seated on the lowest point where the Susquehanna is safely navigable... [and it] is therefore the grand depot of all the produce of this vast river which waters a great part of Pennsylvania.... [T]he produce now brought consists of about 100,000 bbls flour, 300,000 bus of wheat; an immense quantity of lumber—a great deal of coal, iron, whiskey, beef, pork, and in fact all the productions of the country....

The produce is brought down from the extreme waters of the river, in two kinds of boat, one called keel boats are long flat bottomed vessels built regularly like a barge but very flat and carry 50 to 60 tons. [See fig. 25, p. 75.] Arks are vessels of the rudest

Sec. 1, fig. 2 "View of the Susquehannah from the foot of the falls above Columbia." [29 October 1801.] View to west. River mile 41.1/kilometer 66.4. (Latrobe Sketchbook VIII.)

Sec. 1, fig. 3 "View of Chickisalunga Rocks looking to the Southward." East side of river. [29 October 1801.] River mile 42.2/kilometer 67.9. (Latrobe Sketchbook VIII.)

and strongest construction, a flat frame of timber is laid for the bottom very strong on which rough parts are placed upright and boards fastened on the outside, which are roughly tho securely caulked—they are in fact nothing more, than a vast, rough, and unwieldy box of this shape, being flat bottomed and perpendicular at the sides, about 60 to 90 feet [18 to 27 meters] long, from 15 to 20 [4.5 to 6 meters] wide and about 5 feet [1.5 meters] deep—they are so rough as to be put together only with wooden pins or dowells. They carry an immense quantity, and draw about 2 feet [0.6 meters] water—these are built at the places of embarkation on the upper parts of the river and loaded at the time of freshes only, as they cannot be navigated in any other [season]—they have neither oars or sails but depend solely on the velocity of the current and are guided by a long oar at each end—thus strong they bounce and tumble over the falls and rapids, it being only necessary to keep them from running aground: as they are then knocked to pieces and the cargo with difficulty saved—the boatmen are so expert however that these accidents do not often happen, when they arrive at Columbia, they are unloaded, broke up, and sold as lumber—several we saw at Columbia, had a vast quantity of coal on board. [See fig. 26, p. 76.]

... some of the Arks venture down the river for the Baltimore markett—but below Columbia there is such a succession of falls and obstructions, that accidents are very frequent and the risque too great to be often attempted—[but] vast numbers of rafts of boards, and timber go down.

Source: Joseph E. Walker, ed., *Pleasure and Business in Western Pennsylvania: The Journal of Joshua Gilpin, 1809* (Harrisburg, Pa.: The Pennsylvania Historical and Museum Commission, 1975), pp. 11–13.

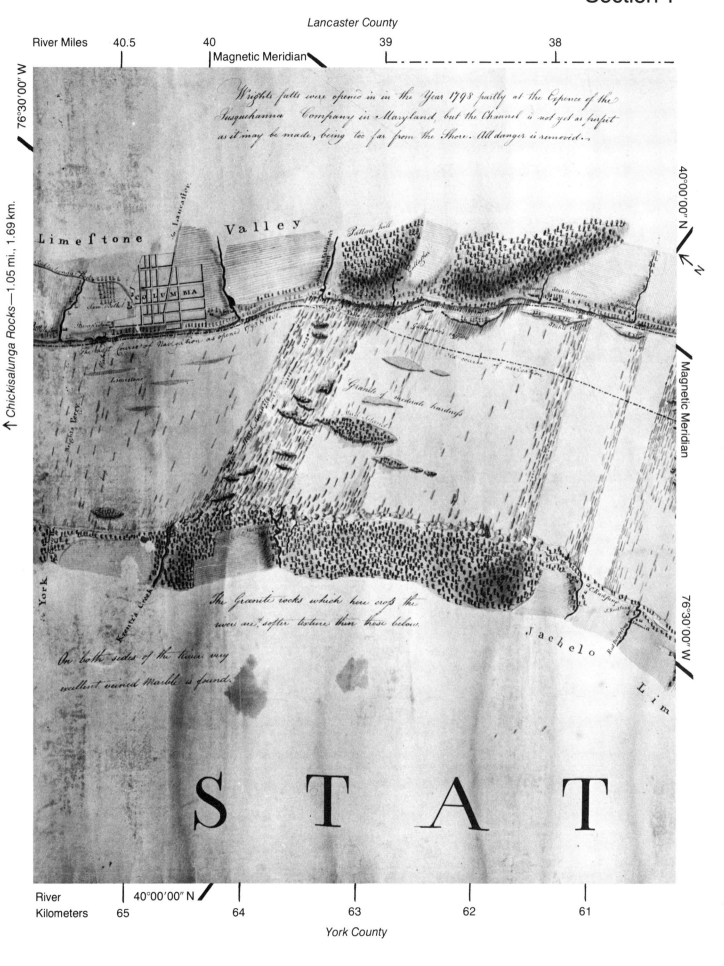

Susquehanna River Survey Map: Section 2

Sec. 2, fig. 1 Panorama, looking east and south from a point just to the north of Section 2. Turkey Hill is partially obscured by large trees on the right. 2 November 1801. (Field Book 2.)

Sec. 2, fig. 2 "Anderson's Mill, below Wrights Ferry, drawn in a Fog." This drawing shows the head of Wright's fall looking down the River. Looking downstream on east shore. Last week of September 1801. River mile 39.5/kilometer 63.5. (Latrobe Sketchbook VIII.)

Sec. 2, fig. 3 Looking north from Turkey Hill on east shore. Last week of September 1801. River mile 34.7/kilometer 55.8. (Latrobe Sketchbook VIII.)

The second section of the map has two contrasting features. On the left are several large islands formed by the sediment deposited in the relatively still water behind the Turkey Hill Falls. These islands are of exceptional interest to the study of river phenomena because their sizes, shapes, and locations change constantly under the influence of currents, floods, and ice jams.

On the right-hand portion of the section is the beginning of the dangerous rapids which made it necessary to undertake navigational improvements if the Susquehanna was to be a route of commerce. Latrobe described this scene in his report to Governor McKean of Pennsylvania.

> ... The most formidable obstruction in Pennsylvania, below Columbia, was perhaps Turkey Hill Falls. [Turkey Hill is the bulge on the east shore at river mile 34.5/kilometer 55.6.]

Section 2

The river after spreading to the width of nearly two miles [3.2 kilometers], suddenly contracts itself on breaking through the mountains to the width of 60 chains, or 3/4th of a mile [1.2 kilometers]. The whole bed is obstructed by high ridges of rocks, extending in regular lines directly across from shore to shore, while at the same time it is choaked by rocky islands; this state of the river continues about a mile, still contracting its width; a natural channel on the Eastern Shore, generally from 50 to 100 feet deep [15.2 meters to 30.5 meters], and from 60 to 100 feet wide [18.2 to 38.5 meters] then commences, through which, in the autumn, the whole river is discharged with astonishing rapidity.

(Latrobe to McKean, 24 November 1801, in Susquehanna Canal Company, *Report of the Governor and Directors to the Proprietors of the Susquehanna Canal* [Baltimore, 1802], pp. 15–16.)

Susquehanna River Survey Map: Section 3

Although Latrobe surveyed this section of the river thoroughly, he did not make a significant pictorial record of it other than the map itself, and the annotative illustrations lie slightly outside of this section.

Latrobe's two watercolors represent opposite views from the same point on the farm of George Stoner, one of the contractors on the Susquehanna improvement. Presumably Latrobe and his surveying party were welcomed to the farm and spent some time there. The scene on the left of the typical Pennsylvania German barn nestled in the hills is one of Latrobe's most serene and well-balanced watercolors. In emotional contrast to it is the unfinished view of the rapid descent to the rocky, foaming Susquehanna River.

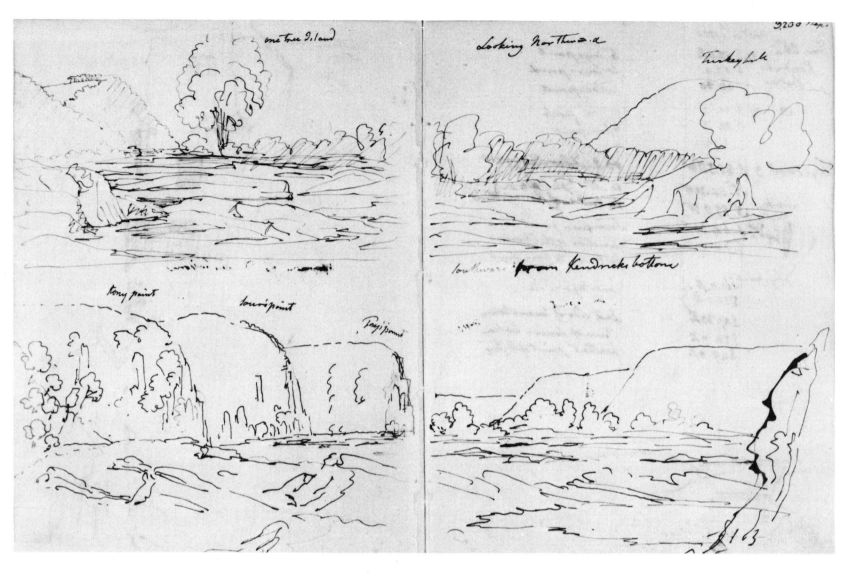

Sec. 3, fig. 1 Two panoramic views from Kendrick's Bottom. Top view, looking to the north. Note cultivated land on top of bluff on left bank. Bottom view, looking to the south. Fry's Point, the most distant bluff on the left bank, lies just within section 3. 4 November 1801. River mile 34.2/kilometer 55.1. (Field Book 1.)

Sec. 3, fig. 2 "Geo. Stoners on Pequai [Pequea] Creek, Burkhalter's ferry, Susquehannah." Last week of September 1801. River mile 27.3/kilometer 44. (Latrobe Sketchbook VIII.)

Sec. 3, fig. 3 "View of the Susquehannah, from the top of Geo. Stoners land on the road to Burkhalters ferry." Last week of September 1801. Looking west through Pequea Creek valley. Unfinished. River mile 27.3/kilometer 44. (Latrobe Sketchbook VIII.)

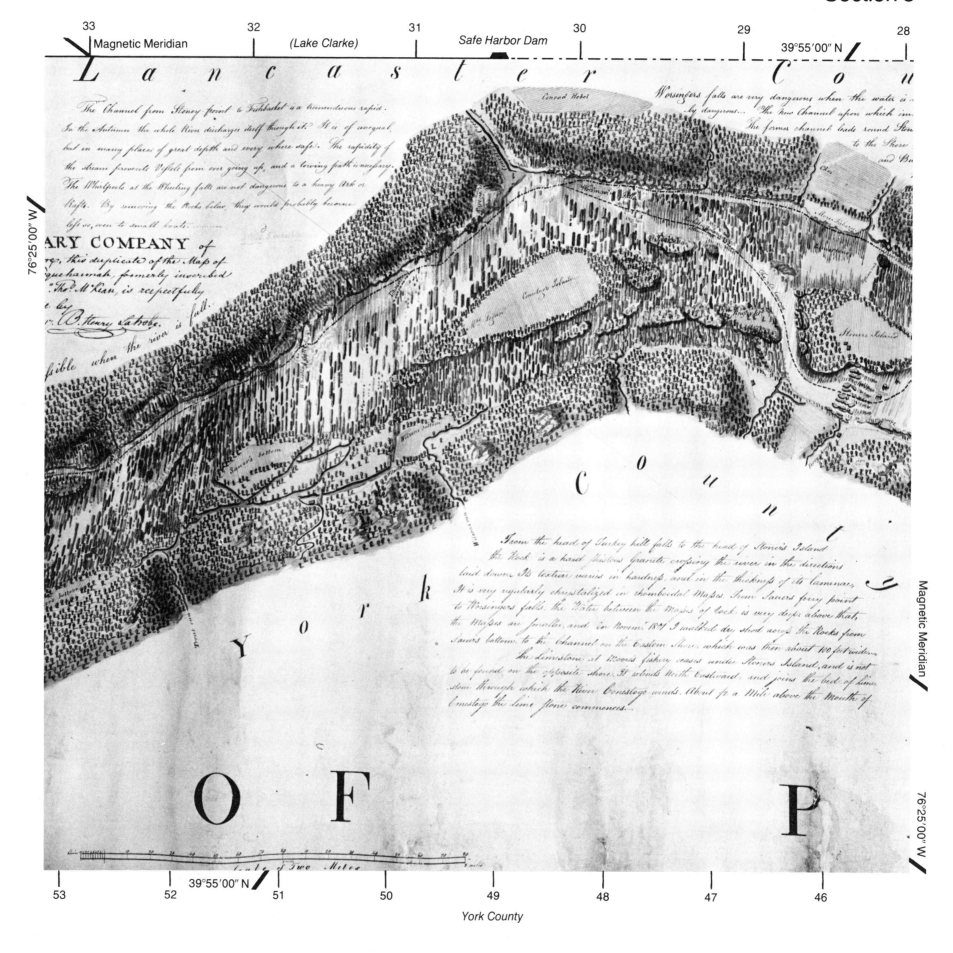

Section 5

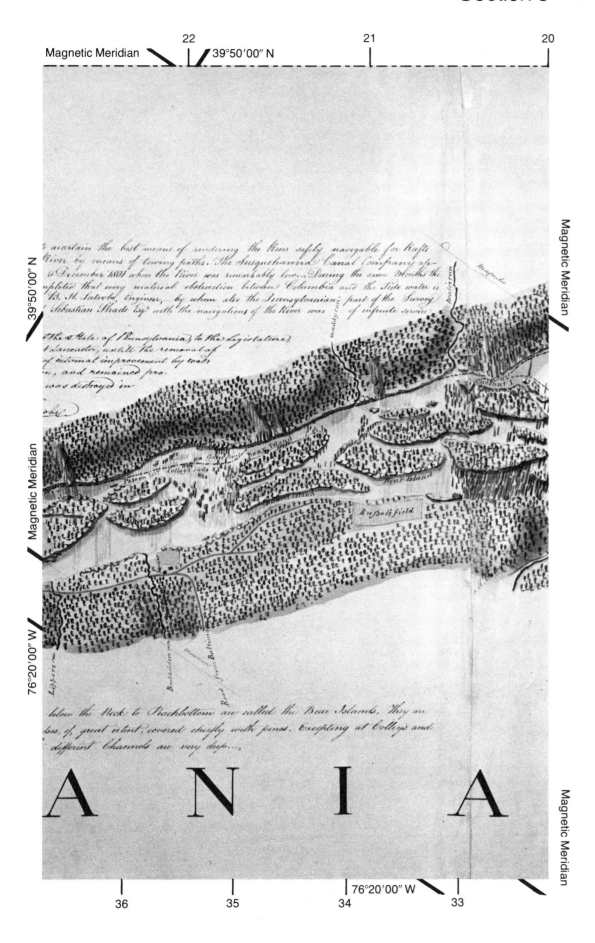

Secs. 4 & 5, fig. 1 "Fulton's ferry, on the East shore, below Burkhalter's ferry." [11 November 1801.] River mile 27/ kilometer 43.5. (Latrobe Sketchbook VIII.)

Secs. 4 & 5, fig. 2 A page of BHL's field notes includes the preliminary sketch for the watercolor above. Some of the writing on this page is in the hand of an unidentified member of the survey party, probably Robert Hazlehurst. 11 November 1801. (Field Book 2.)

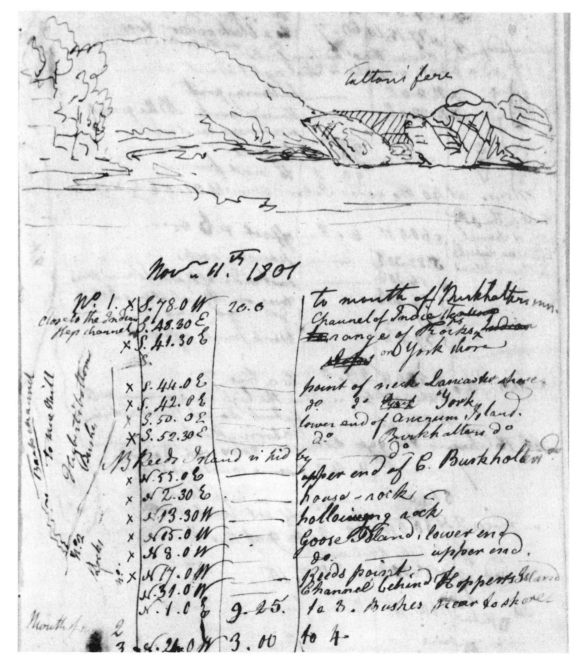

Susquehanna River Survey Map: Sections 4 and 5

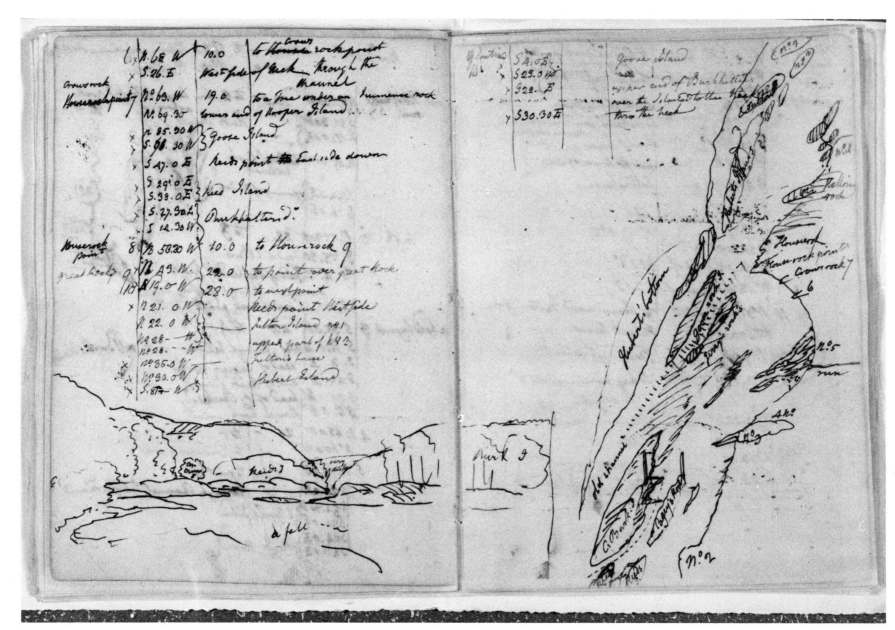

Secs. 4 & 5, fig. 3 This two-page section from BHL's field notes for 10 November 1801 includes survey data, a panoramic view (bottom left), and a topographic sketch, and illustrates the variety of records which BHL employed in making the map. Note how the sketch on the right (river mile 24.6/kilometer 39.6 at the bottom, river mile 26.8/kilometer 43.1 at the top) differs markedly from the final rendering in section 4. (Field Book 2.)

Susquehanna River Survey Map: Sections 4 and 5

Secs. 4 & 5, fig. 4 "Stoney point." Looking south. Although this view and the one below are not in these sections, the scenery is typical of the river between Turkey Hill and the Mason and Dixon Line. Unfinished. Last week of September 1801. River mile 33.8/kilometer 54.3. (Latrobe Sketchbook VIII.)

The area covered by Latrobe's survey party ends at the boundary between sections 4 and 5. Although Latrobe traveled south along the rest of the river and made some further watercolors and sketches, from this point he relied upon Christian Hauducoeur's survey data in making the map.

These two sections contained some of the most remote locations within the map area, and Latrobe was impressed by their beauty. He wrote to his wife: "... as to the natural Scenery in which we have been engaged, it is so Savage in many instances, and so beautiful in others, that I could not fail to find in that alone matter enough for twenty letters." (BHL to Mary E. Latrobe, 10 November 1801, Mrs. Gamble Latrobe Collection, MdHS)

Secs. 4 & 5, fig. 5 "Susquehannah at Stoney point." Looking south. In the right foreground is the "whirlpool" (probably a large circular eddy) which Latrobe noted on the map. Last week of September 1801. River mile 33.7/kilometer 54.1. (Latrobe Sketchbook VIII.)

Secs. 4 & 5, fig. 6 View of Lancaster Neck Mountain, east shore, from the Pinnacles, west shore. Some spots from subsequent discoloration on left side. [12 November 1801.] River mile 24.8/kilometer 40.0. (Latrobe Sketchbook VIII.)

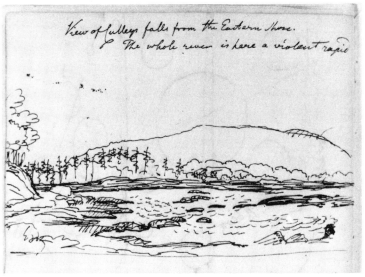

Secs. 4 & 5, fig. 7 "View of Culleys falls from the Eastern Shore. The whole river is here a violent rapid." [13 November 1801.] River mile 21.9/kilometer 35.3. (Field Book 2.)

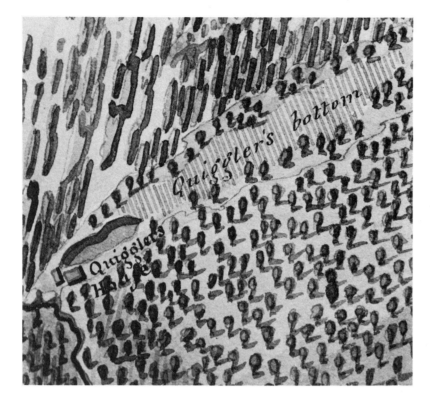

Sec. 6, fig. 1 Detail of map section 2, at Quiggler's Bottom. The river's shoreline runs from the upper right corner to the lower left. Above the shoreline the river is full of rocky ledges. Quiggler's Bottom is largely a cultivated field, symbolized by even rows of brush strokes, although to the left are a house, a barn, and a small pond or meadow. The remainder of the detail is a forest. River mile 33.5/ kilometer 54.0.

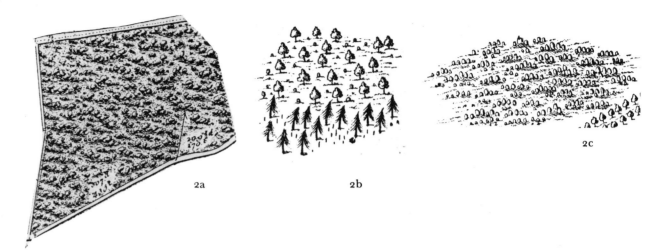

Sec. 6, fig. 2a From Dupain de Montesson, *La science de l'arpenteur* (Paris, 1766), planche II. Sec. 6, figs. 2b and 2c. From J. C. G. Hayne, *Deutliche und ausführliche Anweisung, wie man das militairische Aufnehmen nach dem Augenmaas ohne Lehrmeister erlernen könne* (Leipzig, 1794), tabelle II.

The heavily forested banks of the Susquehanna called for extensive use of vegetation symbols on the map. Relying on European cartographic tradition, represented here by figures from contemporary handbooks (figs. 2a, 2b, 2c), Latrobe drew thousands of symbols for broadleaf trees, giving each a shadow.

He gave attention to cultural features as well. At right are sketches of a riverbank cabin. Latrobe's field study (fig. 3) suggests that he was intrigued by the enigmatic quality of the woman in the doorway looking out over the river. He transferred the scene to his sketchbook (fig. 4) but never completed it.

Sec. 6, fig. 3 Pen and ink study of a log cabin on the river. 4 November 1801. (Field Book 2.)

Sec. 6, fig. 4 Pencil sketch of a log cabin on the river, showing lines of perspective. [4 November 1801.] (Latrobe Sketchbook VIII.)

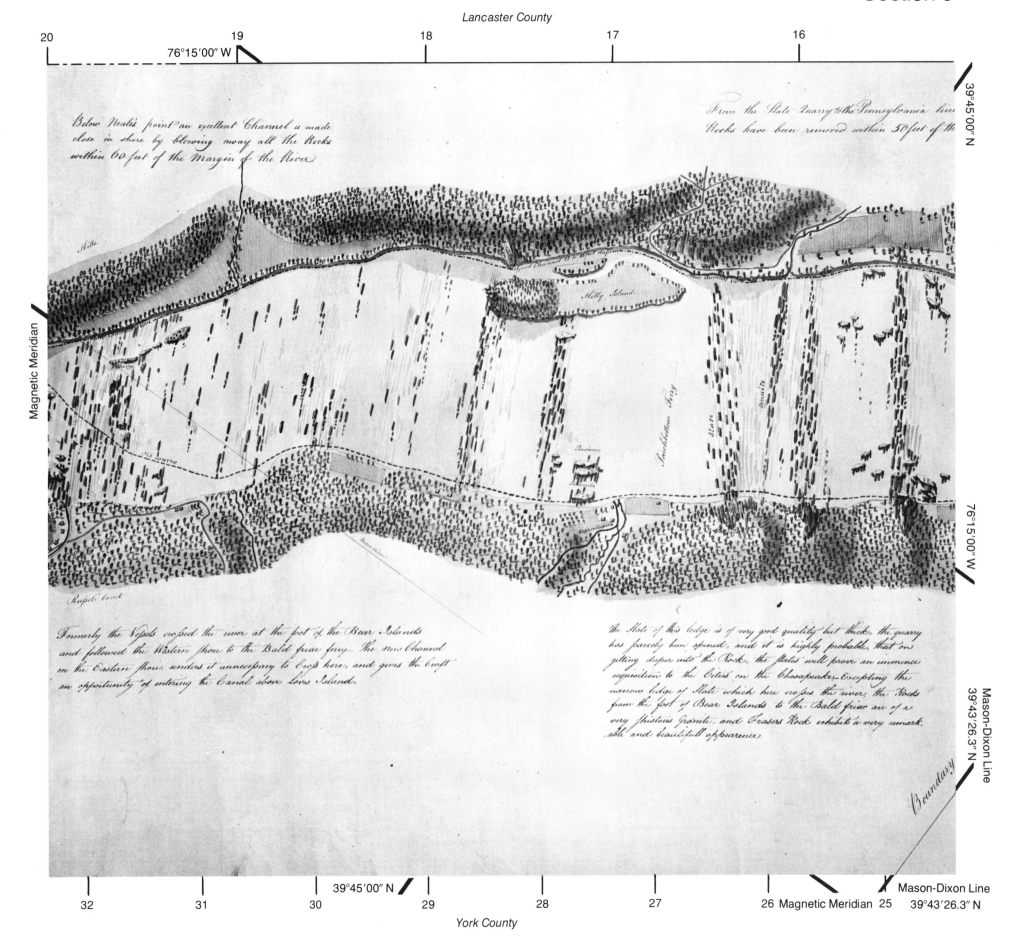

Susquehanna River Survey Map: Section 7

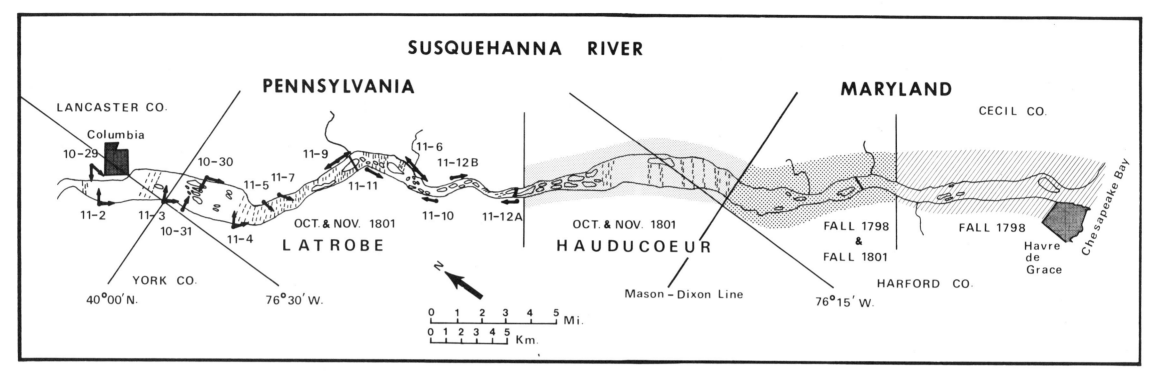

Sec. 7, fig. 1 BHL's and Hauducoeur's surveys of the Susquehanna River, 1798–1801. The arrows represent the approximate points of BHL's survey stations and the general direction of his sightings. Each point is dated. The right-hand side of the map shows the extent of Christian Hauducoeur's two surveys, which overlapped in the portion of the river just south of the Mason-Dixon line.

Sec. 7, fig. 2 *A Map of the Head of Chesapeake Bay and Susquehanna River Shewing the Navigation of the same with a Topographical description of the surrounding Country from an actual Survey by C. P. Hauducoeur, Engineer* (engraved, 1799). Detail corresponding to section 7.

In section 7 the map changes from a detailed watercolor representation to a simpler pencil and ink version. Since the map was meant for the governor of Pennsylvania, there was little reason for Latrobe to embellish it below the Mason and Dixon line as he had done above it.

The entire Maryland portion relies upon Christian Hauducoeur's 1799 map of the region (fig. 2) and his subsequent survey in 1801. Being primarily a navigation chart south of the state line, Latrobe's map gives little information about the shores, but it does revise and refine some details of Hauducoeur's published map, in light of the new channel for rivercraft.

The map at the top of the page indicates the areas covered by the Latrobe and Hauducoeur surveys, and thus their contributions to the basic map data.

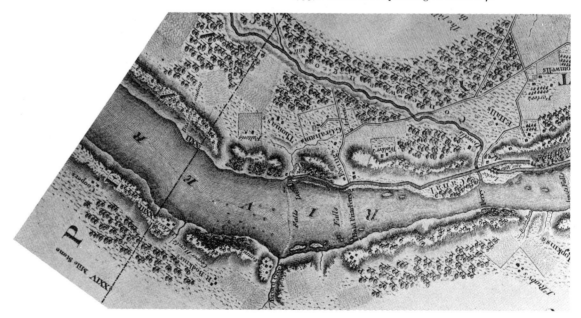

Section 7

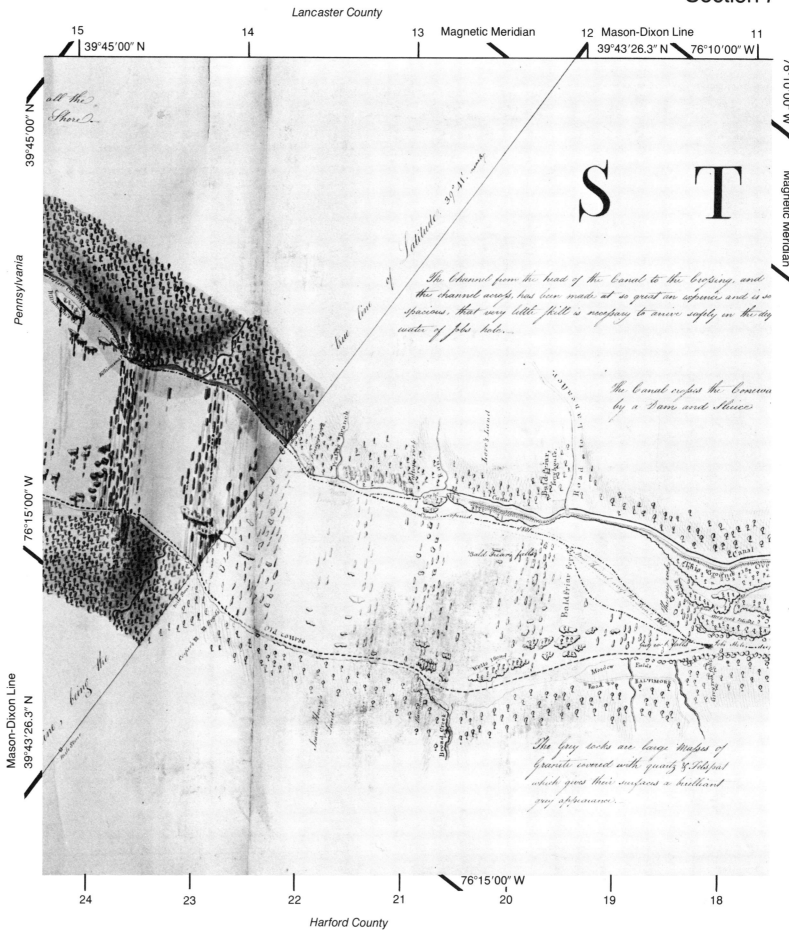

Latrobe's Instructions to Navigators

For about 150 Yards [137 meters] above the Baldfriar ferry house [river mile 12.1/kilometer 19.5], the Channel has been opened in a most capital manner, so as to give ample room to prepare early for crossing the river. The Rocks that are left on the right will occasion breakers that will point out how far westward you may steer, and in crossing you will meet with no obstructions that do not shew themselves, all sharp low rocks being taken out of the Way. It will be perfectly safe to steer between a white flint rock and a long black rock lying across the stream, and to the westward of the long rock, keeping over to the bushy Island, the passage is safe. On reaching the Western shore, one simple direction is sufficient to the tide. *Keep in shore.*

(Latrobe to McKean, appendix, 24 November 1801, Susquehanna River Improvement File, Pennsylvania Historical and Museum Commission, Harrisburg, Pennsylvania.)

Susquehanna River Survey Map: Section 8

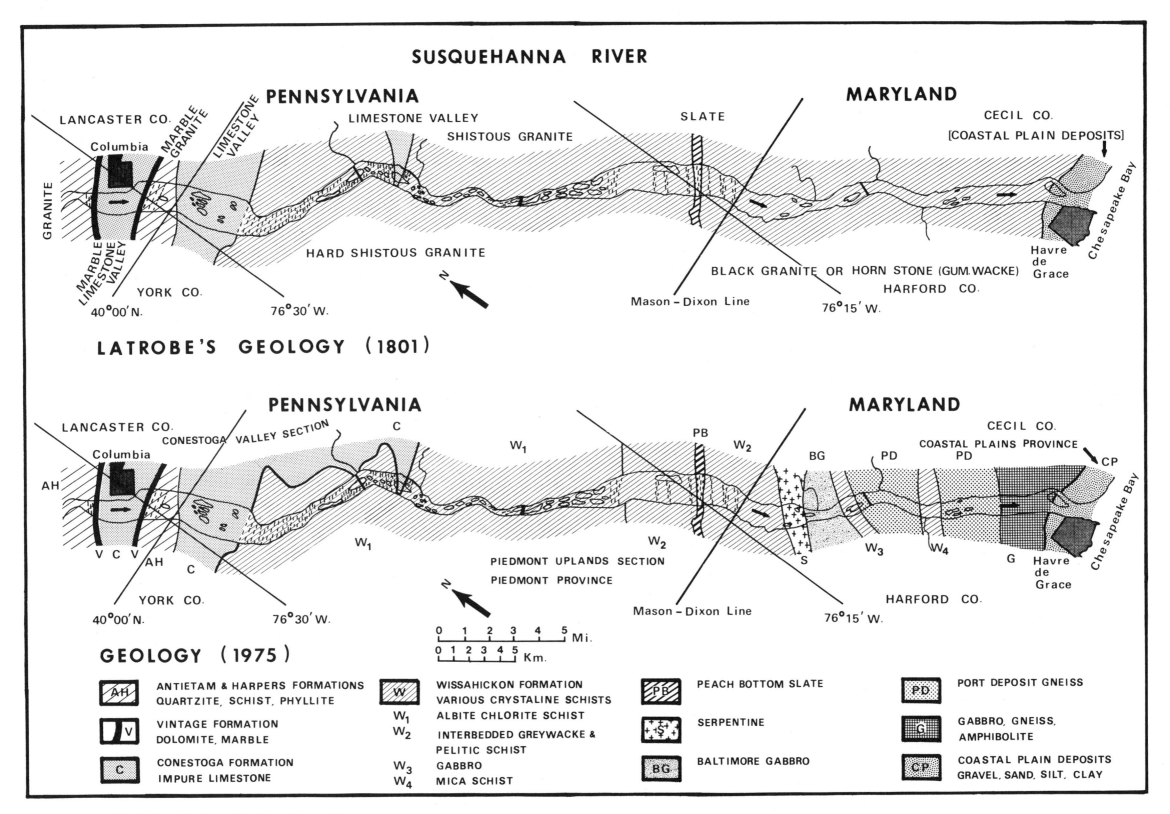

Sec. 8, fig. 1 Geology of the Susquehanna River.

Throughout the Susquehanna map Latrobe's annotations demonstrate his interest in the geological features of the river and adjacent land. On this section he noted the presence of "black Granite or Horn Stone (Gum Wacke)," a feature which other cartographers, such as Hauducoeur (fig. 2) considered unimportant to indicate on a map.

Although he generalized, Latrobe identified the geological features of the lower Susquehanna basin in a manner that is remarkably consistent with modern surveys (fig. 1).

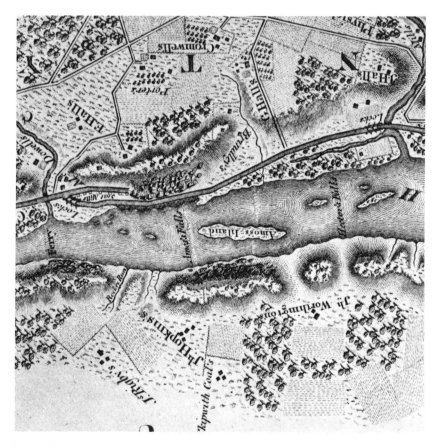

Sec. 8, fig. 2 *A Map of the Head of Chesapeake Bay and Susquehanna River Shewing the Navigation of the same with a Topographical description of the surrounding Country from an actual Survey by C. P. Hauducoeur, Engineer* (engraved, 1799). Detail corresponding to section 8.

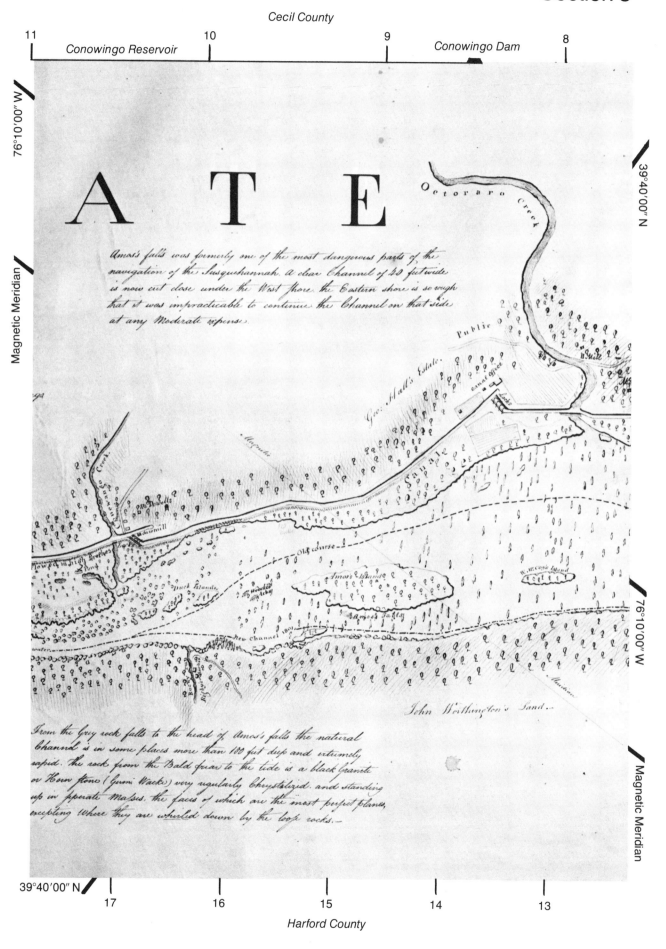

Susquehanna River Survey Map: Section 9

Latrobe employed two systems to illustrate changes in elevation (cartographic relief). On the Pennsylvania portion there is shaded relief produced through the use of colored hachure lines and wash (fig. 1), with the illusion of depth coming from the differences in the strength of color. On the Maryland portion there is rudimentary hachuring (fig. 2), a method of drawing fine lines to simulate slopes. A contemporary manual indicates the variety of hachuring forms (fig. 3). Hauducoeur's map (fig. 4) used hachuring more effectively than Latrobe's.

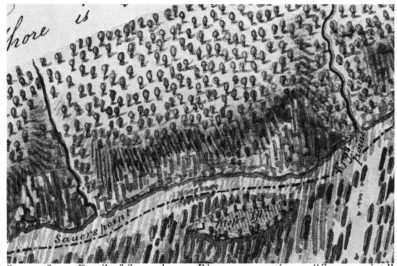

Sec. 9, fig. 1 Detail of Susquehanna River map, section 2, "Sauers point." River mile 33.5/kilometer 54.0.

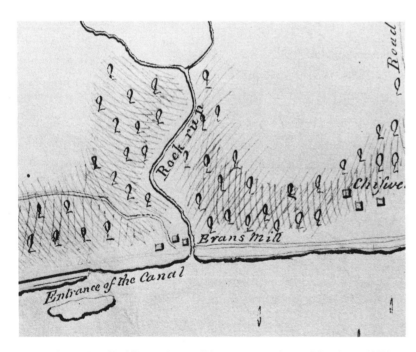

Sec. 9, fig. 2 Detail of Susquehanna River map, section 9, "Rock run." River mile 4.4/kilometer 7.1.

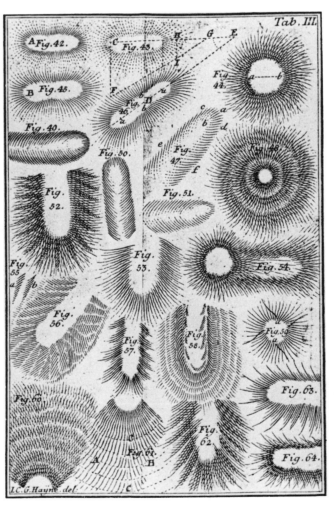

Sec. 9, fig. 3 From J. C. G. Hayne, *Deutliche und ausfürliche Anweisung, wie man das militairische Aufnehmen nach dem Augenmaas ohne Lehrmeister erlernen könne* (Leipzig, 1794), tabelle III.

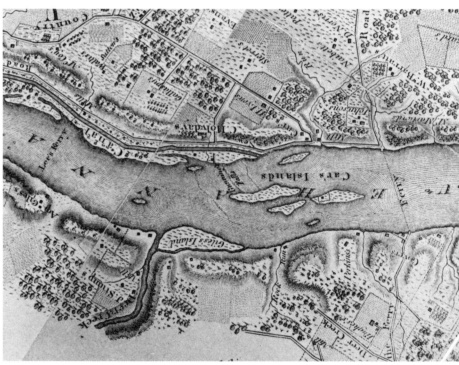

Sec. 9, fig. 4 *A Map of the Head of Chesapeake Bay and Susquehanna River Shewing the Navigation of the same with a Topographical description of the surrounding Country from an actual Survey by C. P. Hauducoeur, Engineer* (engraved, 1799). Detail corresponding to section 9.

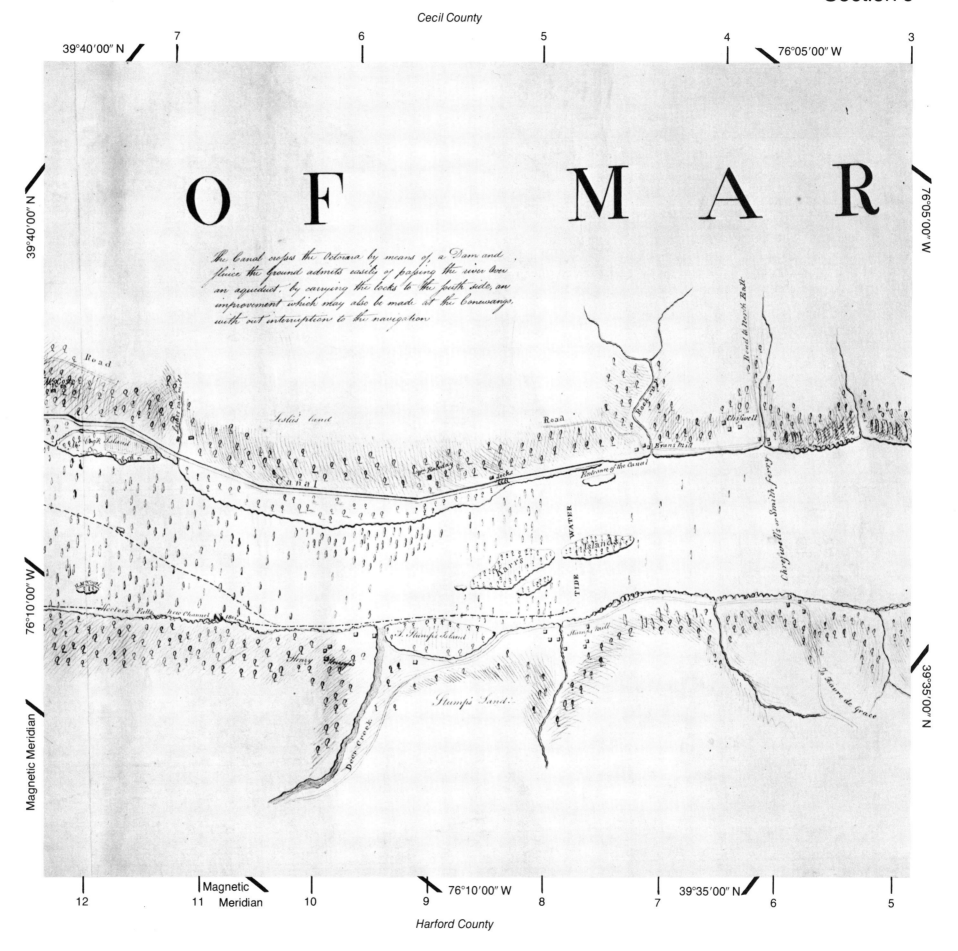

Susquehanna River Survey Map: Section 10

On the last section at the head of the Chesapeake Bay is Havre de Grace, a harbor and commercial town similar to Columbia on the first section of the map. Havre de Grace was the destination for the boatmen who risked the passage of the lower Susquehanna, and on arrival they sold their goods and boats there. Almost without exception they walked or rode a horse home, since it was extremely difficult to tow a vessel upstream. Not until the Susquehanna and Tidewater Canal from Wrightsville (opposite Columbia) to Havre de Grace was completed in 1840 was there a satisfactory water route for those who wished to ascend the river.

An important economic feature of this area was the ferry crossings at Havre de Grace for the main road between Baltimore and Philadelphia, and all points north and south. Latrobe crossed the river here many times, and he sometimes stayed at the tavern on the east side of the river at the end of the "Post Road to Philadelphia."

Hauducoeur's map (fig. 3) also includes the tavern and ferries. A reduced copy of the entire Hauducoeur map (fig. 2) shows that it was only partly concerned with depicting the Susquehanna River, and that much of it charted the upper Chesapeake Bay and its shoreline.

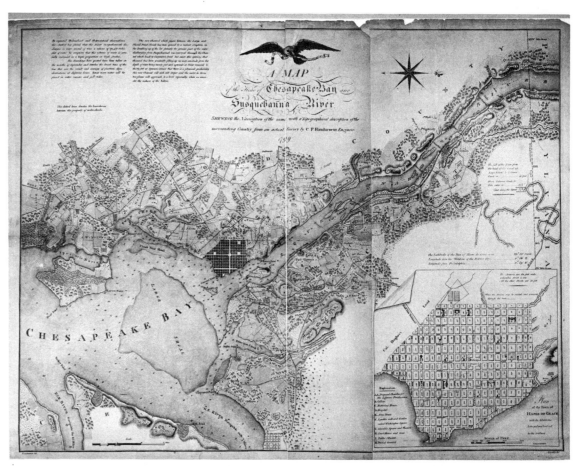

Sec. 10, fig. 2 *A Map of the Head of Chesapeake Bay and Susquehanna River Shewing the Navigation of the same with a Topographical description of the surrounding Country from an actual Survey by C. P. Hauducoeur, Engineer* (engraved, 1799). The map is oriented with north to the right. (Courtesy of Maryland Historical Society.)

Sec. 10, fig. 1 "From the ferry at Havre de Grace." Looking to the east. Palmer's Island (present-day Garrett Island) is on the left, and has a house near the border of the sketch. [4 or 5 October 1801.] River mile 0.0/kilometer 0.0. (Latrobe Sketchbook VIII.)

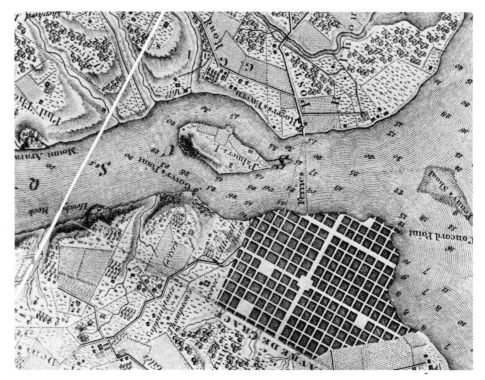

Sec. 10, fig. 3 *A Map of the Head of Chesapeake Bay and Susquehanna River Shewing the Navigation of the same with a Topographical description of the surrounding Country from an actual Survey by C. P. Hauducoeur, Engineer* (engraved, 1799). Detail corresponding to section 10.

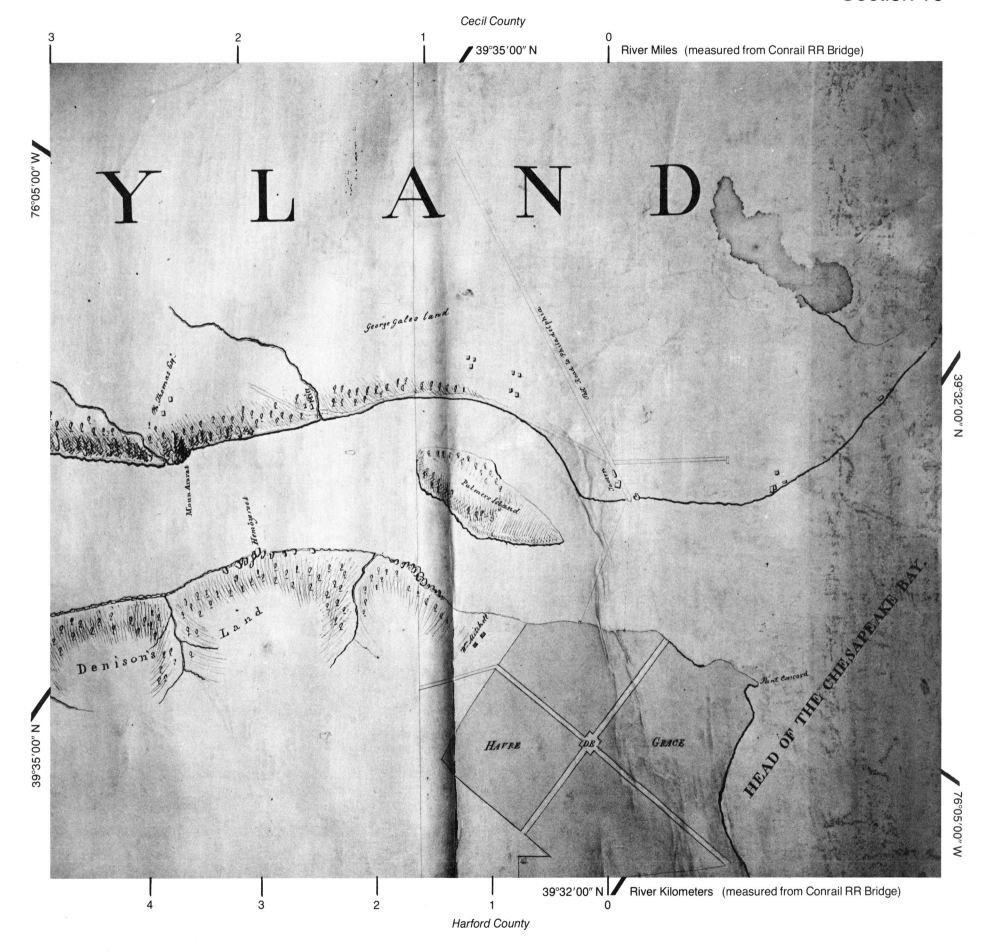

The Engineering Drawings

CHELMER AND BLACKWATER NAVIGATION

1 "A Survey of the Course of the River Blackwater in the County of Essex"

2 "A Map or Plan of the intended Improvement of the Navigation of the Rivers Blackwater and Chelmer in the County of Essex"

These two maps are the only Latrobe engineering drawings which remain from his career in England, and they are among his last works there. They indicate the high level of his surveying and cartographic skill at the time he left for America.

The occasion for these two drawings was the construction of a navigation from Chelmsford in Essex, northeast of London, to the tidal estuary of the Chelmer and Blackwater rivers on the coast. The town of Maldon was located at the shallow upper part of the estuary, and the coal ships which were Maldon's major traffic had to transfer their cargo to lighters for it to be brought up. With the aim of eliminating that middle step and allowing coal to be transferred directly from ships to canal boats the Act of Parliament which created the Chelmer and Blackwater Navigation in 1793 provided for the main route to bypass Maldon and terminate at Collier's Reach, where the transfer of cargoes took place. A Maldon branch of the navigation was planned, but it would not have the depth to accommodate coal ships.

The businessmen of Maldon who depended upon the river traffic soon petitioned Parliament to change the Act to require the navigation's proprietors to provide Maldon with as deep a channel as the bypass. Two men of the Strutt family gave the protest financial support and hired a London attorney to lobby for Maldon in Parliament. One of the attorney's first acts was to hire Benjamin Henry Latrobe to make a survey and report on the Maldon branch.

Latrobe worked on the Maldon project from about May 1793 to April 1795. Although he conducted surveys, wrote two reports, drew two maps, made at least two appearances in Parliament, and had numerous consultations with the attorney and Maldon partisans, his work seems to have had little effect on developments, and the proprietors completed the navigation on the original plan.

Latrobe's 1793 map (no. 1) shows the town of Maldon on the left (west) and Colliers' Reach in the Chelmer–Blackwater estuary to the right (east). His accompanying report discussed his proposed improvement of the Chelmer to allow coal ships to pass up to Fullbridge in Maldon. Latrobe believed that a channel shorter and better than the old one could be created by cutting across shoals at bends in the river and by deepening the remaining parts of the old channel. The old channel he colored a light shade, and the new channel he lettered "B" and made a darker shade. From Fullbridge to the new Chelmer and Blackwater Navigation Latrobe proposed cutting a short canal with one lock. The entire length of the improvement was about two miles (3.2 kilometers).

Essentially, Latrobe attempted to create the same result that the Act of 1793 aimed at: eliminating the lightering of coal and permitting direct loading of the canal boats. In his report he even claimed that his proposal was cheaper than the bypass canal to Colliers' Reach.

The 1794 map (no. 2) was prepared for a second Parliamentary petition after the first petition failed. It shows all of the features contained in the previous map but extends further to the west and north, giving a view of the route and projected terminus of the Chelmer and Blackwater Navigation (labeled "Chelmsford Canal"). (The navigation, having followed the Chelmer River easterly from Chelmsford, appears on the left side of the map [west of Maldon], where it crosses to the Blackwater River for a short distance, and from Heybridge goes straight to Colliers' Reach.)

Since his original plan apparently met with considerable opposition from affected landowners, Latrobe's second map contained a revision of his proposal for the connection between Fullbridge and the navigation. He altered the segment above Fullbridge to follow the course of the Chelmer River to the Beleigh Mill bridge, and proposed dredging that section to create an adequate depth for canal boats. Just west of the bridge he planned a lock that would attain the level of the navigation and proposed a short canal from the lock to make a junction with the navigation.

Overall, Latrobe's proposals demonstrate his familiarity with waterway engineering, since, although the Maldon navigation was short, it required planning both river improvements and a canal. Additionally, the two maps confirm that Latrobe was already an accomplished draftsman because both are clear and beautiful renderings of his plans.

REFERENCES

Manuscript

Chelmsford, England. Essex Record Office. BHL, "Report upon the practicability and advantage of making the River Blackwater navigable," 30 December 1793; BHL, "Report upon the plan of the intended improvement of the rivers Blackwater and Chelmer," 28 October 1794; BHL to Strutt, 24 February 1794, 4 October 1794, 15 April 1795; Birch to Strutt, bill, October 1793–May 1796.

Printed

Booker, John. *Essex in the Industrial Revolution.* Chelmsford: Essex County Council, 1974.
Brown, A. F. J. *Essex at Work, 1700–1815.* Chelmsford: Essex County Council, 1969.
"Canal." *The Cyclopaedia.* Edited by Abraham Rees. 1st American ed. 40 vols. Philadelphia, 1805–24.

Chelmer and Blackwater Navigation

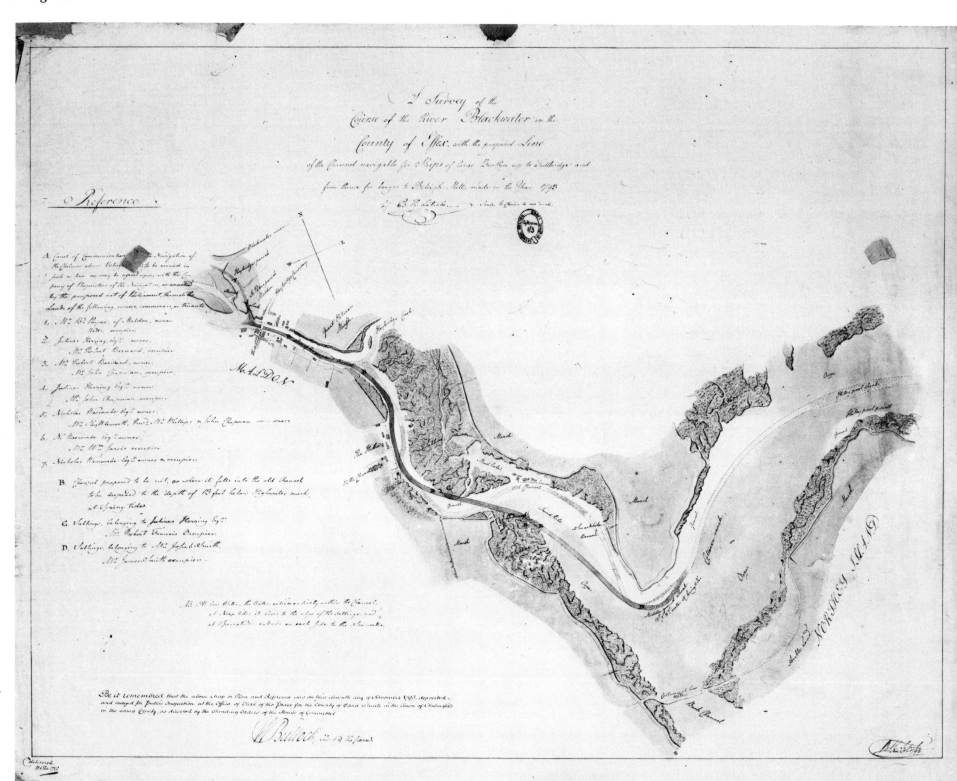

1 "A Survey of the Course of the River Blackwater in the County of Essex"

size: 27″ × 37″ (68.6 cm × 94.0 cm)
media: pencil, pen and ink, watercolor
date: delivered 11 November 1793
location: Essex Record Office,
 Chelmsford, England

Chelmer and Blackwater Navigation

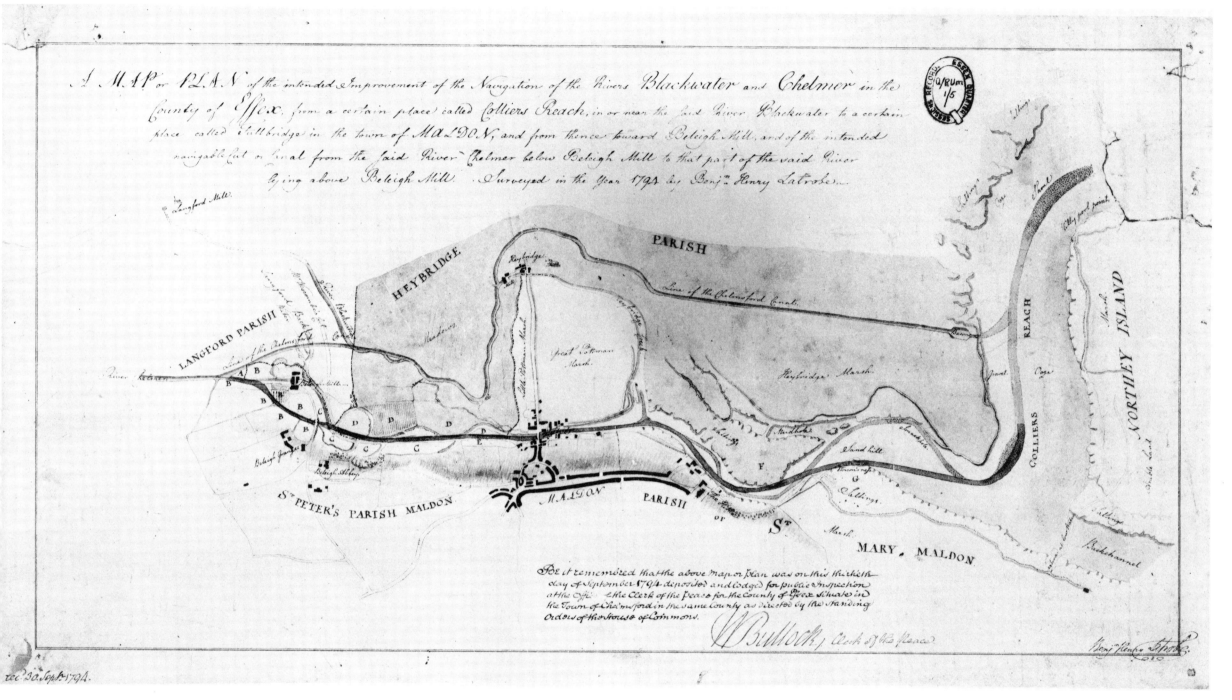

2 "A Map or Plan of the intended Improvement of the Navigation of the Rivers Blackwater and Chelmer in the County of Essex"

size: 12″ × 23″ (30.5 cm × 58.4 cm)
media: pencil, pen and ink, watercolor
date: [September] 1794
location: Essex Record Office, Chelmsford, England

NAVAL DRYDOCK AND POTOMAC CANAL EXTENSION

3 "No. I. Section of Locks; Plan of Locks necessary to elevate the ships into the arsenal"

4 "No. II. [Plan, sections and elevations of the drydock]"

5 Detail from Potomac Canal Extension No. I: including "Mean Section of the Canal" and start of the canal

6 Detail from Potomac Canal Extension No. II: including two aqueducts and Mr. Foxall's works

7 Detail from Potomac Canal Extension No. IIII: plan of basin, lock, and aqueduct over Rock Creek

8 Detail from Potomac Canal Extension No. IIII: section of the aqueduct over Rock Creek

Latrobe's designs for a naval drydock and the Potomac Canal extension in Washington, D.C., have their origin in the politics of the Jeffersonian Republicans who came to power after the election of 1800. The Republicans had opposed the previous Federalist administrations as extravagant and warmongering, and intended to reduce government spending drastically, especially in the military. The navy, which had been enlarged because of an anticipated war with France, was a logical target of economy-minded Republicans. The new administration sold some ships, dismantled others, and suspended construction of those on the stocks. Some of those remaining were to be kept out of service but available when needed.

Thomas Jefferson, in his first term as president, conceived the idea of building a huge covered drydock, or "naval arsenal," in which several vessels could be stored and kept dry, yet could be quickly readied for service; he had even decided on a particular type of roof. Late in 1802 he called on Latrobe to complete the plans and make estimates for Congress.

Latrobe was limited in his approaches to the task by the broad outlines of Jefferson's conception, but it is likely that he drew liberally on his knowledge of European canals, docks, and harbor improvements in developing his final plan.

Latrobe had to select a location for the drydock on the Eastern (Anacostia) Branch of the Potomac River at Washington because that was where the new Navy Yard was located. He decided on a spot at the end of 9th Street S.E., where a gravel shoal promised a good foundation for locks, and where a valley extending inland from the shore provided a natural excavation for the drydock. Latrobe's drawing of the locks (no. 3) does not indicate how far he expected to extend the lower lock into the Eastern Branch, other than indicating that the mouth of it was to intersect the "Line of Wharves" which already existed. The lower lock, on the right-hand side of the drawing in plan and section, is to be protected by rounded masonry piers on either side of its entrance. Extending slightly further than the piers is the wooden sill (shown only in section) of the lower lock's entrance doors.

Several features of the locks are of interest. Latrobe's drawing shows them made of dressed stone, with only the doors and sills made of wood. The doors open into rounded recesses rather than rectangular ones such as were common in British locks. The sluices for carrying water around the doors are circular tunnels in the masonry lock wall, and apparently have wickets for opening and closing them located directly beneath the sluice houses. The houses, one for each sluice, have Neoclassical domed roofs, an expression of Latrobe's architectural style.

If the drydock had been put in use, a ship would have been towed by horses or mules[1] into the lower lock through open entrance doors (shown closed). The lock's doors would then be closed, possibly by some mechanism in the sluice houses, since no balance beams are shown, and the sluices at the entrance to the upper lock would be opened to let in water. The ship would be raised twelve feet above high water in the Eastern Branch; the water level in the two locks would then be equal. The doors to the upper lock would then be opened, the ship towed into the upper lock and raised another twelve feet. The ship would proceed into the drydock at twenty-four feet above high water in the Eastern Branch. At the entrance to the drydock there was an additional pair of doors to allow it to be sealed off.

The turning dock, which Latrobe located between the upper lock and the drydock, was necessary if the government did not purchase the land directly north of the locks for the drydock. The vessels would then be rotated ninety degrees and towed into a drydock located in the Navy Yard.

The drydock (no. 4) would have been the largest covered structure in the United States by a large margin. Designed by Latrobe to hold twelve frigates, three abreast, it would have enclosed an area about 175 feet (53 meters) wide and 800 feet (244 meters) long, or more than three acres. Latrobe generally

---

1. BHL does not name the motive power, but horses and mules, often operating through pulleys or capstans, were commonly used for such tasks.

followed Jefferson's ideas by designing a roof after that of the grain market in Paris (*Halle au Blé*): a single span of timber arches covered with painted sheet metal. (This is one of Latrobe's earliest proposals for sheet metal roofing; he was to suggest it frequently in the future.[2]) The roof was to be supported on both sides by a series of masonry arches and piers interrupted by central colonnaded entrances. At the north end would be the guardhouse. Within this massive shed drydocked ships would rest on blocks, possibly with some planking removed from their hulls to improve ventilation of their interiors.

A drydock and locks of the size envisioned by Jefferson and Latrobe would have required a vast amount of water—Latrobe estimated their capacity at 3,515,200 cubic feet. Jefferson thought there were only two sources of so much water: the Tiber Creek and associated springs about two miles to the north of the Navy Yard, and the canal around the Little Falls of the Potomac some eight miles to the northwest. In the summer of 1802 he instructed Captain Thomas Tingey, commandant of the Navy Yard, to examine those sources of water in regard to their usefulness for supplying the proposed drydock. With the assistance of Nicholas King, surveyor of the city of Washington, Tingey found that both sources were adequate. The water from the Tiber and the springs, however, was insufficient to fill the drydock rapidly, and Tingey recommended that a reservoir be created to hold at least twice the contents of the drydock so that a supply would always be on hand. Latrobe agreed with Tingey's and King's findings, but he preferred bringing water from Little Falls.

> There cannot be one moments hesitation as to the abstract merits of each of these methods of supply. The Potomac canal may be brought by continuation from the locks at the little falls through George Town to Rockey Creek and through the City to the navy yard. This canal would not only fill all the works in twelve hours, but convey to the navy yard all the timber, stores and provisions which the whole range of the upper navigation of the Potomac could supply, comprehending not only a great part of Virginia and Maryland, but also a very considerable portion of the most fertile western counties of Pennsylvania.[3]

Latrobe did not write a detailed report on his proposal for the Potomac Canal extension, but he did draw four plans which traced the proposed route from Little Falls to an aqueduct over Rock Creek. They are the earliest of Latrobe's American canal drawings and are suggestive of his upcoming work on the Chesapeake and Delaware Canal. Some details from the drawings are reproduced here.[4]

The first detail (no. 5) shows that the cross section, or prism, of the canal was small: the width 12 feet on the bottom and 22 feet at the waterline, and the water 3 feet deep. Less than two years later Latrobe adopted almost the same dimensions for the Chesapeake and Delaware Canal feeder. (Both canals had the same order of purposes: primarily to supply water to another work some distance away, and, secondarily, to be navigable.) The drawing also indicates that earth excavated from the canal to form its trough would be used to make the towpath and bermside (or offside) embankments.

The first and second details (nos. 5 and 6) display Latrobe's plans for overcoming various obstacles along the canal's route. Small watercourses he would cross with embankments pierced by culverts ("tunnels") of stone. Most of the culverts are circular in section, although a few are square. Larger creeks required arched stone aqueducts. Since a road ran along the same shore of the Potomac River, Latrobe planned an alternate route for the road where it occupied ground needed for the canal.

The third and fourth details (nos. 7 and 8) show Latrobe's solutions to several problems which he faced in taking the canal into Washington: (1) how to cross Rock Creek, (2) how to traverse the hill on the opposite side of Rock Creek, and (3) how to carry the canal into the city without disrupting its intricate street pattern. He solved the last problem first by turning the canal onto the axis of Pennsylvania Avenue, the grandest street of the city, and directing the canal down its center. He approached the avenue with a lock which brought the canal down to the correct level, a splendid aqueduct with circular openings in its masonry, and a tunnel through the hill on the east side of the creek. The aqueduct was similar to contemporary British designs, especially Benjamin Outram's Marple Aqueduct on the Peak Forest Canal, which was also pierced by round openings to lighten the load on the abutments.

Since there are no final reports or further drawings of Latrobe's proposal for the Potomac Canal extension his plan for the route of the canal in Washington is uncertain. Given the topography and the necessity of keeping the canal

---

2. The Schuylkill rolling mill (at the Philadelphia Waterworks), in which BHL had a financial interest, was just beginning to produce sheet metal in the fall of 1802.

3. [Thomas Jefferson], *Message from the President of the United States, Transmitting Plans and Estimates of a Dry Dock, for the Preservation of our Ships of War* (Washington, 1802), pp. 10–11.

4. The complete drawings are in Prints and Photographs Division, LC.

Naval Drydock and Potomac Canal Extension

level, it is likely that he wanted to continue down the center of Pennsylvania Avenue to the Capitol, tunnel through Capitol Hill, then proceed south to the Navy Yard on the Eastern Branch.[5]

But these plans for the naval arsenal and Potomac Canal extension were too grand, and Congress hesitated to commit money to a scheme both unprecedented and expensive. A potential testimony to Latrobe's virtuosity and a major engineering accomplishment remained rather a glimpse of things to come. The project was, however, the first activity in Latrobe's long association with public works in Washington, and, from a technical standpoint, the design of the Potomac Canal extension foreshadowed his plans for the Chesapeake and Delaware Canal feeder.

5. See fig. 3, p. 10 above.

REFERENCES

Manuscript

Washington, D.C. Library of Congress. Thomas Jefferson Papers. BHL to Jefferson, 22 and 28 November 1802.

Printed

Bacon-Foster, Corra. *Early Chapters in the Development of the Patowmac Route to the West*. Washington: Columbia Historical Society, 1912.

Ferguson, Eugene S. "Mr. Jefferson's Dry Docks." *The American Neptune* 11 (1951): 108–14.

Harris, Robert. *Canals and Their Architecture*. New York: Frederich A. Praeger, 1969.

[Jefferson, Thomas], *Message from the President of the United States, Transmitting Plans and Estimates of a Dry Dock, for the Preservation of our Ships of War*. Washington, 1802.

Norton, Paul F. "Jefferson's Plans for Mothballing the Frigates." *U.S. Naval Institute Proceedings* 82 (1956): 736–41.

Sanderlin, Walter S. *The Great National Project: A History of the Chesapeake and Ohio Canal*. The Johns Hopkins University Studies in Historical and Political Science, series 64, no. 1. Baltimore: Johns Hopkins Press, 1946.

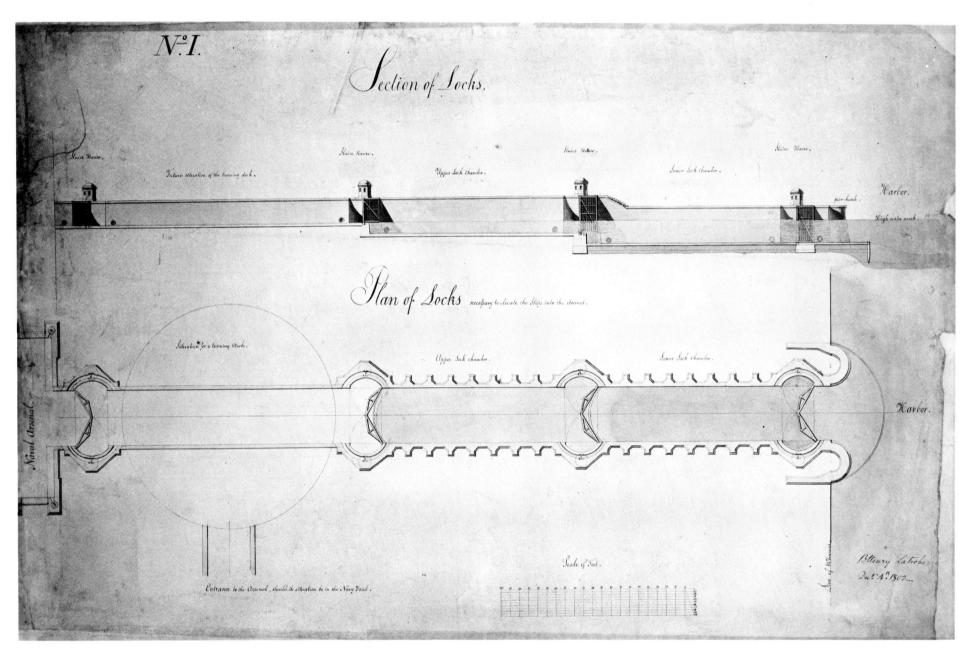

3 "No. I. Section of Locks; Plan of Locks necessary to elevate the ships into the arsenal"

size: $20\frac{1}{16}'' \times 31\frac{15}{16}''$ (50.9 cm × 81.1 cm)
media: pencil, pen and ink, watercolor
date: 4 December 1802
location: Library of Congress, Washington, D.C.

Naval Drydock and Potomac Canal Extension

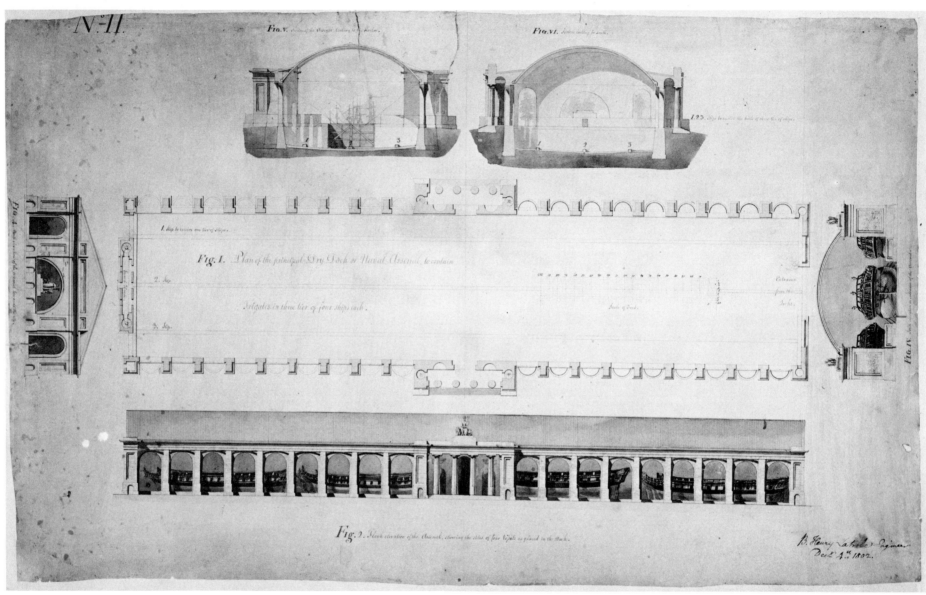

4 "No. II. [Plan, sections and elevations of the drydock]"

size: 19 15/16" × 33 3/16" (50.6 cm × 84.3 cm)
media: pencil, pen and ink, watercolor
date: 4 December 1802
location: Library of Congress, Washington, D.C.

Naval Drydock and Potomac Canal Extension

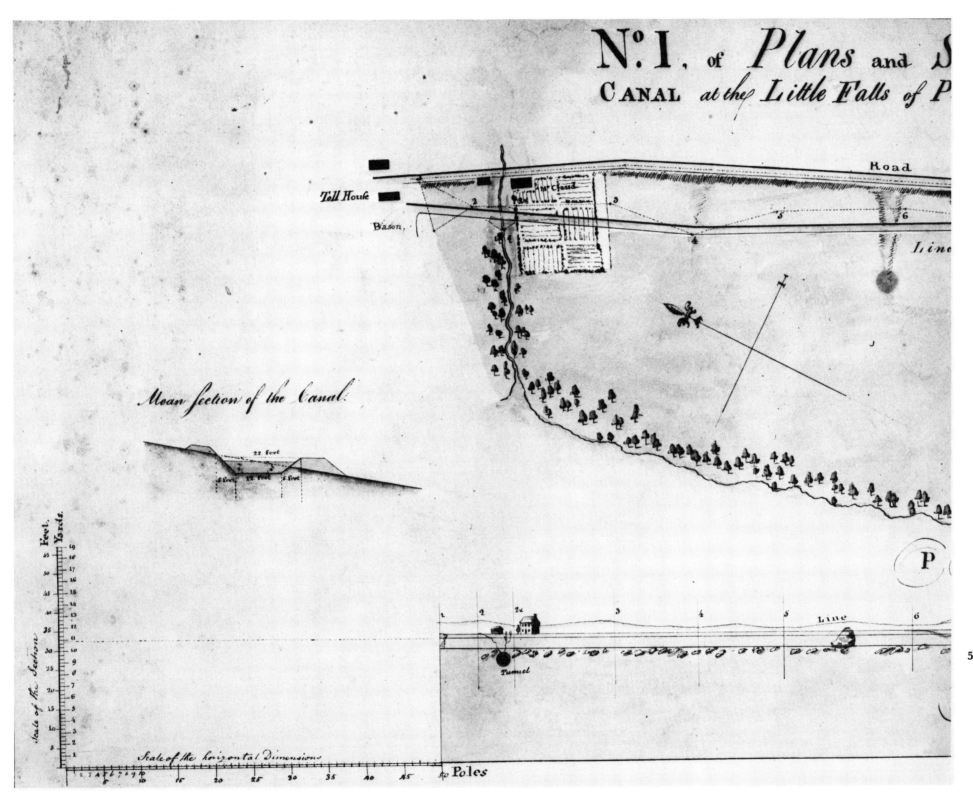

5 Detail from Potomac Canal Extension No. I: including "Mean Section of the Canal" and start of the canal

size: from 22″ × 32¼″ (55.9 cm × 81.9 cm) original
media: pencil, pen and ink, watercolor
date: [December 1802]
location: Library of Congress, Washington, D.C.

Naval Drydock and Potomac Canal Extension

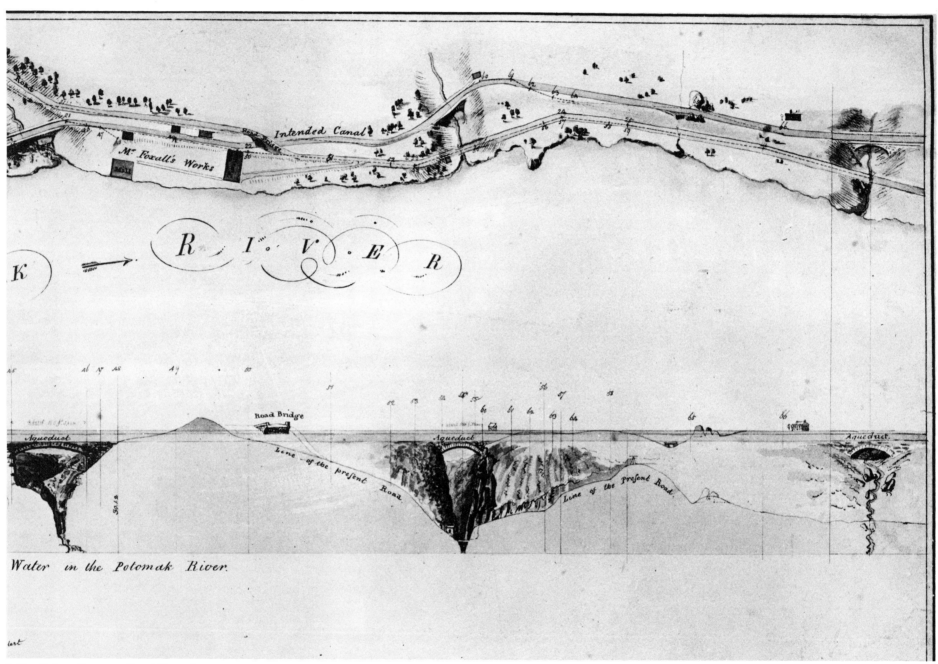

6 Detail from Potomac Canal Extension
No. II: including two aqueducts and
Mr. Foxall's works

size: from 21½″ × 32½″ (54.6 cm × 82.6 cm) original
media: pencil, pen and ink, watercolor
date: December 1802
location: Library of Congress, Washington, D.C.

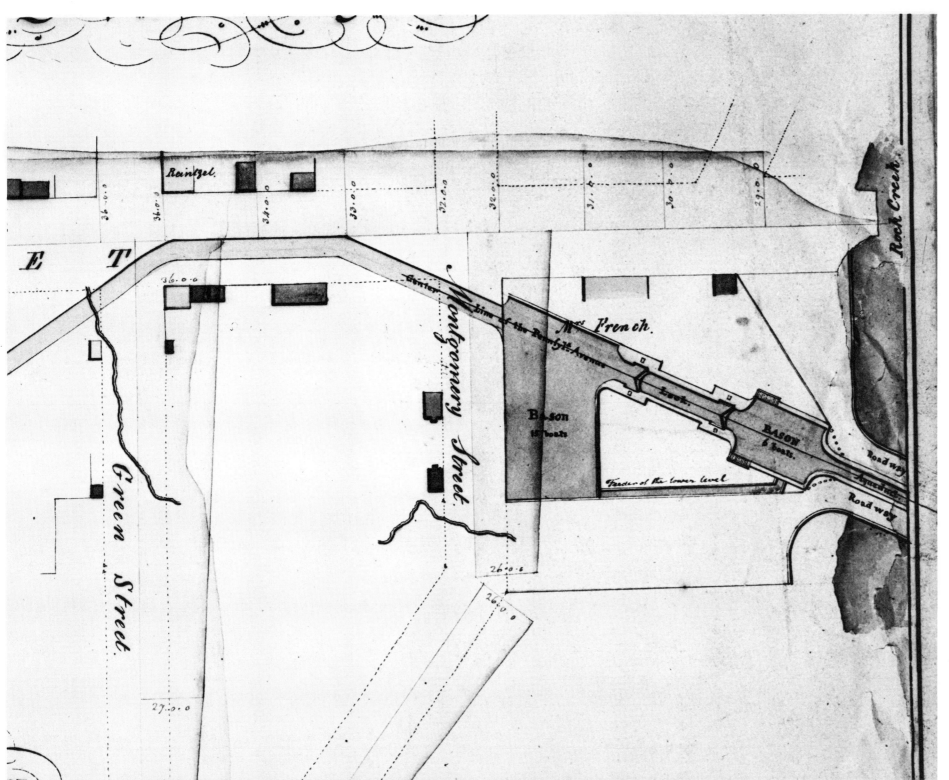

7 Detail from Potomac Canal Extension No. IIII: plan of basin, lock, and aqueduct over Rock Creek

size: from 21½″ × 32¼″ (54.6 cm × 81.9 cm) original
media: pencil, pen and ink, watercolor
date: December 1802
location: Library of Congress, Washington, D.C.

Naval Drydock and Potomac Canal Extension

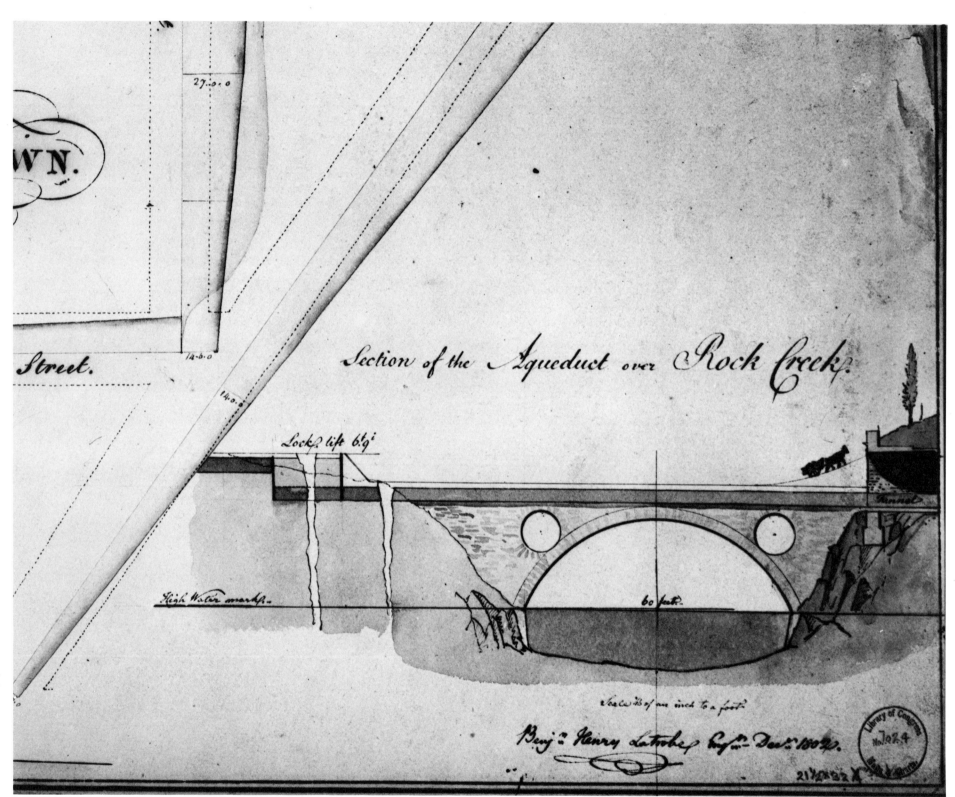

8 Detail from Potomac Canal Extension No. IIII: section of the aqueduct over Rock Creek

size: from 21¼″ × 32¼″ (54.6 cm × 81.9 cm) original
media: pencil, pen and ink, watercolor
date: December 1802
location: Library of Congress, Washington, D.C.

CHESAPEAKE AND DELAWARE CANAL

9 Plan of the Elk Creek aqueduct

10 "South, or downstream Elevation of the Aqueduct over Elk creek"

11 "Transverse Section of the Aqueduct over Elk Creek"

12 "Elevation of the Aqueduct over the head of Cow run"

Benjamin Henry Latrobe surveyed the Chesapeake and Delaware Canal of Maryland and Delaware in 1803, and in 1804–5 directed the construction of a feeder to supply water to the summit level of the proposed main canal. But work stopped late in 1805 before the feeder was complete because the canal company failed to attract enough capital for the project. Latrobe left the company in 1806.

The aqueduct over the Elk River on the Chesapeake and Delaware Canal feeder was to be located near Upper Forge, now the village of Elk Mills, Maryland. Latrobe planned for the aqueduct to receive water from a mill race on the west bank of the river and to carry it across the river to the beginning of the feeder on the east bank. The aqueduct was not built, although the foundation of the eastern abutment was laid and raised to the springing course in 1804.

Latrobe's drawings of the proposed aqueduct show that the masonry is rubble, with some ashlar facing for the piers and the indented date ornament (no. 10). The canal trough is carried over the river by three segmental arches which at their highest are only eight feet above river level. The aqueduct's plan and section (nos. 9 and 11) reveal that a large amount of earth and clay puddling (for watertightness) fills the interior, occupying almost two-thirds of the aqueduct's cross section at the waterline.

It is possible to imagine what the completed aqueduct would have looked like by considering the road bridge built in 1804 by Latrobe's mason, Thomas Vickers, over the feeder south of the aqueduct site. (See fig. 9a, p. 17.) The bridge has a single segmental arch of 15-foot span, or half the span intended for one arch of the aqueduct. Its rubble construction consists of large flat stones laid in mortar to form the arch, and irregularly shaped stones above to the level of the roadbed. That is clearly the pattern which Latrobe depicts in his elevation of the Elk River aqueduct.

The Cow Run aqueduct (no. 12) was to be located near the source of Cow Run, a tributary of the Elk River. It was never constructed, although the feeder was completed to both of its approaches and some stone was gathered at the site. Latrobe planned this aqueduct to be of rubble masonry, except for ashlar in the string course and coping at the top. The arch is circular, like the existing culverts on the feeder south of Elk Mills (see fig. 9b, p. 18), rather than segmental like the road bridge or the proposed Elk River aqueduct. The section of the Cow Run aqueduct is almost identical to the section of the Elk River aqueduct (no. 11).

Latrobe's plans for the Elk River and Cow Run aqueducts are testimonials to his British engineering principles: the necessity for permanence and architectural soundness in canal construction.[1]

1. Compare BHL's aqueduct designs with those of James Brindley and John Rennie: Cyril T. G. Boucher, *James Brindley, Engineer, 1716–1772* (Norwich: Goose and Son, 1968), p. 74; Cyril T. G. Boucher, *John Rennie, 1761–1821* (Manchester: Manchester University Press, 1963), plate X, facing p. 101.

REFERENCES

Manuscript

Baltimore, Maryland. Maryland Historical Society. Letterbooks of Benjamin Henry Latrobe. BHL to Cooch, 5 August 1804; BHL to Gilpin, 26 August 1804; BHL to Johns, 1 November 1804.

Wilmington, Delaware. Historical Society of Delaware. Chesapeake and Delaware Canal Company Papers. Reports of the Engineer.

Printed

Harris, Robert. *Canals and Their Architecture.* New York: Frederich A. Praeger, 1969.

Chesapeake and Delaware Canal

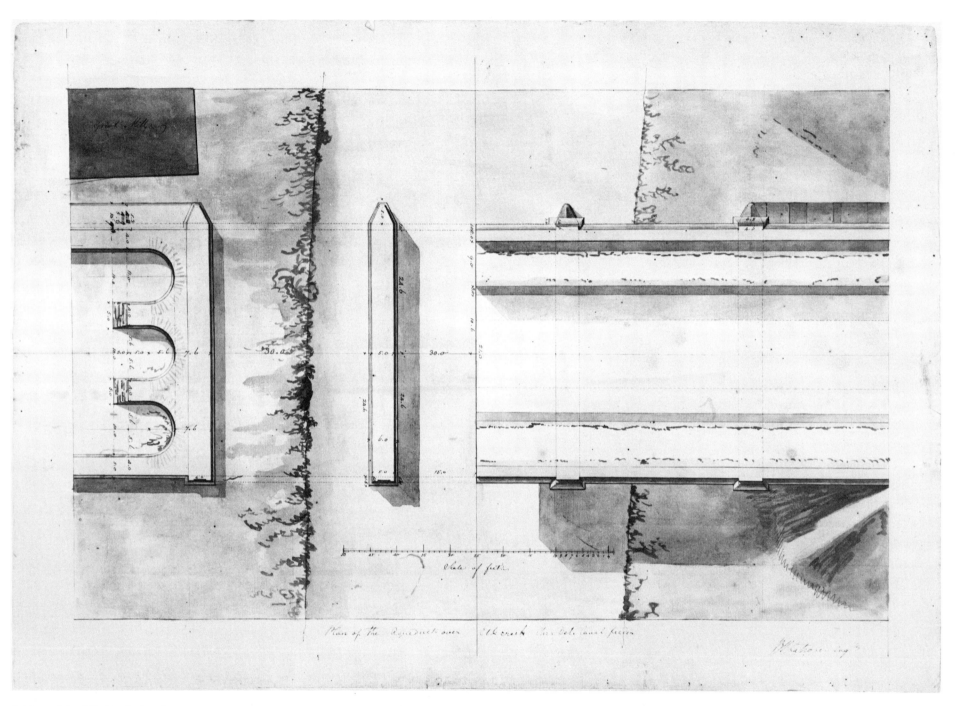

9 Plan of the Elk Creek aqueduct

size: 14 15/16" × 21 3/8" (37.9 cm × 54.3 cm)
media: pencil, pen and ink, watercolor
date: [1804]
location: Robert Brooke Papers, The
  New York Public Library

Chesapeake and Delaware Canal

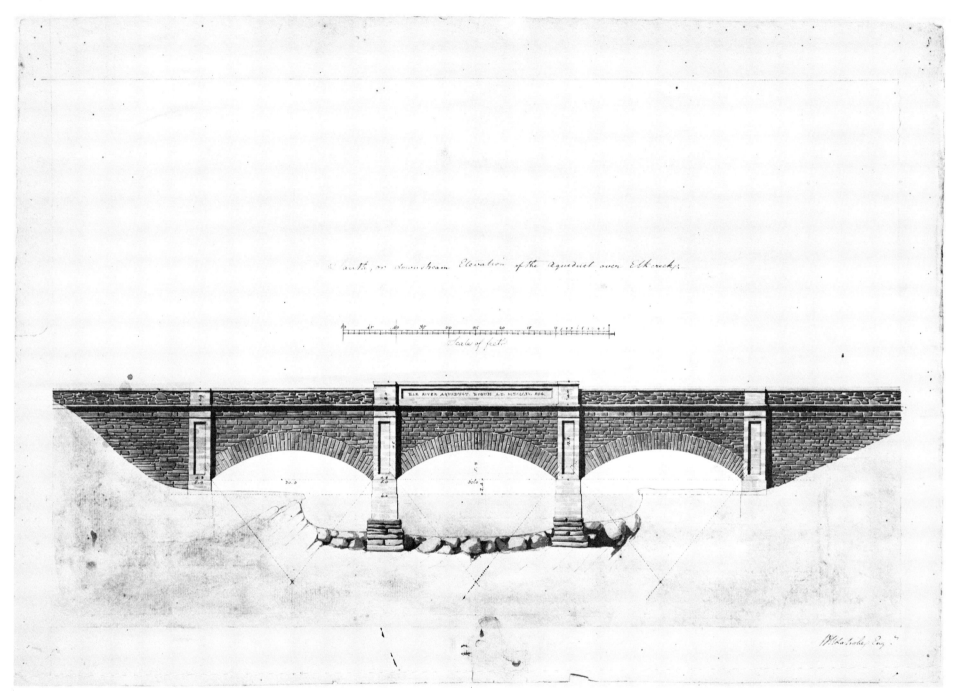

10 "South, or downstream Elevation of the Aqueduct over Elk creek"

size: 14⅞" × 21 7/16" (37.8 cm × 54.4 cm)
media: pencil, pen and ink, watercolor
date: [1804]
location: Robert Brooke Papers, The New York Public Library

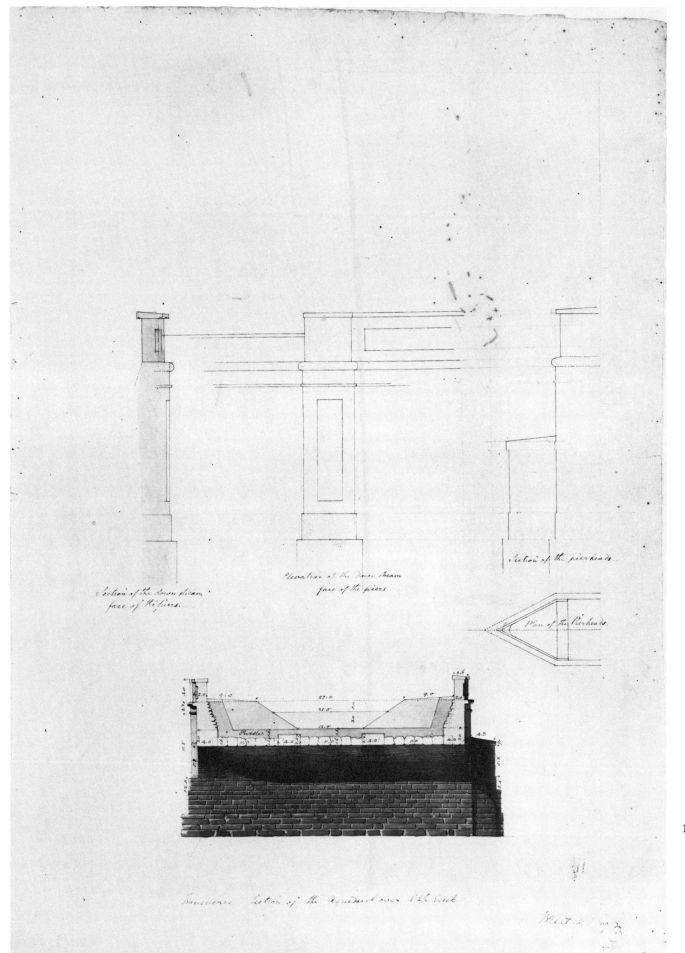

11 "Transverse Section of the Aqueduct over Elk Creek"

size: 21¼" × 14⅞" (53.9 cm × 37.8 cm)
media: pencil, pen and ink, watercolor
date: [1804]
location: Robert Brooke Papers, The New York Public Library

Chesapeake and Delaware Canal

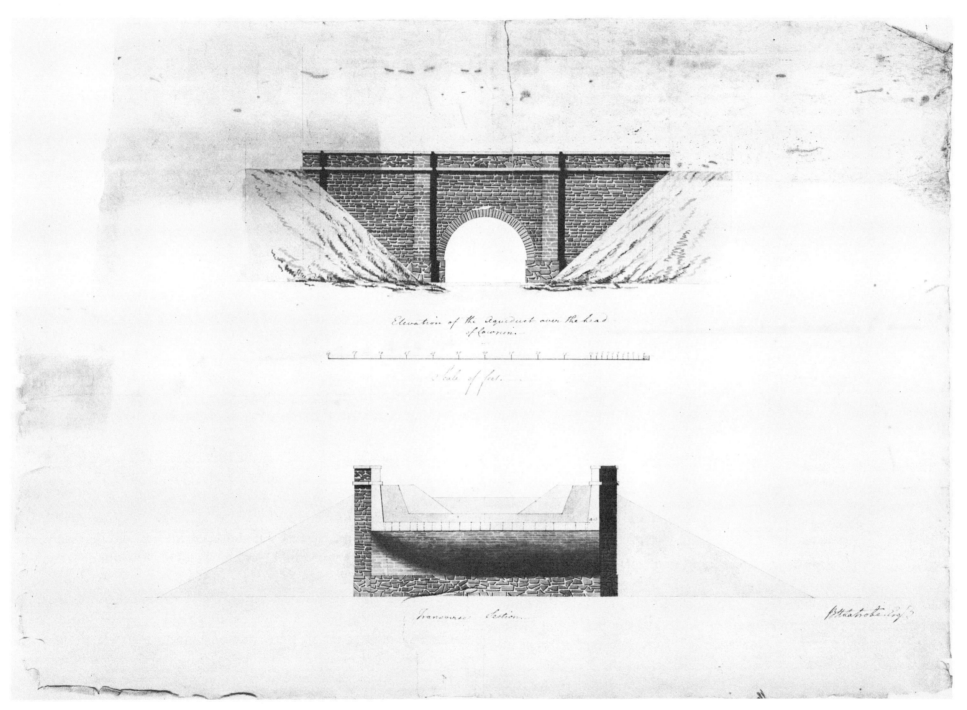

12 "Elevation of the Aqueduct over the head of Cow run"

size: 15 5/16" × 21 3/4" (38.8 cm × 55.2 cm)
media: pencil, pen and ink, watercolor
date: [1804]
location: Robert Brooke Papers, The New York Public Library

WASHINGTON CANAL

13 "Plan of the Groundwork; Plan of the piling"

14 "Longitudinal Section of the upper Gate; Plan of the Lock Floor" (See also colorplate 1.)

15 "Transverse Section of the Lock" (See also colorplate 2.)

16 Plan and sections of a wooden lock (See also colorplate 3.)

17 "Details of the Gates of the Locks" (See also colorplate 4.)

These five drawings, apparently working drawings for the carpenters, show different perspectives of the two wooden locks Latrobe built for the Washington (D.C.) Canal in 1810 and 1811. They were exact duplicates, and the plans represent both locks. One was located just south of N Street near the terminus of the canal at the Eastern Branch and the other was in the dredged channel of the Tiber Creek at 6th Street.[1] They raised the middle level of the canal about three feet above high tide.

Wooden locks were common in Britain in the eighteenth century, and were erected on the Middlesex Canal in Massachusetts which was opened in 1803. Latrobe believed wooden locks to be a temporary expedient, and he recommended that they be replaced with masonry locks in the future. They did cause problems on the Washington Canal because they were damaged by high water during storms and by movements in the loose soil adjacent to them.

Latrobe's plan of the upper gate of the lock (no. 13) is divided into a plan of the piling on the right side and a plan of the structure at the floor level on the left side. The lock opens toward the top of the drawing. On the right the pilings are spaced about three feet apart and form an outline of the lock's shape. Notable is the string of notched piles running from right to left which are connected by sheet piling, probably to provide strength and rigidity to the lock foundation. The piles were pine logs and were set by a floating pile driver.

On the left is a plan of the floor level of the lock. At the top of the drawing, the diagonal opening in the entrance wall of the lock is the beginning of the sluice. The sluice carried water across the sill (Latrobe calls it the "Selle"), under the floor where the lock doors stood, and then into the lock itself. The location of the heel post, on which the left door hung, is shown by a circle abutting a large square pile with a curved inset (the quoin). The door is shown in outline slanting from the heel post to the upper right where it would have met the door on the right side.

The section of the upper gates (upper half of no. 14) shows the vertical bars of the sluice grate to the left of the door. The grate trapped floating debris so it could not clog the sluice. The opening of the sluice into the lock chamber is directly below the door, and from that opening water flowed under the entrance flooring and into the lock.

The door is shown in a closed position against the mitre sill. The sill, which is shown beveled upward to the right, made a good seal when the door was closed. When open for a boat to enter the lock, the door fit into a recessed portion of the wall between the heel post and the sluice grate. The beam going from the end of the door to the heel post and beyond is the balance beam which was used to open and close the door. It is not in correct perspective, and appears much thicker than it should be. The heel post is fastened to the top of the lock wall by strips of iron known in a later era as *goon necks*. The inside wall of the lock chamber is shown to be a combination of horizontal planking and exposed upright studs. The floor of the lock is made up of stringers laid on piles, with a thin wooden deck covering the stringers. A feature of the section is the waterline which Latrobe drew in at 3 feet 6 inches (1.1 meters) above the sill of the upper gate.

The "Plan of the Lock Floor" in the same drawing (no. 14) is similar to the first plan (no. 13), since it is also for the upper gates. However, it does include dimensions showing the width of the lock to be 18 feet (5 meters), the width of the sluice to be 4 feet 6 inches (1.4 meters) wide, and the width of the door recess (and therefore the door) to be 12 feet (3.7 meters).[2]

The "Transverse Section of the Lock" (no. 15) also gives a number of dimensions, including 10 feet 6 inches (3.2 meters) as the height of the lock, and 7 feet (2.1 meters) as the height of the door. Latrobe has rendered different sections in the two halves of the drawing, but they are both near the heel posts. On the left side Latrobe again shows the sheet piling which intersects the quoin and heel post. A heavy timber runs from the top of the heel post across both outer rows of piles, and a slanting member ties the quoin to a horizontal timber attached to the piles. All of these stiffening elements insure that the lock doors will continue to be close fitting and well balanced.

On the right side of the drawing is a full view of a closed lock door, with its

---

1. BHL to President and Directors of the Washington Canal Company, 5 April 1817, Letterbooks; BHL, "Washington Canal. Memorandum... April 26th 1811 &c.," Letterbooks. See fig. 10, p. 20 above.

2. The dimensions noted in the lock recess are 4.0 and 8.0, or a total of 12 feet (3.7 meters).

balance beam extending to the right. An iron strap holds the beam to the meeting edge of the door, and the heel post is also attached to the quoin by ironwork.

The full plan of the lock (bottom of no. 16) contains the only view of a lower gate (*right*). It shows the recessed area inside the lock for the doors when open, and indicates that sheet piling ran between the heel posts of the lower gate as well as the upper. Wickets allowing the lock to be emptied of water when boats locked down to tide level may have been in the doors of the lower gate, but none are indicated. Although no dimensions are noted on this plan it can be estimated (by comparing it to the plan of the upper gate [no. 13]) that the lock was about 127 feet (38.7 meters) long from end to end, and the chamber between the doors was about 93 feet (28.4 meters) long.

The five sketches above the full plan of the lock are differing views of the sluice mechanism at the upper gate. The two to the left are similar to those in drawing no. 14. On the upper right is a section of the sluice from the grate to the lock door. It shows a slanting wicket with a crank-operated drop gate; the gate is up and water is pouring through. Below that section is another parallel to and in front of the drop gate which shows that the crank and gate were joined by a rack and pinion. The gate is down in that view. The uncolored sketch in the group of sketches is similar to the "Transverse Section of the Lock" (no. 15) which depicts the vent of the sluice under the sill.

The most complicated of Latrobe's drawings of the locks is titled "Details of the Gates of the Locks" (no. 17). ("Washington City Canal 1810–1811" is in a later and unidentified hand.) In the upper left, lower left, and lower right are sections and a plan of the upper gate. They are similar to previous drawings. The other views are details of the lower part of a lock door and of the mitre sill.

In the upper right of the drawing is a series of views depicting the lower pivot of the heel post. The "Section of the foot of the Upper Gates Showing the Gate opened back" shows a corner of the door in the recess of the wall, and the attachment of the iron pivot to the heel post and the place of its socket in the gate sill. To the right is a plan of the same part ("Plan of the foot of the Upper Gates") which incidentally reveals how closely the heel post and quoin were expected to fit. Further to the right are sketches of the iron socket and pivot ("details of the hanging pivot"). The pivot fixture is held to the bottom of the heel post by a long bolt passing through the heel post and running the length of the bottom rail of the door. A second much shorter bolt perpendicular to the first holds the fixture to the door by passing through the width of the rail. The iron socket appears to be dovetailed into the gate sill.

In the center of the drawing are sketches of the construction of the bottom rail. To the left, with vertical inscription, is an "exploded" view of how the bottom rail is keyed into the meeting style (the opposite end of the door from the heel post). Latrobe shows the door closed and flush at the bottom with the mitre sill. To the right is a "Plan of the bottom rail of the Gates," and at the bottom of the drawing is a "Section of the Gate Selle [mitre sill] and bottom rail." Each of these sketches indicates the location of the long bolt, and the recessed area in the meeting style for the nut.

In the center of this group of sketches is a broken "Plan of the Hollow coin [quoin] posts—heelpost, and meeting style, of the lock gate." To the left is the massive wooden quoin with the curved recess in the upper right corner for the heel post. The door has the heel post on the left and the meeting style on the right, and the long bolt passing throughout its length. The center section of the door is not a solid thickness, but is covered with planking between the heel post and meeting style. A glance at the elevation of the door in the upper left corner of the drawing shows how the planking is reinforced with two horizontal timbers and one angled timber.

Latrobe's drawings of the wooden locks for the Washington Canal are among the most detailed of his drawings, and are suggestive of the care he took to instruct craftsmen in how he wanted work to be done.

REFERENCES

Manuscript

Baltimore, Maryland. Maryland Historical Society. Letterbooks of Benjamin Henry Latrobe. "Washington Canal. Memorandum . . . April 26th 1811 &c."; BHL to Law, 27 June 1811; BHL to Fulton, 31 July 1811.

Printed

Hadfield, Charles. *British Canals*. 4th ed. New York: Augustus M. Kelley, 1969.
Roberts, Christopher. *The Middlesex Canal, 1793–1860*. Harvard Economic Studies, no. 61. Cambridge, Mass.: Harvard University Press, 1938.

Washington Canal

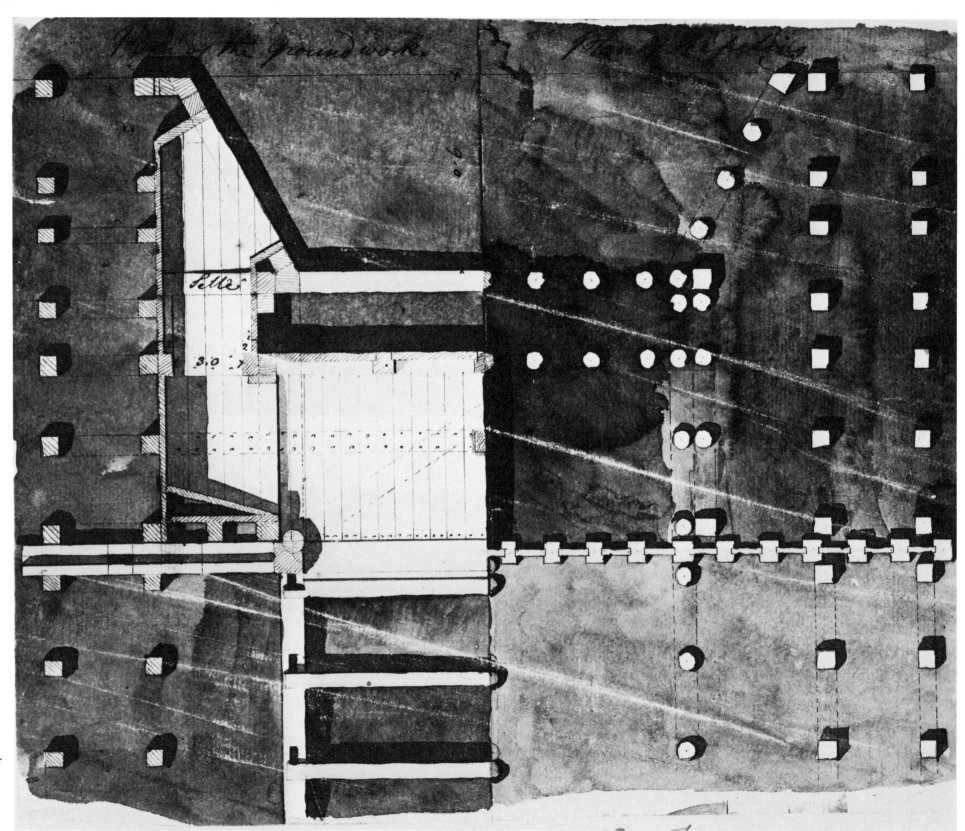

13 "Plan of the Groundwork; Plan of the piling"

size: 9¼" × 11⅞" (23.5 cm × 30.2 cm)
media: pencil, pen and ink, watercolor
date: c. 1810
location: Library of Congress, Washington, D.C.

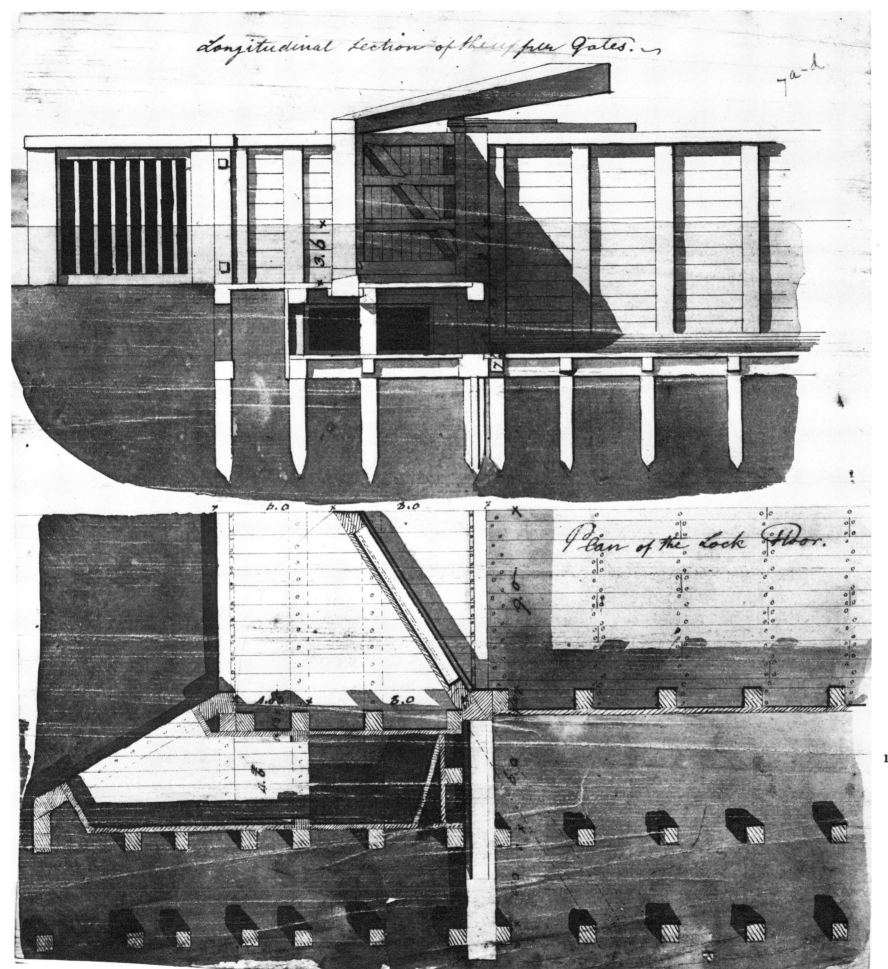

14 "Longitudinal Section of the upper Gate; Plan of the Lock Floor" (See also colorplate 1.)

size: $11\frac{7}{8}'' \times 10\frac{13}{16}''$ (30.1 cm × 27.4 cm)
media: pencil, pen and ink, watercolor
date: c. 1810
location: Library of Congress, Washington, D.C.

Washington Canal

15 "Transverse Section of the Lock"
(See also colorplate 2.)

size: 7 15/16" × 11 3/4" (20.1 cm × 29.9 cm)
media: pencil, pen and ink, watercolor
date: c. 1810
location: Library of Congress,
Washington, D.C.

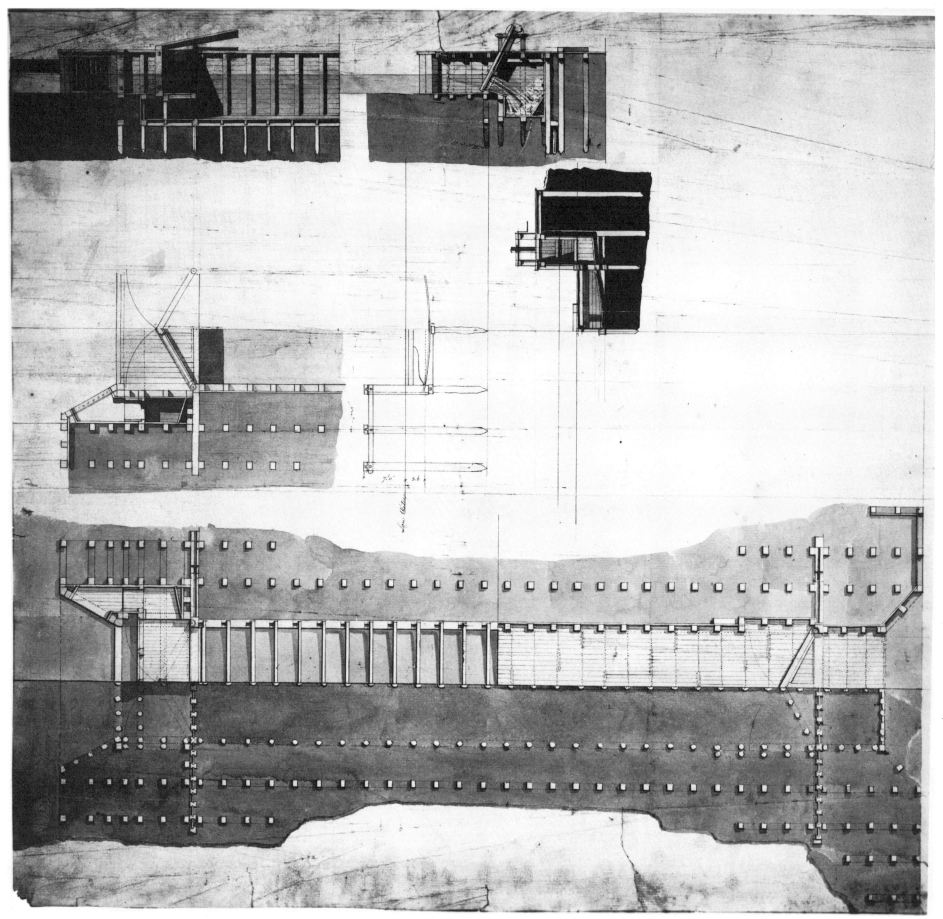

16 Plan and sections of a wooden lock
(See also colorplate 3.)

size: 18 5/16" × 21 1/4" (46.5 cm × 54.0 cm)
media: pencil, pen and ink, watercolor
date: c. 1810
location: Library of Congress, Washington, D.C.

135

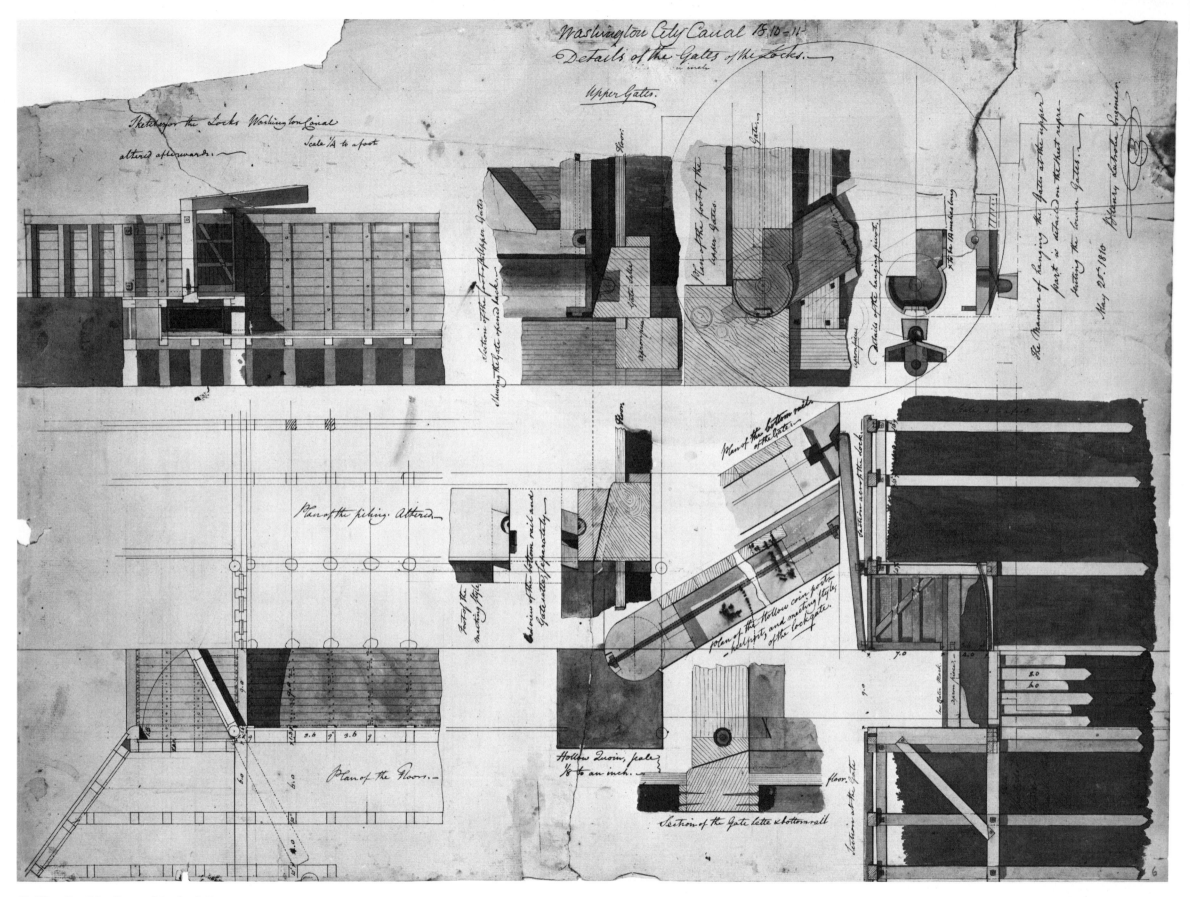

17 "Details of the Gates of the Locks"
(See also colorplate 4.)

size: 27 3/16" × 19 7/16" (69.0 cm × 49.3 cm)
media: pencil, pen and ink, watercolor
date: 20 May 1810
location: Library of Congress, Washington, D.C.

Plate 1. Washington Canal—"Longitudinal Section of the upper Gate; Plan of the Lock Floor," c. 1810.

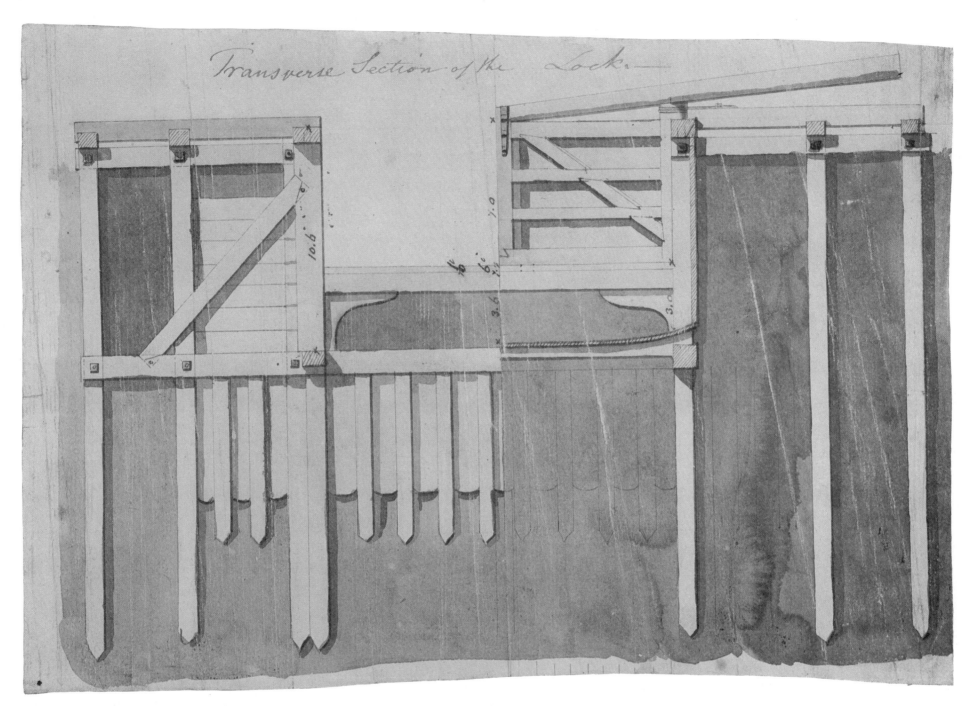

Plate 2. Washington Canal—
"Transverse Section of the Lock," c. 1810.

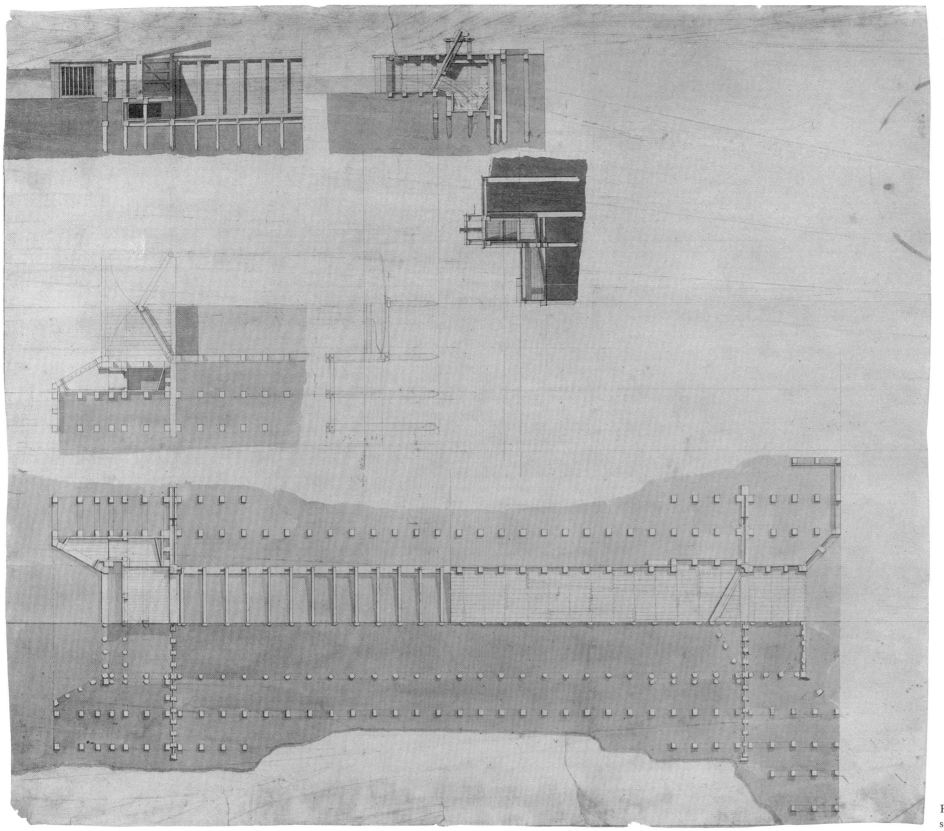

Plate 3. Washington Canal—Plan and sections of a wooden lock, c. 1810.

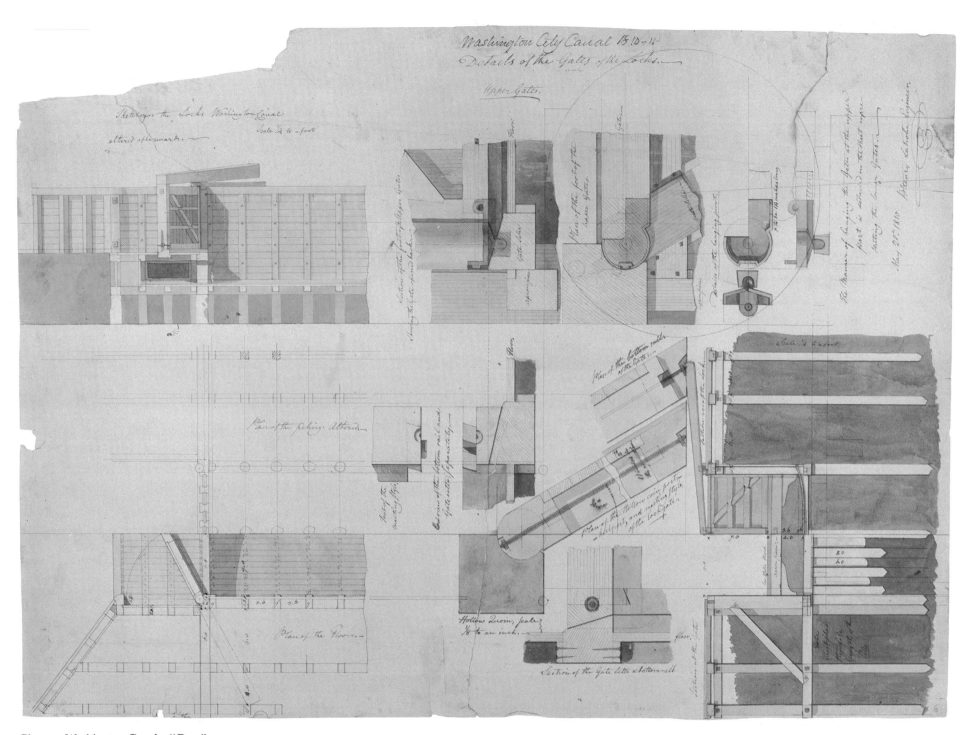

Plate 4. Washington Canal—"Details of the Gates of the Locks," 20 May 1810.

DISMAL SWAMP CANAL

18 "Sketch of Canals *executed* or *proposed* near Norfolk—the latter by B. H. Latrobe. Philad. 1803"

This is Latrobe's sketch of a better route for the Dismal Swamp Canal. In 1803 the canal ran from the Pasquotank River in North Carolina through the Dismal Swamp to the Deep Run estuary near Norfolk, Virginia.

The Dismal Swamp Canal Company was incorporated in 1790, but stock subscriptions lagged and digging did not begin until 1793. After sporadic construction activity the canal was opened for boating in 1805 even though it was unfinished. The canal depended upon natural seepage from the adjacent swamplands for a water supply and had frequent periods of low water. Since no accurate survey of the canal was taken in these early years, the canal company did not know that nearby Lake Drummond was higher than the canal and could have been used for a water supply.

Latrobe was familiar with the Dismal Swamp Canal from a brief visit to the area just after his arrival at Norfolk in March 1796, and from a professional trip in June 1797. The latter time he went at the request of the Dismal Swamp Land Company, which engaged him to "make a survey of all their Land in the Swamps, to cut a compleat Lane round it, and to lay off canals for the supply of Jerico and Smiths Mills."[1] He actually saw only the northern end of the canal on 12 June 1797.

Latrobe's sketch of the canal and proposed alternate route has north at the bottom and south at the top. On the right side is a list of his objections to the existing route of the canal. The first objection indicates that the canal was expected to carry to the north a heavy traffic of goods coming from the interior of North Carolina via the Roanoke River. Latrobe's proposed alternate route would have brought the southern end of the canal much closer to the mouth of the Roanoke, thereby eliminating difficult passages for the small rivercraft over the open waters of Albemarle Sound. He also suggested two possible northern termini, one of which would have bypassed the Deep Run estuary, where Latrobe believed a mill was creating a sandbar which would eventually block the estuary. Finally, he commented on the uncertain water supply and suggested that either the canal be puddled (lined with impermeable clay) or be supplied from the lake. The sketch has a feeder marked from Lake Drummond to a creek which intersects the canal.

The sketch of a railway car in the middle of the drawing seems out of place, as there is no reference to a railway in the sketch or in other papers. It is probable that the car is actually meant to be an addition to the drawing on the other side of the sheet, which illustrates the route of a railway proposed by Latrobe (no. 19).

The handwriting on this sketch is not Latrobe's (it is probably that of John Vaughan, the secretary of the American Philosophical Society), but the wording of the "objections" closely parallels part of a letter of Latrobe's dated 16 December 1803.[2] The sketch itself is apparently his.

---

1. BHL to Wood, 14 February 1798, Executive Papers of Governor James Wood, Virginia State Library, Richmond, Va., printed in BHL, *Journals*, 2:362–65.

2. BHL to Wheeler, 16 December 1803, Letterbooks.

REFERENCES

Brown, Alexander Crosby. *The Dismal Swamp Canal*. Chesapeake, Va.: Norfolk County Historical Society, 1970.

[Gallatin, Albert]. *Report of the Secretary of the Treasury on the Subject of Public Roads & Canals*. Reprint. Washington, 1808. New York: Augustus M. Kelley, 1968.

Latrobe, Benjamin Henry. *The Virginia Journals of Benjamin Henry Latrobe, 1795–1798*. Edited by Edward C. Carter II et al. 2 vols. New Haven: Yale University Press, 1977.

Dismal Swamp Canal

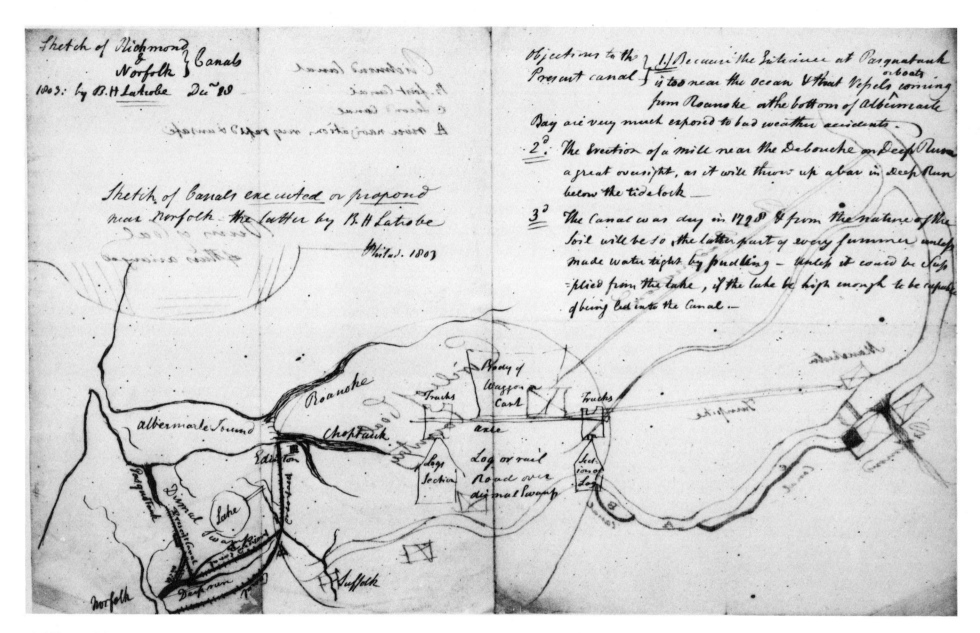

18 "Sketch of Canals *executed* or *proposed* near Norfolk—the latter by B. H. Latrobe. Philad. 1803"

size: 9″ × 15″ (22.9 cm × 38.1 cm)
medium: pen and ink
date: 19 December 1803
location: American Philosophical Society, Philadelphia, Pennsylvania

JAMES RIVER CANAL AND RAILROAD

19 Map of James River Canal and proposed railroad

The purpose for which this map was made is unknown, although it is the earliest expression of Latrobe's proposal for the construction of a railroad from the coal mines west of Richmond, Virginia, to tidewater on the James River. It is oriented with south at the top and north at the bottom. The handwriting is probably that of John Vaughan, the secretary of the American Philosophical Society.

This map is interesting because it shows three different types of transportation. The James River Canal was built in the latter 1790s, partly while Latrobe lived in Richmond. Its two sections extended only a short distance west of the city, but they were a first step in opening a navigation for boats traveling from the interior of the state to Richmond. A note on the map reminds the reader, however, that the section of the river between the two canal sections was still "very rapid and unsafe." A few years later Latrobe also commented that boats in the canal could carry only two hundred bushels of coal, but "if the canal were well constructed, 1000 bushels might be as easily and cheaply conveyed."[1] To provide an alternate route, a turnpike from the mines to Manchester was under construction at about the time Latrobe made this map, and was completed in 1807. In 1808 Latrobe believed its transportation costs about equal to the canal's.

Latrobe's idea for a better means of taking the coal to tidewater, which he first expressed in writing six years after making this map, was to build a railway to follow the valley of the Falling Creek from the vicinity of the southernmost mines to the James River below Richmond. He believed that the route could be made to go downhill all the way, and if it was, that a horse or mule could easily pull forty tons of coal over the railroad.

The kind of railroad Latrobe wanted to build is not certain. He sketched a completely wooden railroad on his map of the Dismal Swamp Canal (no. 18, p. 138), probably the same kind that Albert Gallatin in later years remembered discussing with Latrobe.[2] (The Dismal Swamp Canal sketch is on the reverse of this one.) Later Latrobe advocated iron railways. "Rail roads may be constructed of iron or of timber. The most durable (but also the most expensive) railroads, consist of cast iron *rails* let down on stone foundations; such roads will last for ages. Cast iron rails secured on beds of timber, are sufficiently durable for our country, and of moderate expense. Rail roads entirely of timber, are fit only for temporary purposes."[3] Motive power was to be provided by horses or mules—Latrobe did not anticipate the development of locomotives.

The construction of a coal railway at the Virginia mines did not materialize for Latrobe (nor was one built in his lifetime), although in 1809 he attempted to persuade mineowner Henry Heth to build one. Not until 1831 was a railway opened between the mines and the James River.

1. [Albert Gallatin], *Report of the Secretary of the Treasury on the Subject of Public Roads & Canals* (Washington, 1808; reprint ed., New York: Augustus M. Kelley, 1968), p. 89.
2. Hamlin, *Latrobe*, p. 555.
3. Gallatin, *Report on Public Roads & Canals*, pp. 104–5.

REFERENCES

Manuscript

Baltimore, Maryland. Maryland Historical Society. Letterbooks of Benjamin Henry Latrobe. BHL to C. I. Latrobe, 5 June 1805; BHL to Gilpin, 19 August 1805; BHL to Bollman, 21 May 1809, 30 May 1809, 8 December 1809; BHL to Heth, 3 July 1809, 25 September 1809, 21 November 1809; BHL to Gamble, 19 September 1810.

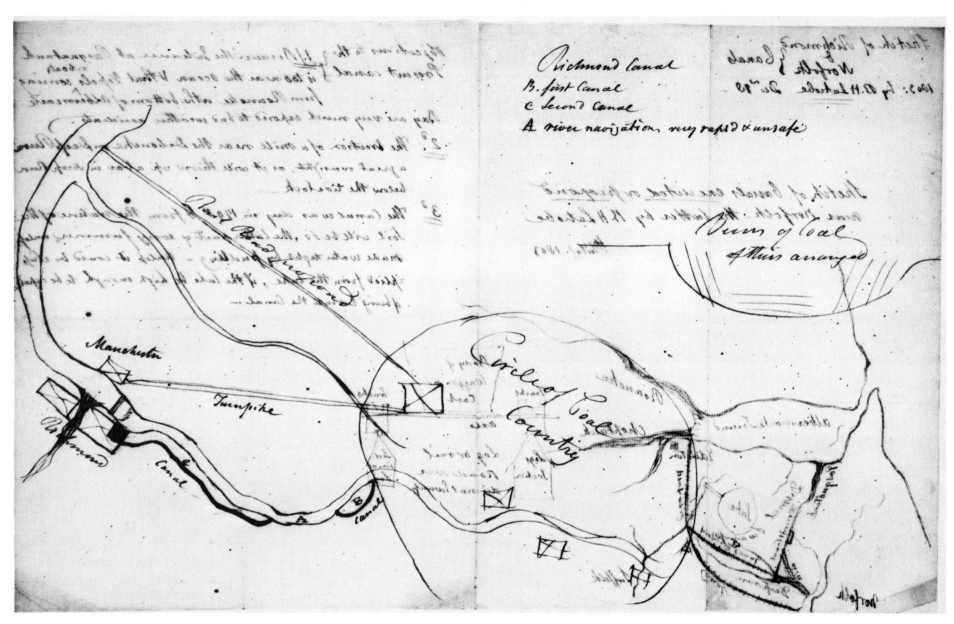

19 Map of James River Canal and
proposed railroad

size: 9″ × 15″ (22.9 cm × 38.1 cm)
medium: pen and ink
date: 19 December 1803
location: American Philosophical
Society, Philadelphia, Pennsylvania

NATIONAL ROAD (CUMBERLAND ROAD)

20 Plan and sections for the National Road

Benjamin Henry Latrobe knew about turnpikes from his extensive travels in England and the United States. During his years in England turnpikes there had been improved extensively under trusts organized by wealthy persons. In addition to his knowledge derived from personal observation, Latrobe also stated that he had learned a great deal from "Mr. Wilkes, an opulent banker in London, who possessed large estates in the county of Derby in England— ... [and who was] engaged largely in the making of turnpike roads in that county."[1] Latrobe's arrival in the United States in 1796 coincided with the beginning of the first turnpike era here. The Lancaster Turnpike was under construction in Pennsylvania and Latrobe was able to observe it in the process of completion.

This engraving illustrated Latrobe's published letter of 1813 to Albert Gallatin concerning the instructions given to the contractors for the first miles of the National Road. This federal highway, often called the Cumberland Road, was first surveyed in 1806, but not begun until 1811. As secretary of the treasury, Gallatin had overall direction of the project. Apparently he had asked for Latrobe's comment because they were friends and because Latrobe, by virtue of being surveyor of the public buildings and engineer to the Navy, was the closest thing to a chief engineer in the government. In addition, Gallatin knew that two of the three contractors for the first segment of the road were James Cochran and Charles Randall, men who had worked for Latrobe on the Philadelphia Waterworks, the Chesapeake and Delaware Canal feeder, and the Washington Canal.

Figure 1 is Latrobe's rendering of a typical section of the National Road as required by the contractor's specifications. It is located on a slope because, Latrobe noted, most of the road was in elevated terrain, and was or would be cut out of mountainsides. The roadway is 30 feet wide, as indicated by the scale at the top of the figure, with a hardened road in the central 20 feet (6.1 meters) and shoulders of 5 feet (1.5 meters) on each side. The road is composed of a bottom layer of stones 7 inches (18 cm) in diameter and less, and an upper layer of stones 3 inches (8 cm) in diameter and less. The two layers together are about 12 inches (30 cm) thick where the road meets the shoulders and 18 inches (46 cm) thick in the middle. The roadway has a ditch in the inner shoulder next to the hillside.

Latrobe objected to the ditch in the shoulder because he believed that the quantity of water coming off the hillside would be so great that the ditch would soon erode and undermine the hardened roadway. In figure 2 he illustrated his suggestion that ditches be dug on the bank above the road so that water could be diverted into the woods and fields, or directed towards culverts laid under the road at appropriate intervals. In suggesting this, Latrobe was also inveighing against the practice of draining water across the road in shallow oblique gutters which he felt caused serious erosion of the road surface.

His solution to the problem of diverting surface water seems closely related to his canal engineering practice. Present remains of the Chesapeake and Delaware Canal feeder include parallel ditches leading to culverts. Latrobe's ideal road is similar to the canal by having water runoff carefully separated from the transportation route. Apparently he visualized canals and turnpikes as similar forms.

Figure 3 shows Latrobe's suggested design for the surface of the National Road. The roadway is still 30 feet (9.1 meters) wide, but all 10 feet (3 meters) of the shoulder are to the left side. He felt that the shoulder was of sufficient width to be traveled by wagons when the ground was hard and dry in the summer (a "summer road") or frozen in the winter. Latrobe wanted all of the paved portion on the remaining 20 feet of road. He believed that if the pavement abutted the hillside there would be no opportunity for erosion of the road, and if his plan for ditching was followed (figure 2) little water would come down to that spot anyway.

Latrobe stated that the pavement should be made of one size of stones because if two sizes are used the smaller always work down into the crevasses between the larger, leaving the larger on top. Thus the original intent of making a smooth surface with smaller stones is frustrated. Latrobe suggested that only stones of 3-inch diameter or smaller be used and that the pavement could be thinner than the National Road's specifications. He also thought that a level pavement would be better than one that was thicker in the middle, since a convex road made carriage and wagon wheels run on the thinner outside part of the road and caused greater destruction of the surface. These suggestions were similar to those which John Loudon McAdam, the famous Scottish highway builder, first made public before a committee of the House

---

1. BHL, "To the Editor of the Emporium [on the construction of the National Road]," *Emporium of the Arts and Sciences*, new series, 3 (1814):296.

of Commons in 1811, but which were not widely adopted in Britain for a decade.[2]

Latrobe's criticism of the specifications for the National Road represented a professional engineer's understanding of past problems and current best practice. In 1808 one turnpike company had publicly admitted the error it had made in composing its road of two layers of different sized stone, stating that "the larger stones will, in time, work up to the surface by the action of heavy carriages, and prove highly detrimental."[3]

Although the report to Gallatin did not lead to a change in the National Road's specifications, it provided a brief summary of a different method of construction for new turnpike companies. In 1820 Latrobe traveled on a Pennsylvania turnpike which he was told used his ideas as a guide. He, and at least one other traveler who left a record of his passage, found that particular turnpike to be an excellent one.[4] By 1826 the National Road was experiencing the problems Latrobe had predicted: on long sections the upper layer of the roadbed was gone, exposing the larger stones and making the road nearly impassable. Two years later the superintendent of the National Road adopted the McAdam system of covering the surface exclusively with small, well-packed stones.[5]

In spite of such examples, however, it was thirty years before road-building ideas similar to Latrobe's became orthodox. W. M. Gillespie's *A Manual of the Principles and Practice of Road-Making* restated in greater detail all of the arguments Latrobe had made against convexity of the surface, layers of different sized stone, and draining water across the road. Latrobe possessed an advanced theory of road construction for his day.

5. Cochran, *New ASP. Transportation*, 2:195–99, 207–8, 262, 269.

REFERENCES

Manuscript

Baltimore, Maryland. Maryland Historical Society. Letterbooks of Benjamin Henry Latrobe. BHL to Murray, [15 May 1814]; BHL to Madison, 8 April 1816.
Baltimore, Maryland. Maryland Historical Society. Mrs. Gamble Latrobe Collection. BHL to Harper, 22 January 1820.

Printed

Cochran, Thomas, ed. *The New American State Papers. Transportation.* Vols. 1–2. Wilmington, Del.: Scholarly Resources, 1972.
Court, W. H. B. *A Concise Economic History of Britain: From 1750 to Recent Times.* Cambridge: At the University Press, 1964.
Devereaux, Roy (pseud.). *John Loudon McAdam.* London: Oxford University Press, 1936.
Evans, Oliver. "Turnpike Roads." *Emporium of the Arts and Sciences.* New series, 1(1813): 340–41.
Gillespie, W. M. *A Manual of the Principles and Practice of Road-Making.* 9th ed. New York, 1868.
Latrobe, Benjamin Henry. "To the Editor of the Emporium [on the construction of the National Road]." *Emporium of the Arts and Sciences.* New series, 3 (1814): 284–97.
McAdam, John Loudon. *Remarks on the Present System of Road Making.* 9th ed. London, 1827.
Parks, Roger N. *Roads and Travel in New England, 1790–1840.* Sturbridge, Mass.: Old Sturbridge Village, 1967.
Taylor, George Rogers. *The Transportation Revolution, 1815–1860.* New York: Harper & Row, 1968.

---

2. John Loudon McAdam, *Remarks on the Present System of Road Making*, 9th ed. (London, 1827), pp. 33–36.

3. Thomas Cochran, ed., *The New American State Papers. Transportation*, 7 vols. (Wilmington, Del.: Scholarly Resources, 1972), 1:227. See also Robert F. Hunter, "Turnpike Construction in Antebellum Virginia," *Technology and Culture* 4 (1963):195.

4. Adlard Welby, *A Visit to North America* (London, 1821), reprinted in Reuben Gold Thwaites, ed., *Early Western Travels*, 32 vols. (Cleveland: A. H. Clark Co., 1904–7), 12:192–95; BHL to Harper, 22 January 1820, Mrs. Gamble Latrobe Collection, MdHS.

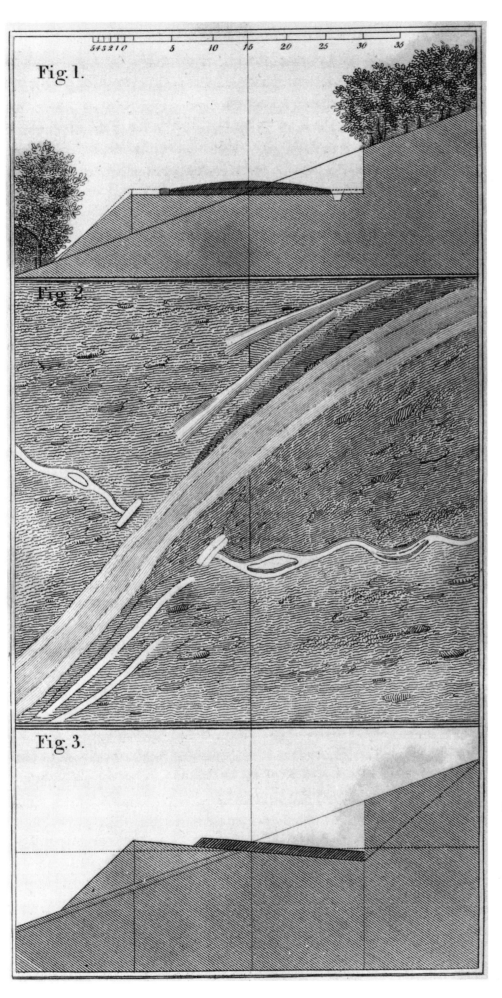

20 Plan and sections for the National Road

size: 7 9/16″ × 3 11/16″ (18.6 cm × 9.3 cm)
medium: engraving
date: c. March 1814
engraver: George Murray
location: *Emporium of the Arts and Sciences*, new series, 3 (1814): facing p. 290

PHILADELPHIA WATERWORKS

21 "No. I. Section of the Works from the Schuylkill to the lower, or Schuylkill Engine house"

This unscaled drawing is a sectional view from the south of the segment of the Philadelphia Waterworks at the Schuylkill River. This segment ran west to east just to the north of Chestnut Street, and had a total length of about 830 feet (253 meters).

Schuylkill water first entered the waterworks by flowing into the basin projecting into the river. In the middle of the basin wall, (*a*) to the extreme left of the drawing, was a tide-lock gate which admitted water on the ebb tide and closed on the flow tide. On the ebb the water of the Schuylkill was less turbid and brackish. (Latrobe was familiar with gates at the Chelsea Waterworks in London which worked the opposite way, allowing water in on the flow and closing on the ebb when sediment settled out of the collected water.[1]) The western wall of the Schuylkill basin was made of masonry, while the northern and southern walls (not seen here) were of earth.

At its eastern end the basin was connected to the canal by a wooden "scouring sluice" (*c*) set in a masonry wall.[2] The sluice was closed when new water was taken into the basin, and opened after settling had occurred. It also had another function according to an official report which stated that "Through the sluice-gate, water can be received or driven out at pleasure, as often as the turbid state of the Schuylkill or the foulness of the canal or tunnels... may require."[3] It seems likely that the cleansing action took place simply by allowing water to flow out of both the canal and basin when the tide was extremely low.[4]

The canal (*d*), which began with the sluice, was "blasted out of the Solid Rock"[5] for a distance of about 200 feet (61 meters). Its bottom was 2½ feet (76 cm) below the low waterline of the river (insuring a water supply in dry seasons), and its total depth was 12 to 16 feet (3.7 to 4.9 meters) below the former ground level. It was 40 feet (12.2 meters) wide at the top.

At the eastern end of the canal the tunnel (*f*) began. Although in this drawing Latrobe marked the tunnel's length at "about 130 yards [119 meters]," in the summer of 1799 he calculated it at 120 yards (110 meters), and several years later said it was 133 yards (122 meters). Approximately the first third of the tunnel was created by cutting down from ground level to the bottom of the tunnel, arching over with brick or masonry at a height of about 6 feet, and covering the new arch with earth. At the end of this cutting a brick-lined shaft (*g*) from the surface provided for inspection and repair of the interior.

Blasting with black powder excavated the last two-thirds of the tunnel horizontally into the rock (*h*). At the end of the tunnel the water flowed into the bottom of the pump well (*k*) of the Schuylkill Engine House. The well (*l*) extended up about 48 feet (14.6 meters) through a treacherous soil which caved in once during digging. Latrobe lined the well with 18 inches (46 cm) of brick to ensure stability.

---

1. BHL, *Answer to the Joint Committee of the Select and Common Council of Philadelphia, on the subject of a plan for supplying the city with water, &c.* ([Philadelphia, 1799]), p. 5; H. W. Dickinson, *Water Supply of Greater London* (London: Newcomen Society, 1954), p. 56.
2. Probably as a measure of economy the sluice was wooden rather than marble, as BHL had planned.
3. *Report to the Select and Common Councils, on the Progress and State of the Water Works, on the 24th of November 1799* (Philadelphia, 1799), p. 26.
4. At the point where the basin entered, the river is influenced by tides in the Delaware River estuary.
5. BHL to Smith, 21 March 1804, Letterbooks.

REFERENCES

Manuscript

Baltimore, Maryland. Maryland Historical Society. Letterbooks of Benjamin Henry Latrobe. BHL to Smith, 21 March 1804.
Philadelphia, Pennsylvania. Franklin Institute Library. Frederick Graff Collection.
Philadelphia, Pennsylvania. Historical Society of Pennsylvania. BHL, "Designs of Buildings erected in the Year 1799 in Philadelphia."

Printed

Gazette (Philadelphia). 20 July 1799.
Latrobe, Benjamin Henry. *Answer to the Joint Committee of the Select and Common Council of Philadelphia, on the subject of a plan for supplying the city with water, &c.* [Philadelphia, 1799].
Poulson's American Daily Advertiser (Philadelphia). 23 October 1801. BHL to Poulson, 22 October 1801.
Report of the Committee Appointed by the Common Council to Enquire into the State of the Water Works. Philadelphia, 1802.
Report to the Select and Common Councils on the Progress and State of the Water Works, on the 24th of November, 1799. Philadelphia, 1799.

Section from the Schuylkill to the Engine House

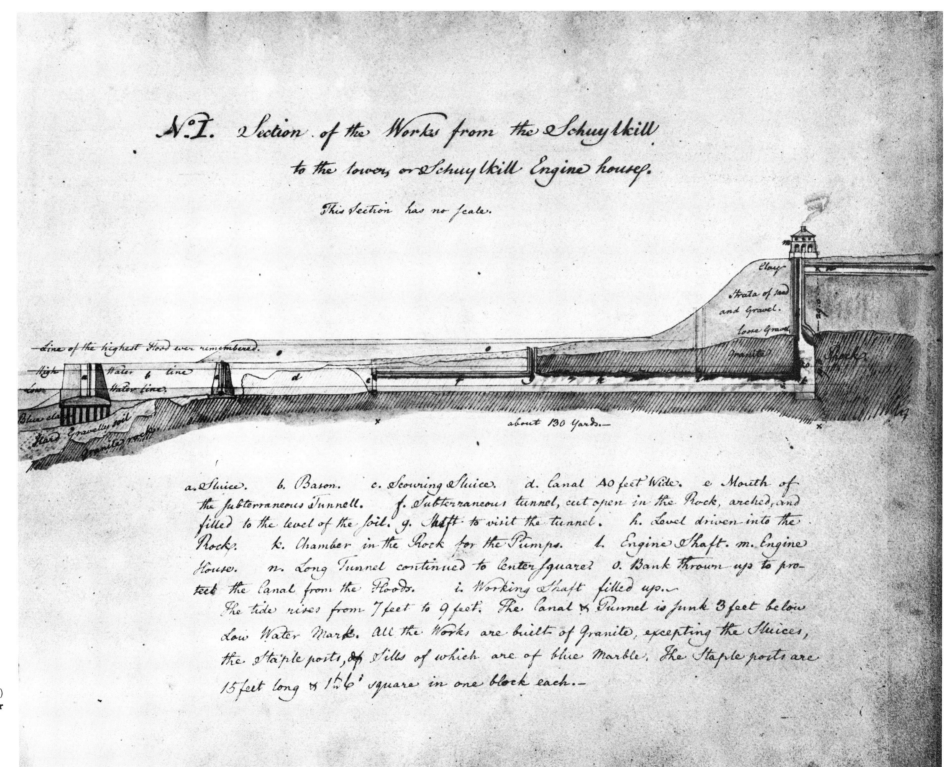

21 "No. I. Section of the Works from the Schuylkill to the lower, or Schuylkill Engine house"

size: 13¼″ × 11⅛″ (33.7 cm × 28.3 cm)
media: pencil, pen and ink, watercolor
date: [1799]
location: The Historical Society of Pennsylvania, Philadelphia

PHILADELPHIA WATERWORKS

The Basin on the Schuylkill River

22 Basin, canal, tunnel, and Schuylkill Engine House

23 "Section of the Basin from North to South; Plan of the Basin Wall & of the Coffre-Dam" (See also colorplate 5.)

24 "Plan of the Platform of the Basin Wall; Plan of the Basin Wall at high Water Mark" (See also colorplate 6.)

25 "West Elevation of the Wall of the Basin; Section of the Basin and of the Coffre Dam thrown up previous to its erection" (See also colorplate 7.)

These four drawings, which focus on the basin of the Philadelphia Waterworks, are among the most beautiful examples of Latrobe's drafting skill. Shaded, symboled, watercolored, and bordered, they appear to have been made for presentation, perhaps to the Watering Committee of the Philadelphia City Councils. The drawings are now preserved in a bound volume.

The plan and section of all the work on the Schuylkill River (no. 22) refine some of the details shown in the previous drawing (no. 21). This section has about the same horizontal scale as the earlier version, but the vertical scale is much reduced and closer to reality. The foundation of the basin wall (on the extreme left) is shown to be random masonry rather than the timber piling which Latrobe depicted in the earlier drawing.

There are several interesting features in the plan. On the left, the masonry wall with its rounded ends stands in the current of the Schuylkill River; the other sides of the basin are sloped earthen banks which, together with the masonry wall, make the basin plan nearly pentagonal. The sluice connecting the basin with the canal is a narrow opening formed by two structures which are made of wood and appear as mirror-image *V*s. (The inscriptions on the plan and at the top of the drawing are in pencil and may be in the hand of John Trautwine, Jr. [1850–1924], an eminent Philadelphia engineer.)

The embankment on the upper (northern) side of the canal is much heavier than that on the lower side, undoubtedly because it faces upstream, the direction from which floodwaters would come. The pair of dotted lines leading from the canal to the engine house represent the tunnel, and have superimposed at intervals three dark circles representing two inspection shafts and the pump well of the engine house. North and south of the canal and tunnel Latrobe symbolically rendered the grassy bank of the river and (by hachuring) the slope of the bluff on which the Schuylkill Engine House stood.

This drawing is the only one extant which exhibits the outline of the Revolutionary War redoubt (a fortification) mentioned by Latrobe in his report to city councils on 2 March 1799. The drawing also precisely locates and orients the engine house and the brick conduit with respect to Chestnut Street on the south.

The remaining drawings exhibit plans and sections of the western wall of the basin. The construction of the wall was difficult because it was set in the channel of the Schuylkill River and required excavation to bedrock to lay a secure foundation. Such work had never been attempted before in Philadelphia, and the strength and permanence of this wall was a matter of concern to the Watering Committee and citizens of Philadelphia. The detailed examination of the western wall in these drawings reflects their anxious interest.

The first drawing of the three (no. 23) contains the "Plan of the Basin Wall & of the Coffre-Dam," one of the most exquisite of the surviving Latrobe drawings. North is to the left. The rendering shows the wall in process of construction, still surrounded by the coffer dam. The dam is made up of an outer layer of three or four rows of wooden pilings connected by wooden stringers which apparently extend just past the midpoint of the wall. The body of the dam is an earthen bank, sloped outward from the top. On the inside, next to the basin wall, is a single row of timber piles connected by planking to make a smooth interior wall for the coffer dam. Each of these piles has a timber prop slanting downward to the base of the wall (see also no. 25). For extra strength, the piles at the corners of the dam are connected to a very large central pile by a web of timbers. Once the coffer dam was in place, water inside was pumped out, and excavation and construction were begun.

The basin wall is shown under construction in masonry, composed of a thick outer wall and a thinner inside wall. There is a small opening for the tide gate in the center.

The "Section of the Basin from North to South," also shown on this drawing, is an example of Latrobe's occasional break with conventional rendering to show different views of a structure within the same drawing.[1] On the left there is a section of the coffer dam and a view of the interconnected pilings

---

1. See drawings for the Harford Run Bridge, 14 and 16 March 1818, Peale Museum, Baltimore, Md., and drawings for the Washington Canal locks, particularly fig. 16, p. 135 (Plan and sections of a wooden lock), c. 1810.

at the north end of the basin wall. The beginnings of the masonry are exhibited at the bottom left of the section. Then Latrobe shows a perspective of the earthen north wall of the basin, extending slightly to the southeast and terminating in the raised bank about 200 yards (183 meters) away which protects the canal.

Abruptly in the center of the section Latrobe returns to the masonry wall, this time showing a completed portion flanking the tide gate. The gate has a toothed rack going upward to mesh with a gear which is turned by a crank. If the gate was self-operating by opening on the ebb and closing on the flow tide, as Latrobe hinted, the machinery for it is not evident. To the right of the gate is merely a dimensioned outline of the earthen bank on the south side of the basin, although Latrobe also indicates the levels of the river's extreme high and low water.

The next drawing (no. 24) exhibits a plan of the masonry wall at the high water mark. The inscribed dimensions make the wall parallel to the river about 169 feet (51.4 meters) long and about 10.6 feet (3.2 meters) wide on its crown. The plan also shows that between the inner and outer parts of the wall there was a gap filled with puddle, a watertight mixture of soft clay.

The upper plan ("Plan of the Platform of the Basin Wall") is remarkable because it shows the top course of masonry on the wall to be composed chiefly of interlocking butterfly headers. That such a rigid pattern of masonry is employed indicates that Latrobe intended the basin to be as sturdy and flood-resistant as possible.

The last drawing of this group (no. 25) has considerable notation in Latrobe's hand describing the geological character of the bed of the Schuylkill River. Latrobe recorded three layers: surface mud; "Hard Gravel and loose Rock"; and bedrock of granite. When Latrobe probed the riverbed prior to construction he was deceived by the middle layer and thought it was firm enough for a foundation. During the excavation he discovered his error, and the base of the wall had to be laid deeper and at greater cost than he had first estimated. The figures on the drawing indicate that for most of the wall he found bedrock about 4 feet (1.2 meters) below the mud, but at the south end it was 9 feet (2.7 meters) down.

The "Section of the Basin" on the lower portion of the drawing includes a section of the coffer dam, confirming the features exhibited in no. 23. The massive earth dam appears trapezoidal in section; extensive systems of piling support it on each side. In the upper drawing ("West Elevation of the Wall of the Basin") we again see the piles arranged around the ends of the masonry wall. The view of the tide gate in the center of the upper drawing, and the section of the wall in the right corner of the lower drawing also verify features which appeared in previous drawings.

REFERENCES

Poulson's American Daily Advertiser (Philadelphia). 27 October 1801. BHL to Poulson, 24 October 1801.
Report to the Select and Common Councils, on the Progress and State of the Water Works, on the 24th of November, 1799. Philadelphia, 1799.

Philadelphia Waterworks

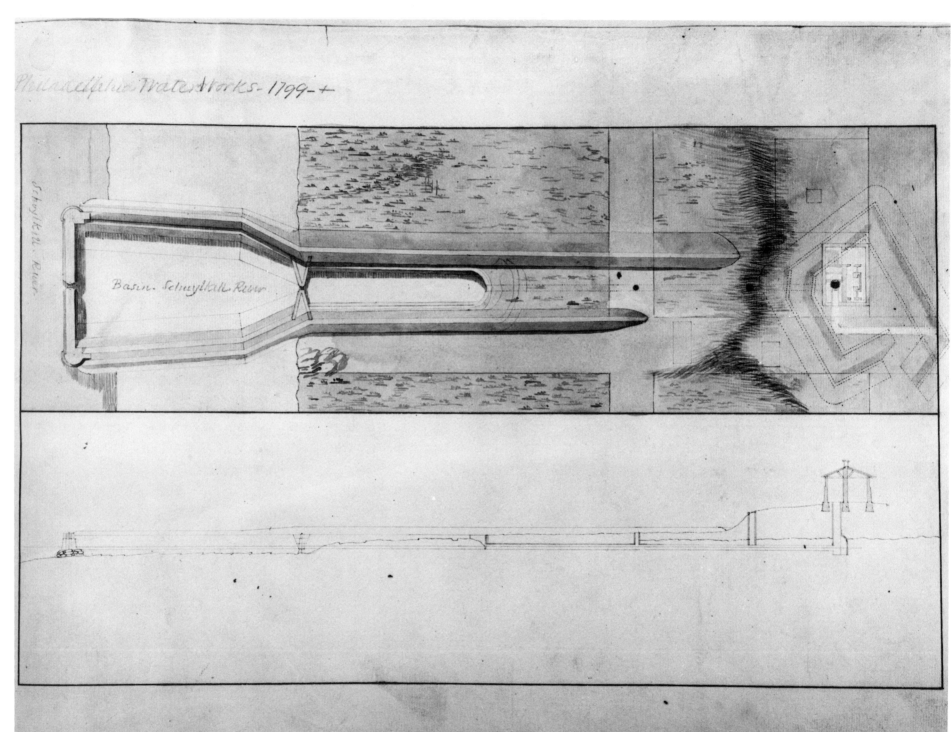

22 Basin, canal, tunnel, and Schuylkill Engine House

size: 14¾" × 21 1/16" (37.4 cm × 53.5 cm)
media: pencil, pen and ink, watercolor
date: c. 1800
location: The Historical Society of Pennsylvania, Philadelphia

The Basin on the Schuylkill River

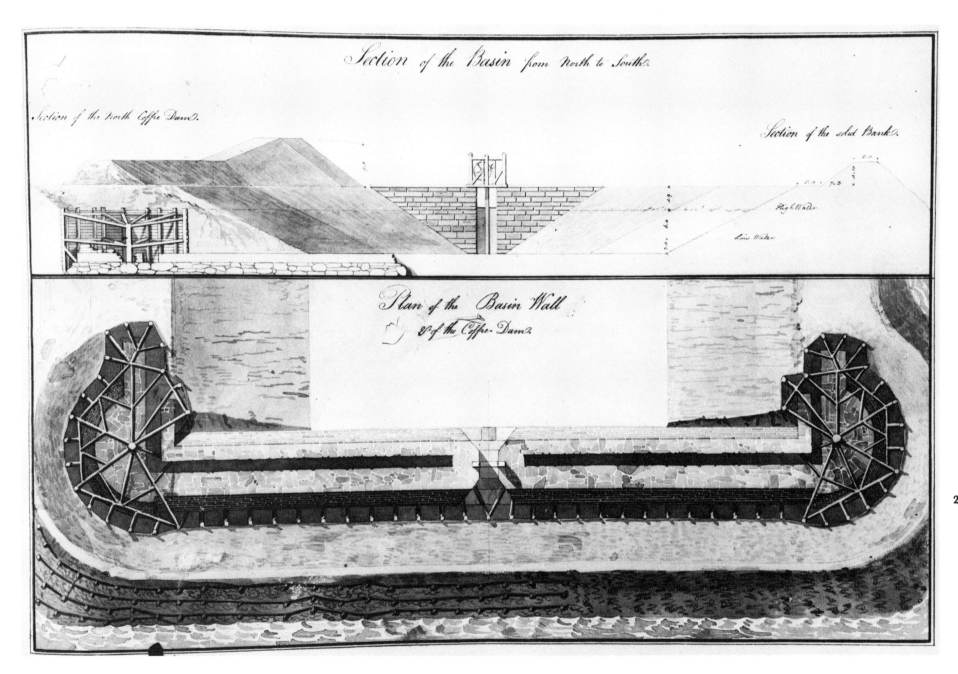

23 "Section of the Basin from North to South; Plan of the Basin Wall & of the Coffre-Dam" (See also colorplate 5.)

size: 14 1/16" × 19 13/16" (35.7 cm × 50.3 cm)
media: pencil, pen and ink, watercolor
date: c. 1800
location: The Historical Society of Pennsylvania, Philadelphia

Philadelphia Waterworks

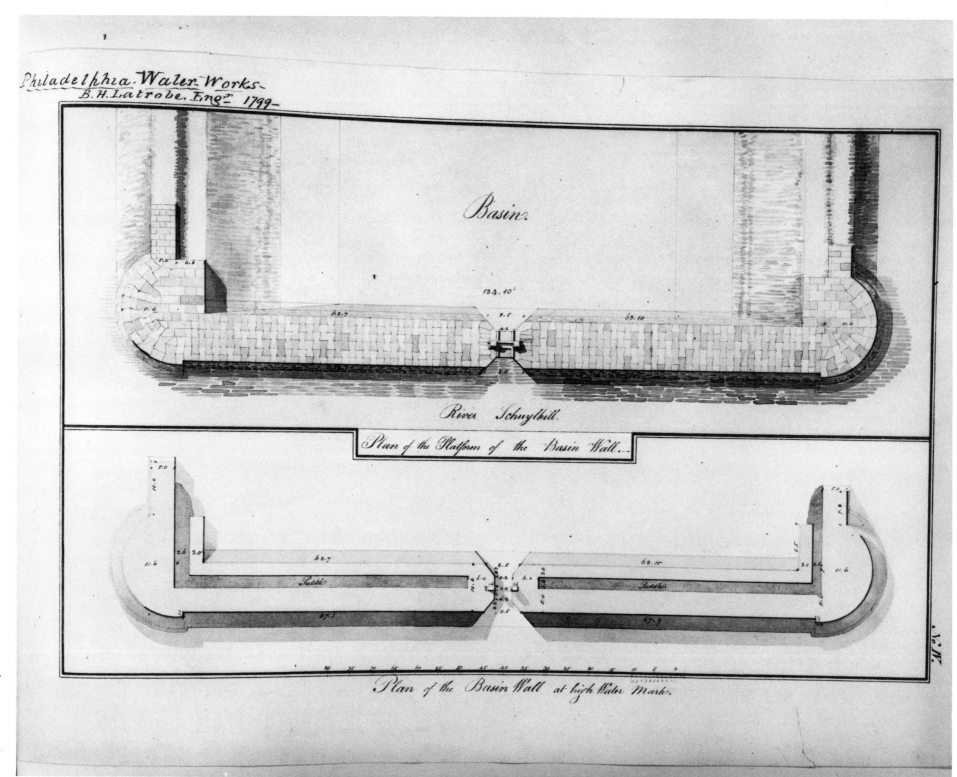

24 "Plan of the Platform of the Basin Wall; Plan of the Basin Wall at high Water Mark" (See also colorplate 6.)

size: $14\frac{1}{8}'' \times 20\frac{9}{16}''$ (36.0 cm × 52.3 cm)
media: pencil, pen and ink, watercolor
date: c. 1800
location: The Historical Society of Pennsylvania, Philadelphia

The Basin on the Schuylkill River

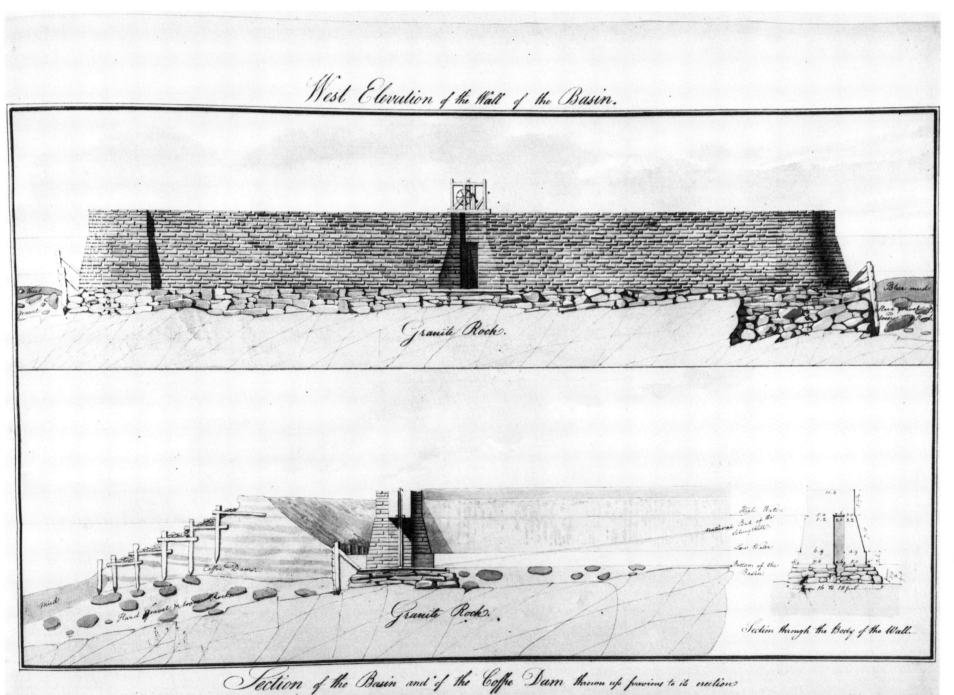

25 "West Elevation of the Wall of the Basin; Section of the Basin and of the Coffre Dam thrown up previous to its erection" (See also colorplate 7.)

size: 14⅜" × 21 1/16" (36.5 cm × 53.5 cm)
media: pencil, pen and ink, watercolor
date: c. 1800
location: The Historical Society of Pennsylvania, Philadelphia

PHILADELPHIA WATERWORKS

26 "Section of the Archimedean Screw"

This drawing is of an Archimedean screw pump used to drain and keep dry the Schuylkill basin while it was under construction. It appears to be part of the previous series of drawings (nos. 22–25), but is now separated from them. The pump is on the inner edge of the southern embankment of the basin (see nos. 22 and 24). The embankment slants away to the left (north), meeting the canal sluice which is just visible at the extreme left. Above the sluice at the top of the distant bluff is the hip-roofed Schuylkill Engine House.

The screw pump itself rests at an angle from lower left to upper right and turns on its long axis. A spiral tube inside it, probably of copper or lead, carries water from the bottom to the top. This tube with its casing of wooden staves was probably fitted with iron pins which rested in iron sockets at each end. At the top of the screw is an iron bevel gear meshing with a much larger gear mounted horizontally under the shed platform. A horse (wearing the collar mounted on the arm above) walks in a circle to drive the larger gear and thus the screw.

When the screw turns, water moves up the spiral tube and empties into a conduit which delivers the water into the river. A drop-gate at the end of the conduit can be closed to prevent water from flowing in during floods.

Archimedean screw pumps have been used in Mediterranean regions since antiquity to raise water continuously over short vertical and horizontal distances. In the eighteenth century they were used for draining the Lincolnshire and Cambridgeshire fens of England. Latrobe may have been familiar with the screw pump through his study of the fens while working for John Smeaton. He thought it to be reliable and used the device not only at the Philadelphia Waterworks basin, but also at the excavation and construction of the locks of the Washington Canal in 1810 and 1811.

REFERENCES

Manuscript

Baltimore, Maryland. Maryland Historical Society. Letterbooks of Benjamin Henry Latrobe. BHL to Roosevelt, 10 November 1804; BHL to President and Directors of the Washington Canal Company, 6 February 1810; BHL to Caldwell, 31 December 1810; BHL to Fulton, 8 July 1811.

Printed

Ewbank, Thomas. *A Descriptive and Historical Account of Hydraulic and Other Machines for Raising Water.* 4th ed. (New York, 1850), pp. 137–41.

The Archimedean Screw

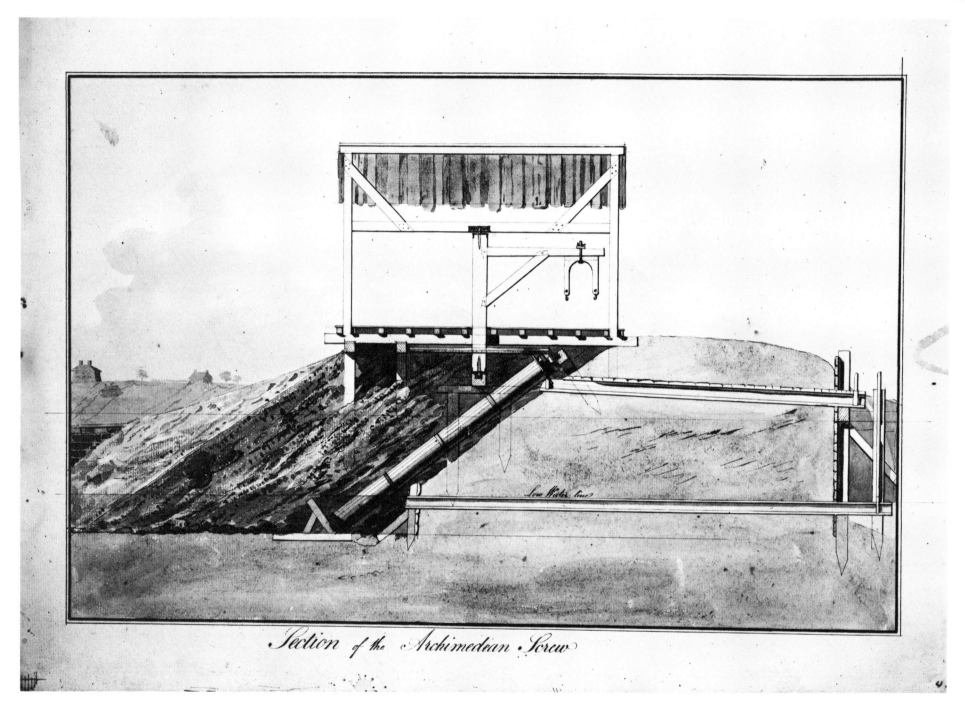

Section of the Archimedean Screw

26 "Section of the Archimedean Screw"

size: 15″ × 21 5/16″ (38.1 cm × 54.1 cm)
media: pencil, pen and ink, watercolor
date: c. 1800
location: Library of Congress, Washington, D.C.

PHILADELPHIA WATERWORKS

Plans of the Schuylkill Engine House

27 "Schuylkill Engine House No. I; Plan of the Foundations"

28 "Schuylkill Engine House No. II; Ground-plan"

These two drawings show the locations of the steam engine and pump in the Schuylkill Engine House. The brief "References" are some of the few remarks by Latrobe concerning the engine and house, but were made before their completion and do not reflect their finished state. Two of Latrobe's figures are thus inaccurate: (1) the engine's steam cylinder had a diameter of 39 inches (99.1 cm) rather than 44 inches, and (2) the pump well was about 48 feet (14.6 meters) deep, rather than 44 feet 6 inches.

The "Plan of the Foundations" (no. 27) indicates the placement of the basic parts of the steam engine, including the boilers, cylinder, condenser, and air pump. The "Lever Beam [or working beam] Wall" was an essential part of steam engine buildings since Newcomen's time. It held the bearing blocks which, in turn, supported the axle of the beam connecting the steam cylinder and the pump. As its thickness suggests, the wall was substantial enough to withstand the strain of the constant working of the powerful engine.

In the upper left of the drawing Latrobe shows the well (heavily shaded) in which the 48-foot pipe to the bottom is located. The three small circles in a row (two within, one outside the well) suggest that Latrobe intended to use a double-acting pump at the Schuylkill like the one at Centre Square,[1] although what kind of pump was installed is unknown. To the left of the engine well the brick conduit (known as the "upper tunnel") proceeds out through the building's foundation on its way to Centre Square.

In the "Ground-plan" of the Schuylkill Engine House (no. 28) Latrobe indicates the location of the boilers and the cold water cistern in which the condenser and air pump were immersed, but renders most completely the flywheel, its shaft, and supports. In this drawing, as in the previous one, the large empty space to the right of the steam engine is apparent: that space was to contain the equipment for the iron-rolling works which would use the "extra power" of the steam engine.

1. See drawings of Centre Square pump, nos. 43, 48, 49, 50.

REFERENCES

Proposals for Establishing a Company, for the purpose of employing the surplus power of the Steam-Engine Erected near the river Schuylkill. [Philadelphia], 1801.

Report to the Select and Common Councils, on the Progress and State of the Water Works, on the 24th of November, 1799. Philadelphia, 1799.

The Schuylkill Engine House

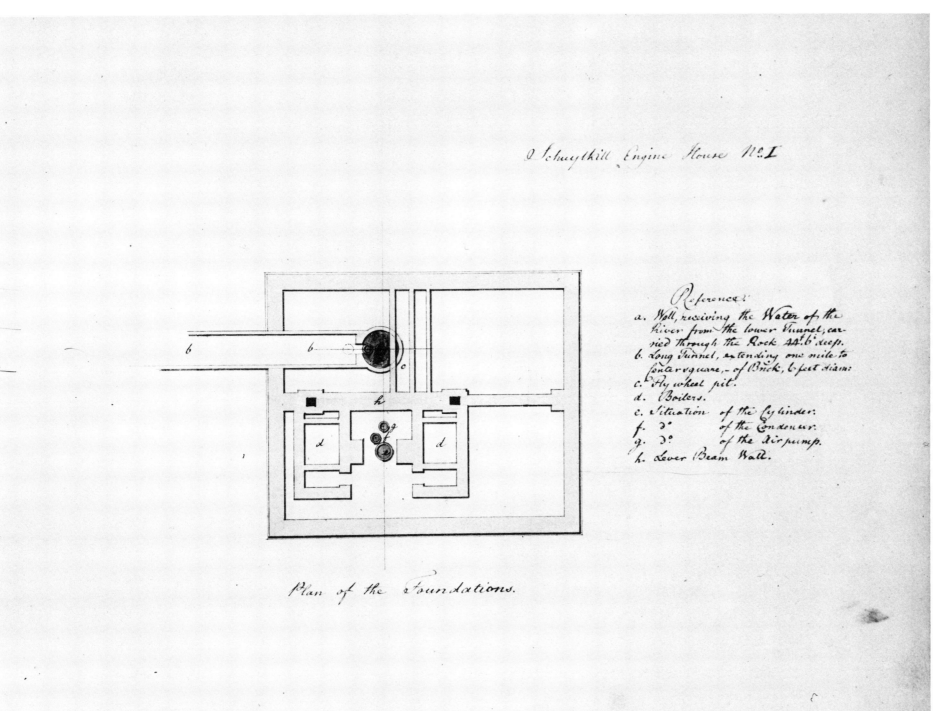

27 "Schuylkill Engine House No. I; Plan of the Foundations"

size: 13¼″ × 11⅛″ (33.7 cm × 28.3 cm)
media: pencil, pen and ink, watercolor
date: [1799]
location: The Historical Society of Pennsylvania, Philadelphia

Philadelphia Waterworks

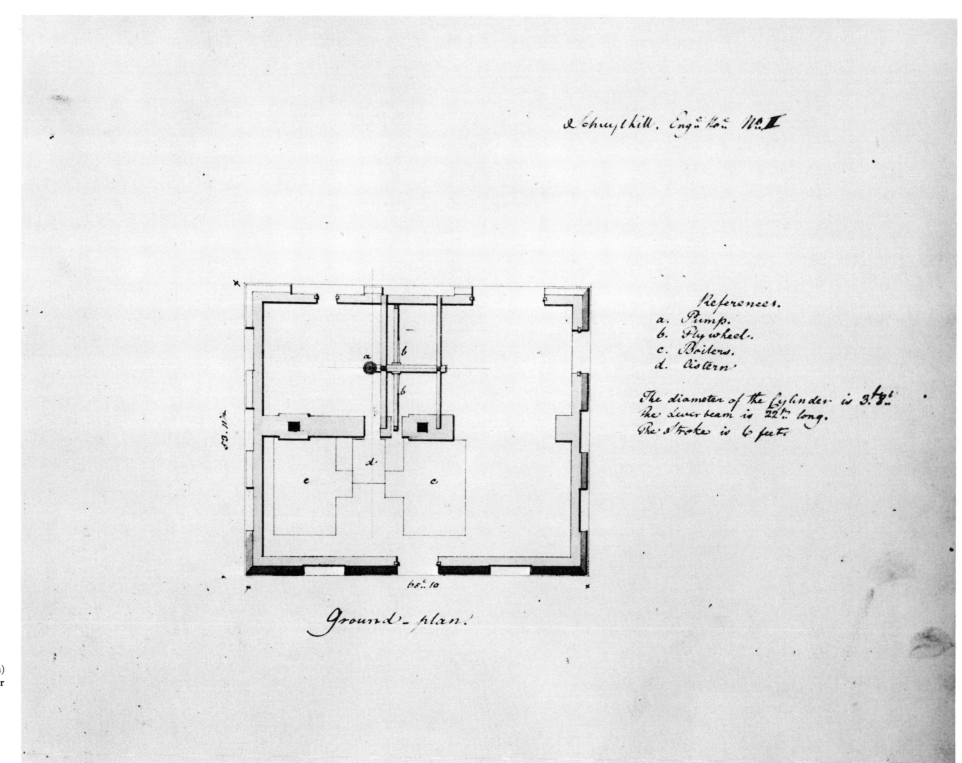

28 "Schuylkill Engine House No. II; Ground-plan"

size: 13¼" × 11⅛" (33.7 cm × 28.3 cm)
media: pencil, pen and ink, watercolor
date: [1799]
location: The Historical Society of Pennsylvania, Philadelphia

PHILADELPHIA WATERWORKS

Drawings of the Schuylkill Engine

29 Plan of the Schuylkill engine

30 Section of the Schuylkill engine (view from the east)

31 Plan and sections of a cast-iron boiler

32 "Section of the Boiler"

33 Section of the Schuylkill engine (view from the south)

The Schuylkill engine was manufactured at the Soho Works near Newark, New Jersey. Since Charles Stoudinger was the chief draftsman at Soho it is probable that the three drawings of the engine in this group are in his hand, since their style is unlike any Latrobe renderings. The designer of the Schuylkill engine was James Smallman, chief engineer at Soho and formerly an employee of Boulton and Watt in England. Smallman copied the very successful Boulton and Watt steam engine of 1787 which admitted steam alternately to each side of the piston (known as a "double-acting" engine). Like Watt engines, the Schuylkill engine employed a separate condenser, but it had three significant variations from the Boulton and Watt pattern: wooden boilers, a double-acting air pump, and a crank attachment of the beam to the flywheel.

Of the three engine drawings, the plan (no. 29) is the simplest, giving little detail besides the horizontal spacing of the major elements of the engine. It exhibits no conflict with Latrobe's plans except that the building is not a rectangle. Presumably Stoudinger's drawings were made with reference to the steam engine, and Latrobe later refined the architectural elements. The two square boilers are located symmetrically on either side of the steam cylinder, the largest circle on the plan. The cylinder's inside diameter was 39 inches (99.1 cm). To the upper left of the cylinder is the condenser, and directly above the cylinder and on the upper right of the condenser is the air pump. The condenser and air pump are within the cold water cistern, which also has a narrow projection containing the hot water pump (shown as a small circle).

Further to the top of the drawing is a circle locating the pump of 17 inches (43.2 cm) diameter which drew the Schuylkill water up from the tunnel. To the right of the pump are the flywheel and flywheel shaft. The flywheel had a diameter of 24 feet (7.3 meters), with a rim of cast iron weighing about 15 to 18 tons (13.6 to 16.3 metric tons). The flywheel shaft was extended to the right to connect with the gearing of the rolling mill. In 1807 the decayed wooden shaft was replaced with an iron one.

The first section (no. 30) shows the boilers and cylinder on the east side of the Schuylkill Engine House. The focus of attention in the drawing is the steam cylinder and its connection to the working beam. The cylinder, which weighed over 4 tons (3.6 metric tons), rests on a solid foundation of wood and bricks, and is secured to it by long bolts. The cylinder was reported to be "rather more than $6\frac{1}{2}$ feet long,"[1] giving a six-foot (1.8 meter) stroke when in operation. The original cylinder of the Schuylkill engine (installed in the fall of 1800) was defective, and was replaced with a new cylinder from Soho in 1803.

The beam, which is shown at the lowest point of the downward piston stroke, was connected to the steam cylinder piston rod with a parallel linkage, part of which appears in this drawing. The axle at the top of the beam rests in bearing blocks bolted to the wooden spring beams.

Behind the linkage of the beam and the piston rod is the connection of the two steam pipes to the valves which admitted steam to the cylinder. The pipes, which were cast in sections with flanges so they could be bolted together, brought low-pressure steam from the boilers.

The boilers shown in the section are wooden ones with their characteristic square cases. The interior flues of cast iron are also outlined. Both boilers rest on brick supports which have an ash pit between them. Only one wooden boiler was built for the Schuylkill engine, and its dimensions were variously reported as "18 feet long, 11 wide and 9 high," "9 feet high, 9 feet wide, and 15 feet long," and "about seventeen feet deep."[2] This boiler soon proved troublesome, as the wooden casing above the waterline in the boiler frequently sprang leaks at the boltholes where the white pine planks were fastened together. Steam pressure was difficult to keep up and the engine had to be stopped when the boiler was repaired.

A cast-iron boiler was constructed for the Schuylkill engine early in 1803, and was an immediate improvement. An engraving of it (no. 31) which accompanied Latrobe's report to the American Philosophical Society on

---

1. 7 July 1800, Thomas P. Cope Diaries, Acc. 975, Quaker Collection, Haverford College Library, Haverford, Pa.

2. Ibid., 9 July 1800; *Gazette* (Philadelphia), 6 April 1801; William H. McFadden, "A Brief History and Review of the Water Supply of Philadelphia," in *Annual Report of the Chief Engineer of the Water Department* (Philadelphia, 1876), p. 16.

steam engines in America shows it had the design of the English "wagon boiler." It was 17 feet (5.2 meters) long and 8 feet (2.4 meters) wide at the bottom, and 19 feet (5.8 meters) long and 10 feet (3.0 meters) wide at the height of 5 feet 7 inches (1.7 meters). Latrobe stated that "the boiler is composed of 70 plates of iron, cast with flanches, and bolted together, so that the flanch and bolts are within the water of the boiler wherever the flame touches it; otherwise they would be burned off in a few days. The pieces are so contrived as to be of only 12 different patterns."[3]

Another boiler of similar construction was built later in 1803 to replace the original wooden boiler at the Schuylkill. A section of a cast-iron boiler apparently drawn by Frederick Graff (no. 32) probably depicts one of the two early cast-iron boilers of the Schuylkill engine. One was replaced about four years later with a third boiler of cast iron.

The second section (no. 33) of the Schuylkill engine reveals the connection of the steam cylinder piston, beam, and flywheel, as well as several other elements. The supporting timbers for the cylinder are indicated clearly, and the cylinder, which is symmetrical, is the same as it is in the first section. The steam valves, hidden from view in the previous section, appear here to the left of the cylinder. They are the necessary valves for intake and exhaust of the steam at each end of the cylinder. The circle shown at the top of the valves is a section of the steam pipe coming in from the boiler, bringing steam to the valves. The valves were operated by levers tripped by pegs on the air pump rod (attached to the beam) which is to the left of the valves. Neither the pegs nor the levers (*working gears*) are shown in this drawing, nor in any other Philadelphia Waterworks drawing.

The piston rod of the cylinder is attached to the parallel linkage at the right end of the beam. The linkage between the beam and the rod forced the rod to work straight up and down, and not at varying angles as it would have done if connected directly to the end of the beam. Vertical working made possible a good seal in the stuffing box (shown in this section) on top of the cylinder.

The rod of the air pump was also connected to the parallel linkage. That rod worked the pump which drew the hot water (condensed steam) out of the condenser, and activated the working gears which operated the valves on

3. BHL, "First Report of Benjamin Henry Latrobe, to the American Philosophical Society, held at Philadelphia; in answer to the enquiry of the Society of Rotterdam, 'Whether any, and what improvements have been made in the construction of Steam-Engines in America?'" *Transactions of the American Philosophical Society* 6 (1809): 96.

the cylinder. Neither the condenser nor the air pump is shown in the sections, but they are in the plan (no. 29) of the engine.

At the center of the beam are the axle and bearing blocks which are shown in the previous section (no. 30). It is notable that the axle is on the top of the beam rather than under it, the latter manner of attachment being typical of English steam engines of this period. The Centre Square engine beam also had an overhead axle.

The rod attached to the left of the axle operated the hot water pump which took the water evacuated from the condenser and raised it to the hot water reservoir above. Water from that reservoir could flow to the boilers (by a pipe shown on this section as a circle superimposed on the reservoir) when the boiler feed valve reacted to the boiler's water falling below a safe level.

There was also a pump which brought water from the pump well into the cold water cistern, but it is not shown in this section. It may have been attached to the main pump. In 1809 a 10-inch (25 cm) diameter cold water pump was installed which not only kept the cold water cistern filled but provided a surplus which flowed into the tunnel to Centre Square.

Connected to the left end of the working beam is the pump rod which worked the 17-inch pump in the well. It was a wooden rod with an iron connection to the beam, and perhaps had iron segments extending down to the pump. The pump raised water into the circular brick conduit leading to Centre Square.

At the left end of the beam there was a crank rod to turn the flywheel, apparently hung on the same axle as that of the pump rod. The flywheel of the Schuylkill engine had two functions. First, it regulated the motion of the engine so that pumping went smoothly. London waterworks engines usually lacked flywheels, but at least one operating and one proposed waterworks engine had them. The flywheel was also necessary for the operation of the rolling mill because the reciprocating motion of the beam had to be converted into rotary motion. The rolling mill machinery was connected to the extended flywheel shaft. The flywheel also helped to give a regular motion to the mill by mitigating the counterforces which its machinery exerted on the steam engine.

The Schuylkill engine operated until 7 September 1815, when the new steam pumping station at Fairmount was started. During its fourteen years of service the Schuylkill engine required virtual reconstruction, needing new boilers, a new cylinder, and numerous other replacements. The engine frequently broke down, giving headaches to all associated with it, including

Latrobe. But it did pump water as required, and provided Philadelphians with the Schuylkill water to which they became ever more accustomed.

REFERENCES

Manuscript

Haverford, Pennsylvania. Haverford College Library. Quaker Collection. Thomas P. Cope Diaries, 1800–1, Acc. 975.

Philadelphia, Pennsylvania. Historical Society of Pennsylvania. George Bollman Collection. Eric Bollman memorandum and letterbook, 1802–3.

Printed

Dickinson, H. W. *Water Supply of Greater London*. London: Newcomen Society, 1954.

Farey, John. *A Treatise on the Steam Engine, Historical, Practical, and Descriptive*. London, 1827.

Latrobe, Benjamin Henry. "First Report of Benjamin Henry Latrobe, to the American Philosophical Society, held at Philadelphia; in answer to the enquiry of the Society of Rotterdam, 'Whether any, and what improvements have been made in the construction of Steam-Engines in America?'" *Transactions of the American Philosophical Society* 6 (1809): 89–98.

Proposals for Establishing a Company, for the purpose of employing the surplus power of the Steam-Engine Erected near the river Schuylkill. [Philadelphia], 1801.

Pursell, Carroll W., Jr. *Early Stationary Steam Engines in America*. Washington: Smithsonian Institution Press, 1969.

Report of the Committee Appointed by the Common Council to Enquire into the State of the Water Works. Philadelphia, 1802.

Watering Committee. *Annual Reports*. Philadelphia, 1803–9.

Philadelphia Waterworks

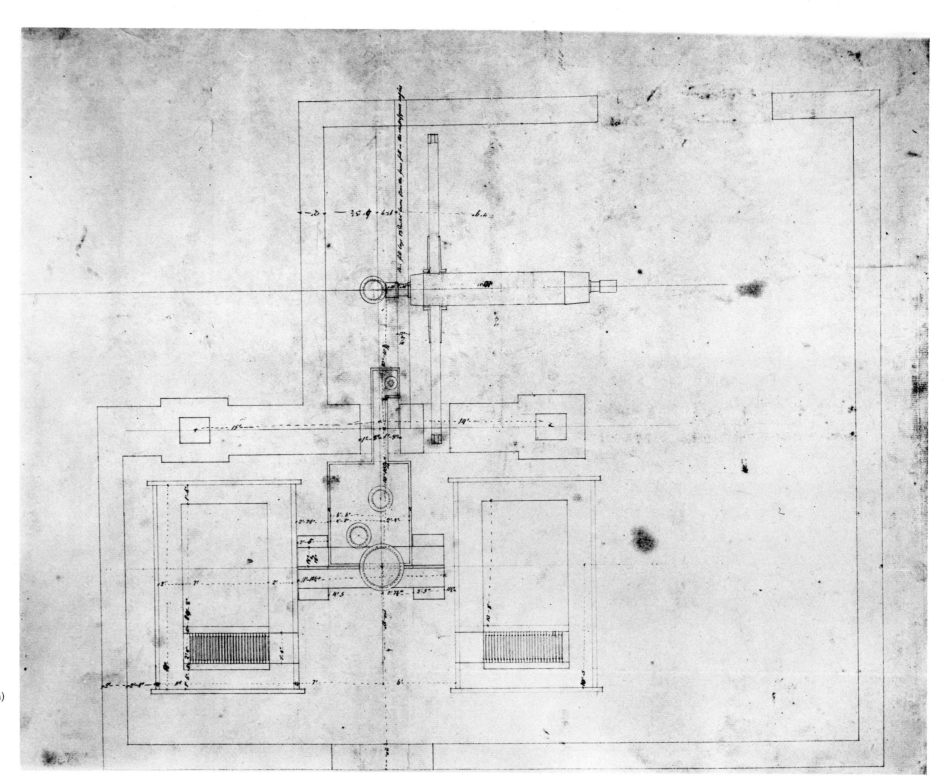

29 Plan of the Schuylkill engine

size: 29″ × 20¾″ (73.6 cm × 51.7 cm)
media: pencil, pen and ink
date: c. 1799
location: The New Jersey Historical
 Society, Newark

The Schuylkill Engine

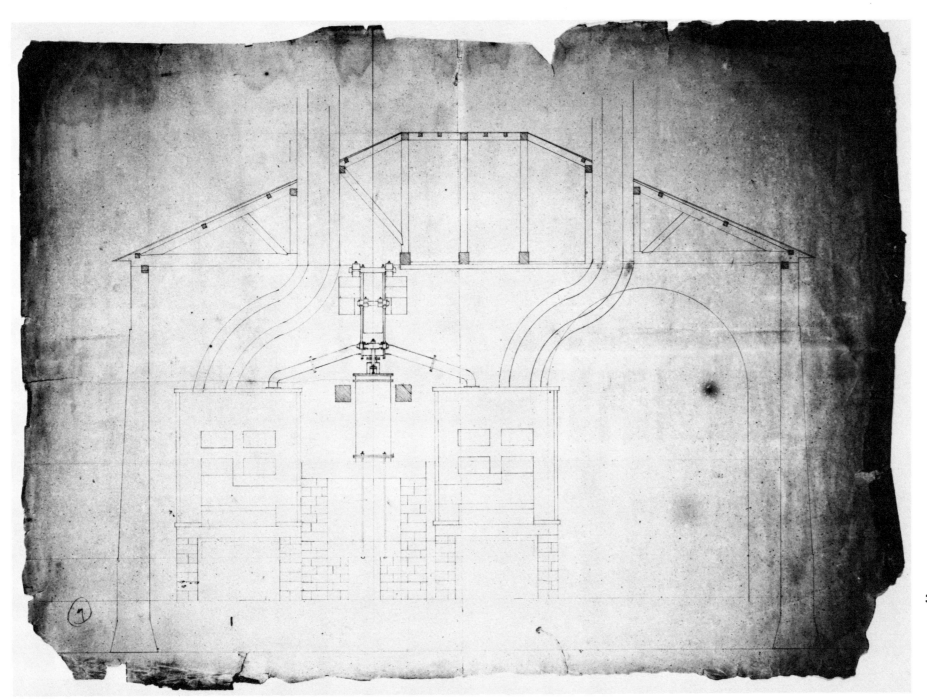

30 Section of the Schuylkill engine (view from the east)

size: 29½" × 42¼" (75 cm × 107.4 cm)
media: pencil, pen and ink
date: c. 1799
location: The New Jersey Historical Society, Newark

31 Plan and sections of a cast-iron boiler

size: detail from 7½″ × 9⅝″ (19.1 cm × 24.4 cm) original
medium: engraving
date: 1809
engraver: Benjamin Tanner
location: *Transactions of the American Philosophical Society* 6 (1809)

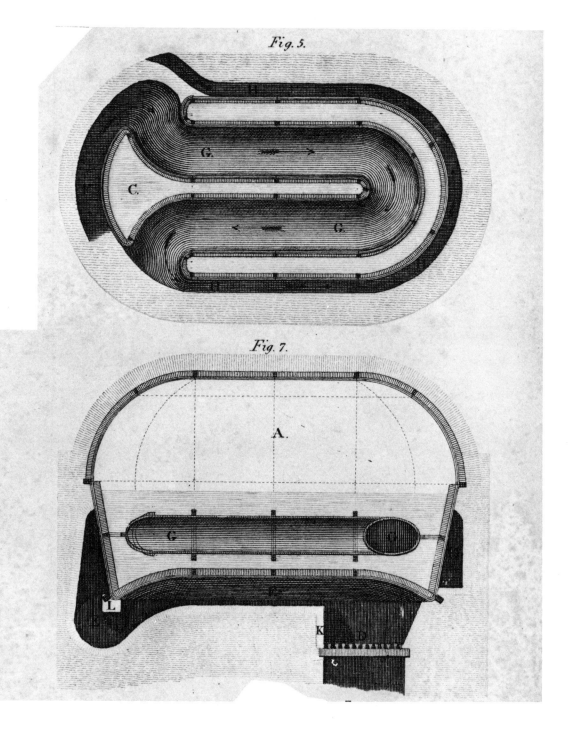

The Schuylkill Engine

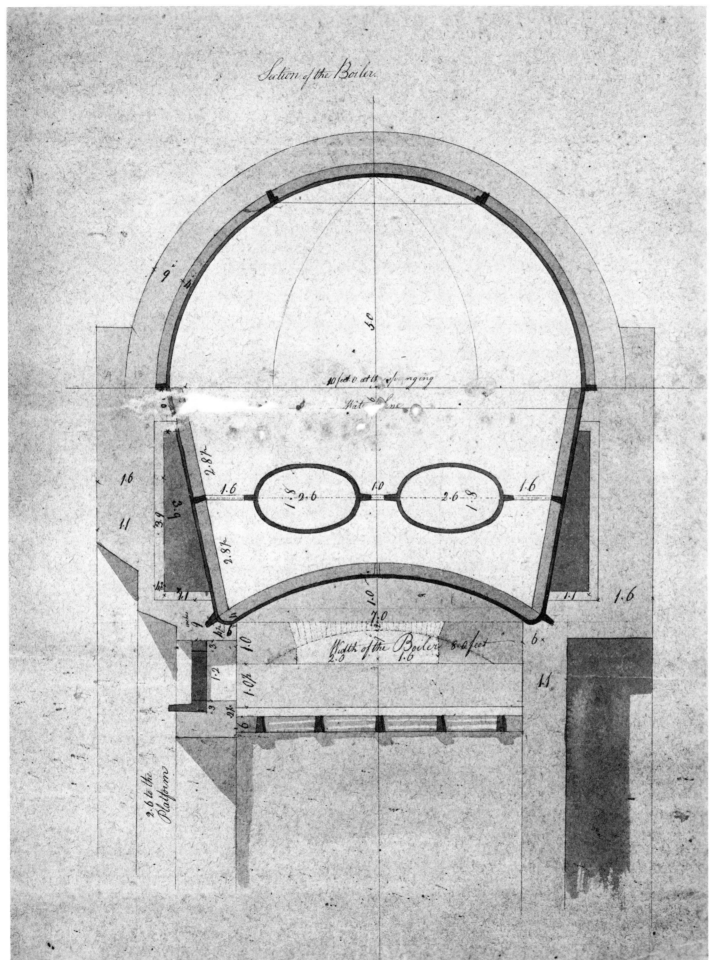

32 "Section of the Boiler"

size: 13⅞″ × 22¼″ (35.2 cm × 56.5 cm)
media: pencil, pen and ink, watercolor
date: c. 1803
location: Frederick Graff Collection,
 Franklin Institute, Philadelphia,
 Pennsylvania

Philadelphia Waterworks

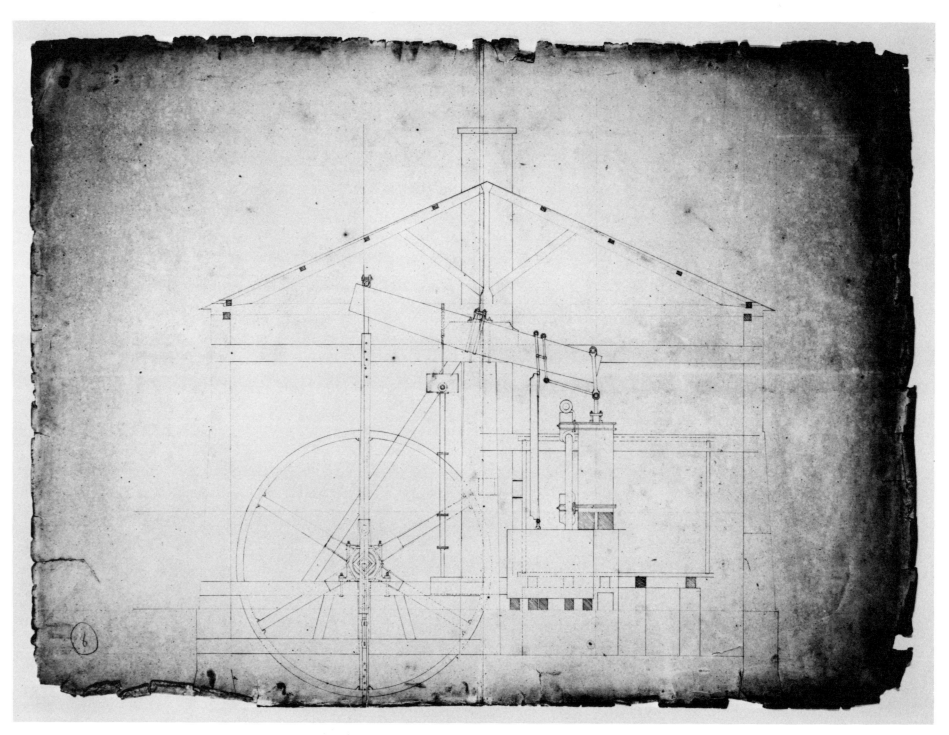

33 Section of the Schuylkill engine (view from the south)

size: 29⅝" × 41¼" (75.3 cm × 104.8 cm)
media: pencil, pen and ink
date: c. 1799
location: The New Jersey Historical Society, Newark

PHILADELPHIA WATERWORKS

34 "Section of the Schuylkill Engine House, Philadelphia, through the Walls from North to South"

This is the only section of the pump well and brick conduit ("tunnel") at the Schuylkill Engine House. In the middle of the drawing are the top of the well and the 17-inch (43.2 cm) (inside diameter) pipe which descended into it. To the right are the "discharging valve" and the conduit to Centre Square. Since the connection between the pipe and the conduit is not shown, the function of the discharging valve is unclear. Its name implies that it was used to empty the conduit when cleaning or repair were necessary. Certainly Latrobe drew it slanting below the bottom of the conduit, and it could have performed that function.

To the left of the pipe and well are the flywheel pit and flywheel supports. Directly above the pipe Latrobe drew in the steam cylinder supports and bolts, although they were located about 20 feet (6.1 meters) to the rear. It was important to show them because the center of the cylinder, indicated by a line equidistant from the bolts, was a reference line for all machinery in the engine house.

Latrobe's marginal notes twice mention "Stoudinger's plan," indicating that he had one of Charles Stoudinger's preliminary drawings for the Schuylkill engine in front of him as he made this one. Latrobe's gloss at the top of the drawing tells us that unlike the pump well at Centre Square, that at the Schuylkill was to be circular, although at a depth of 28 feet (8.5 meters) it also became oval.

REFERENCES

Report of the Watering Committee to the Select & Common Councils, November 13th, 1806. Philadelphia, 1806.
Report to the Select and Common Councils, on the Progress and State of the Water Works, on the 24th of November, 1799. Philadelphia, 1799.

Philadelphia Waterworks

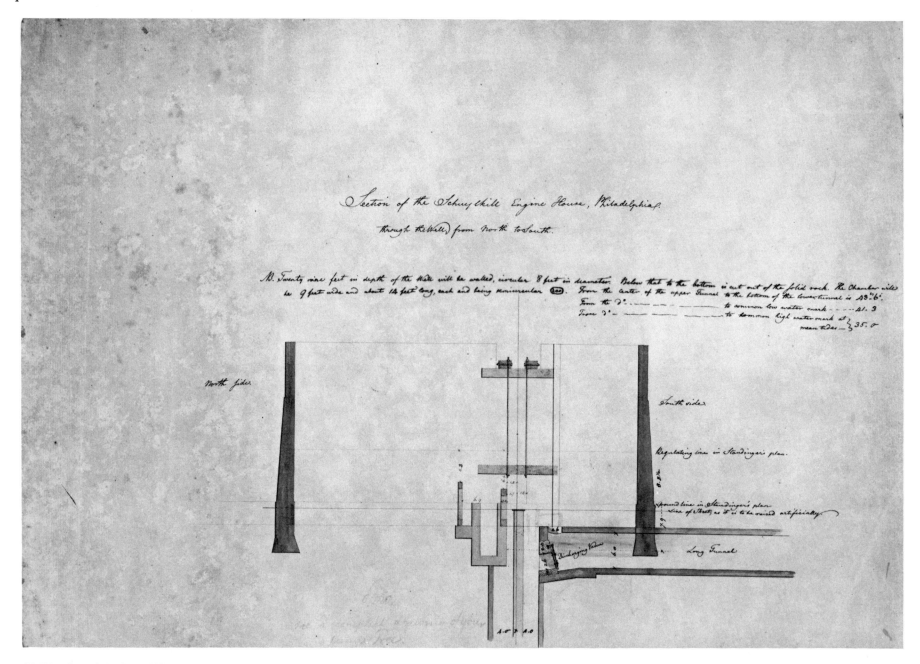

34 "Section of the Schuylkill Engine House, Philadelphia, through the Walls from North to South"

size: 13¼" × 19⅛" (33.7 cm × 48.5 cm)
media: pencil, pen and ink, watercolor
date: c. 1799
location: The New Jersey Historical Society, Newark

PHILADELPHIA WATERWORKS

Drawings of the Machinery for the Rolling Mill at the Schuylkill Engine House

35 "Plan [of the] Schuylkill Mills"

36 "Section from East to West"

37 "Section Showing the Rollers—Schuylkill Mills"

38 "Section parrallel to the Sheer block [of the] Schuylkill Works"

Although Nicholas Roosevelt stated that "Stoudinger was the projector [of the rolling mill], and that L[atrobe] was a mere copyist," these four Latrobe drawings of the Schuylkill rolling mill are the only ones in existence. Latrobe was involved in the planning of the rolling mill with Stoudinger, James Smallman, and Roosevelt from its genesis early in 1799. If he was a "mere copyist" he nonetheless thoroughly understood and approved of the plans.

There were rolling mills in the American colonies by 1750, but Latrobe's drawings are probably the earliest pictorial representations of an American rolling mill. They are certainly the earliest of a steam-powered one, since the Schuylkill mill was the first with steam power in America. (See figs. 19 and 20, p. 49, for views of a contemporary English rolling mill which may be compared with the Schuylkill mill.)

A rolling mill of this era took hammered sheets or bars of wrought iron from a forge and rolled them into smooth sheets for sale. Additionally, a rolling mill usually had slitters for cutting narrow sheets of iron into strips which were rolled into barrel hoops or mechanically sheared into rods. The rods were sold to blacksmiths or other craftsmen who made them into a variety of items.

From 1802 to 1804 the Schuylkill rolling mill occupied the north side of the Schuylkill Engine House and operated when the engine was not pumping water. In a given day it was apparently normal for the rolling mill to run for twelve hours, and for pumping to take six hours. Latrobe's plan of the rolling mill (no. 35) shows the connection of the flywheel with the rollers and slitters and has the dimensions for most parts. The flywheel shaft has a gear $44\frac{1}{2}$ inches (1.13 meters) in diameter at its left end which meshes with the roller cogwheel (bottom) 87 inches (2.21 meters) in diameter, and the slitter cogwheel (top), also $44\frac{1}{2}$ inches in diameter. Each cogwheel or gear is 16 inches (41 cm) wide. These cogwheels probably had cast-iron rims with wooden teeth, or were solid cast iron; in England it was common for the larger cogwheels to have wooden teeth and for the smaller ones to be cast iron. It was also usual English practice in steam rolling mills to drive both the rollers and slitters from the same point on the flywheel shaft, as shown here.

At the middle bottom of the plan Latrobe indicated the location of shears ("Sheers"), but he did not draw them in. On the left side of the drawing (no. 35) are the rollers (bottom) and the slitters (top). The rollers were to be 20 inches (51 cm) in diameter and 36 inches (91 cm) wide, while the slitters were about 16 inches (41 cm) wide. The rollers were a pair of smooth cylinders on horizontal axles; the slitters were ribbed cylinders which meshed together so that a sheet fed into them was cut into strips. Latrobe planned for heavy supporting timbers to underlay the rollers and slitters, as well as the cogwheels. The heavy (14-inch [36 cm] wide) timbers which held the bearing blocks for the cogwheels were called *sills*.

Latrobe's drawing "Section from East to West" (no. 36) is the only other view of the entire mill. His major purpose in making the drawing was to show the dimensions of the frames for the rollers (right), slitters (left), and the flywheel bearing block (center). All three of these have long bolts which anchor them to the heavy bearing timbers below. The thick, threaded tighteners for the roller and slitter frames are an interesting feature of this view. When the machine was operating, it was necessary to make adjustments to both frames according to the original thickness of the iron and the thickness that was to be produced. To roll a half-inch sheet from a three-quarter-inch hammered sheet required several passes through the rollers; after each pass, the frame was tightened slightly in order to bring the rollers closer together. Attempting to roll a sheet that was much thicker than the space between the rollers put a great strain on the machinery, and could break rollers and cogwheels. The slitters were also adjusted to the thickness of the sheet to be divided. Latrobe's "Section Showing the Rollers" (no. 37) has an inset showing the dimensions of the nut on the top of the roller frame tighteners, as well as a better view of them on the frame.

The drawing titled "Section parrallel to the Sheer block" (no. 38) was a correction of the drawings Latrobe had made a few days before, in which he made no provision for the shears. They were standard equipment in a rolling and slitting mill, and were typically made of cast iron and used to cut rods into shorter segments. Latrobe's plan of the shears ("Corrected plan of the Roller Cogwheel & Sheers") shows that he wanted to move the cogwheels

Philadelphia Waterworks

5 inches (12.5 cm) to the right (south) in order to make room for the shears between the roller cogwheel and the roller frame. The shears appear to be awkwardly located because men working them had to use the same working area as men at the rollers and because slit iron had to be carried around the rollers in order to be sheared. However, the section shows that only to the right of the roller cogwheel was there enough room to locate shears of the dimensions which Latrobe planned. Even then he had to extend the shears, as well as the sill which supported them, out one of the doors of the engine house. A wooden shed may have been added to the building to fully accommodate the shears.

The shears in Latrobe's drawing are of a design known in English rolling and slitting mills since at least 1758. The upper cutter is fixed at the right (in Latrobe's section) by the "cheek" and at the left side by a "stake," so that it is set at a slight angle. The lower cutter is hinged at the cheek by the same bolt as the upper arm, so the left end could swing between the shear sill and a stop-bolt on the stake. The left end of the lower cutter was moved by a double cam on the roller cogwheel shaft, and was estimated to work with a strength of close to 500 tons (454 metric tons).

The rolling mill at the Schuylkill Engine House was put in operation by the fall of 1802 under a partnership of Nicholas Roosevelt, Latrobe, and Eric Bollman, a merchant. It produced hoops, rods, and sheet iron, the latter ordered by Latrobe for the roofs of public buildings in Washington and for Thomas Jefferson's residence, Monticello. In January 1806 the City of Philadelphia bought the Schuylkill Engine House, and the mill never operated again.

REFERENCES

Manuscripts

Baltimore, Maryland. Maryland Historical Society. Letterbooks of Benjamin Henry Latrobe. BHL to Mifflin, letters 1803–4; BHL to Peters, 14 February 1807.

Haverford, Pennsylvania. Haverford College Library. Quaker Collection. Thomas P. Cope Diaries, 1800–1, Acc. 975.

Philadelphia, Pennsylvania. Historical Society of Pennsylvania. George Bollman Collection. Eric Bollman memorandum and letterbook, 1802–3.

Printed

Ashton, Thomas Southcliffe. *Iron and Steel in the Industrial Revolution*. 2d ed. Manchester: Manchester University Press, 1951.

Bishop, J. Leander. *A History of American Manufactures from 1608 to 1860*. 2 vols. Philadelphia, 1864.

Farey, John. *A Treatise on the Steam Engine, Historical, Practical, and Descriptive*. London, 1827.

Latrobe, Benjamin Henry. "First Report of Benjamin Henry Latrobe, to the American Philosophical Society, held at Philadelphia; in answer to the enquiry of the Society of Rotterdam, 'Whether any, and what improvements have been made in the construction of Steam-Engines in America?'" *Transactions of the American Philosophical Society* 6 (1809): 89–98.

Proposals for Establishing a Company, for the purpose of employing the surplus power of the Steam-Engine Erected near the river Schuylkill. [Philadelphia], 1801.

Redlich, Fritz. "The Philadelphia Water Works in Relation to the Industrial Revolution in the United States." *Pennsylvania Magazine of History and Biography* 69 (1945): 243–56.

Schubert, H. R. *History of the British Iron and Steel Industry from ca. 450 B.C. to A.D. 1775*. London: Routledge & Kegan Paul, 1957.

The Machinery for the Rolling Mill

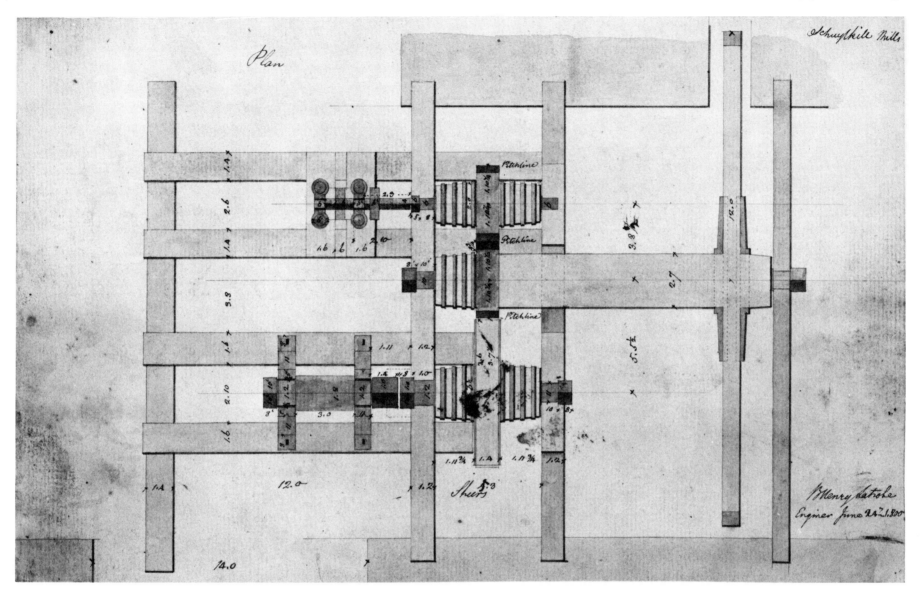

35 "Plan [of the] Schuylkill Mills"

size: 6⅞″ × 11¼″ (17.4 cm × 28.5 cm)
media: pencil, pen and ink, watercolor
date: 24 June 1800
location: The New Jersey Historical
  Society, Newark

Philadelphia Waterworks

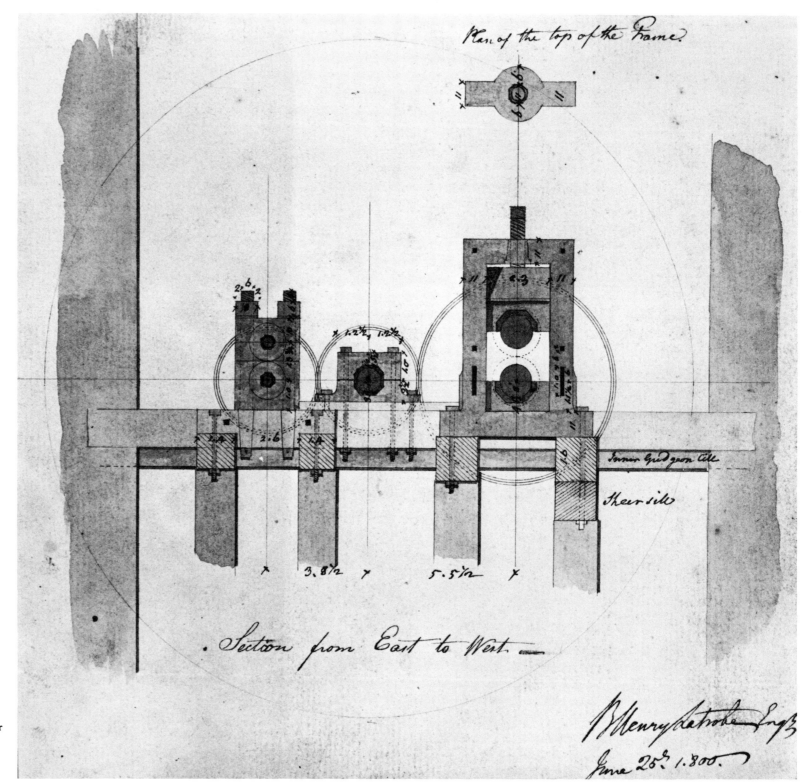

36 "Section from East to West"

size: 6⅞" × 7⅝" (17.5 cm × 19.3 cm)
media: pencil, pen and ink, watercolor
date: 25 June 1800
location: The New Jersey Historical Society, Newark

The Machinery for the Rolling Mill

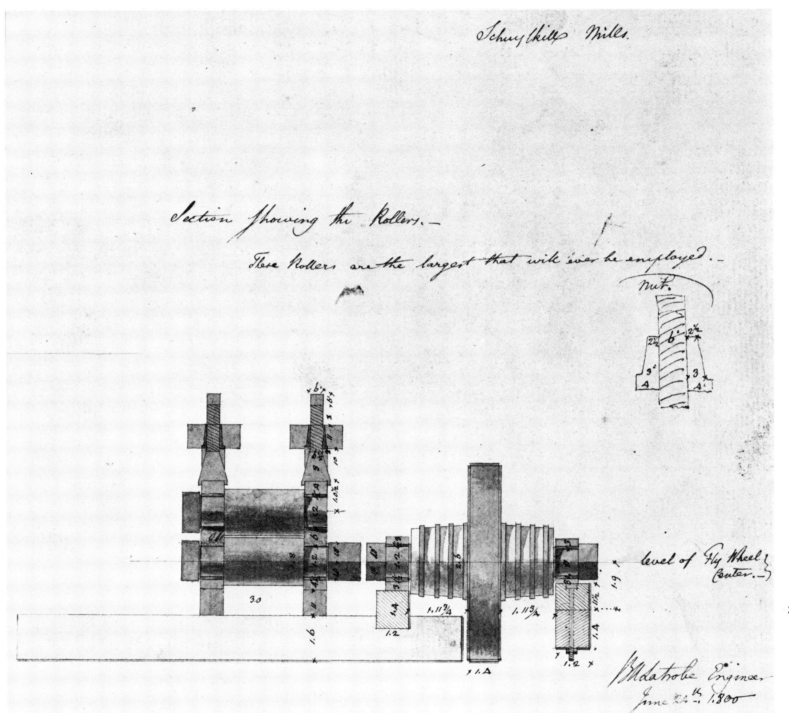

37 "Section Showing the Rollers—
Schuylkill Mills"

size: 6⅝" × 7 11/16" (16.9 cm × 19.5 cm)
media: pencil, pen and ink, watercolor
date: 24 June 1800
location: The New Jersey Historical Society, Newark

Philadelphia Waterworks

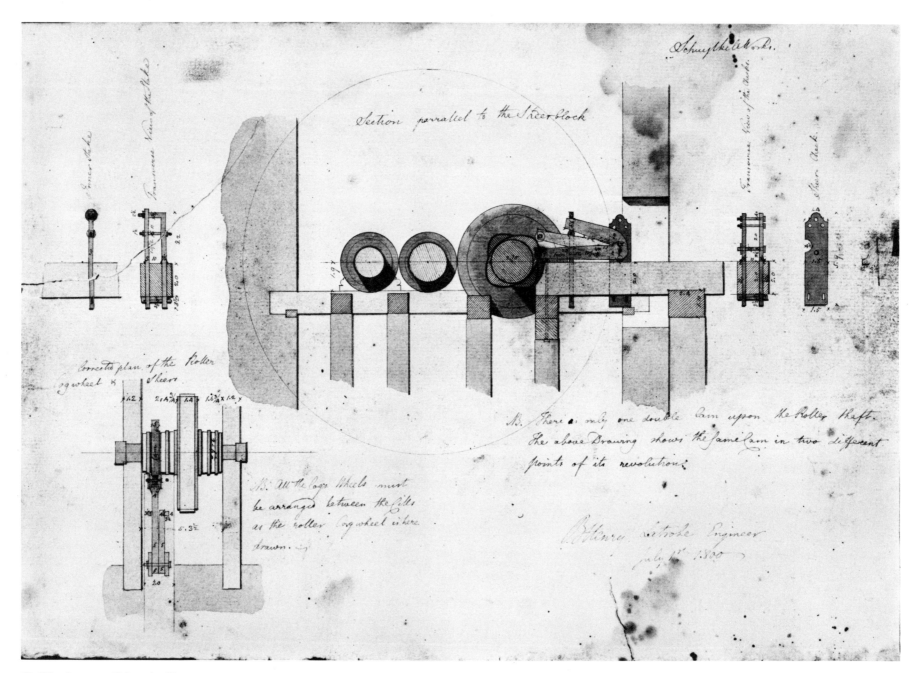

38 "Section parrallel to the Sheer block [of the] Schuylkill Works"

size: 9⅞" × 14¾" (25.1 cm × 37.4 cm)
media: pencil, pen and ink, watercolor
date: 1 July 1800
location: The New Jersey Historical Society, Newark

PHILADELPHIA WATERWORKS

39 "Sketch for a design of an Engine house and Water office in the City of Philadelphia March 1799" (See also colorplate 8.)

Presumably the earliest of the Centre Square drawings is the one by Latrobe titled "Sketch for a design of an Engine house and Water office in the City of Philadelphia March 1799," which shows a well-developed plan for the location and design of the engine. Major distinguishing features of the waterworks engines are present, among them the double-acting air pump, overhead pivot of the beam, flywheel, and wooden boilers. Two parts were different when the Centre Square engine was built, however: the water pump was double-acting instead of single and the reservoirs in the dome were wooden rather than marble.

Latrobe's "Memo" at the bottom of the drawing lists information to be given immediately to Nicholas Roosevelt, who had the contract for the engines. The information concerns the distances water had to be pumped vertically at the Schuylkill and Centre Square, so Latrobe probably wanted to insure that Roosevelt delivered the correct lengths of pipe for the pumps.

Latrobe probably displayed this drawing at the 2 March 1799 meeting of the Philadelphia city councils when his plan for the waterworks was adopted.

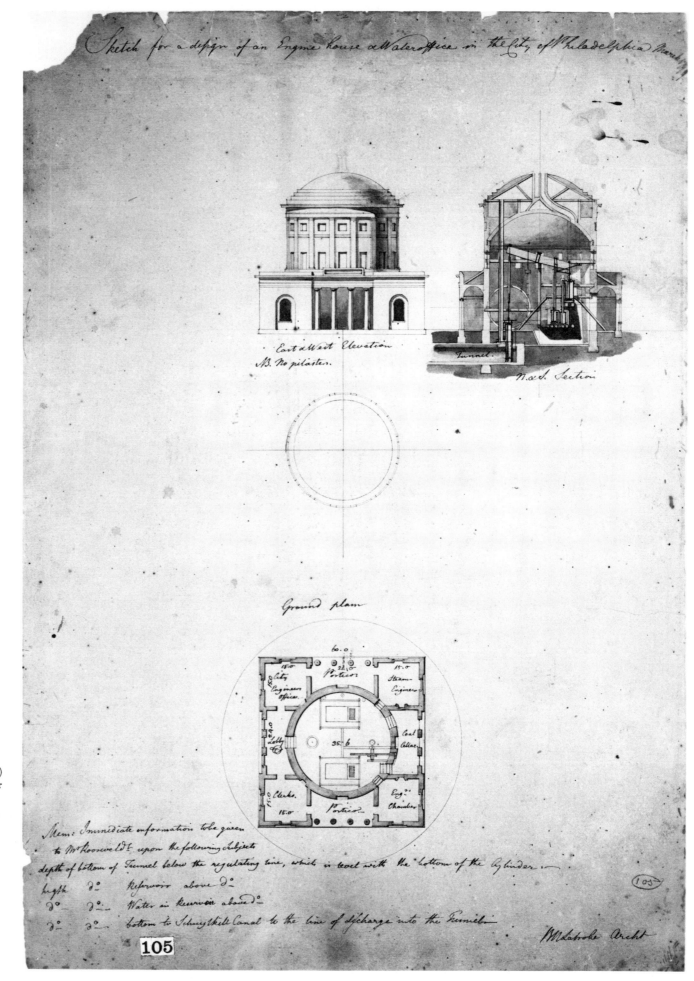

39 "Sketch for a design of an Engine house and Water office in the City of Philadelphia March 1799" (See also colorplate 8.)

size: 21⅜" × 15½" (54.2 cm × 38.1 cm)
media: pencil, pen and ink, watercolor
date: March 1799
location: Maryland Historical Society, Baltimore

PHILADELPHIA WATERWORKS

40 Plan of the foundation of the Centre Square Engine House

South is at the top and north at the bottom of the plan, contrary to the usual map orientation. Thus, the conduit bringing water from the Schuylkill Engine House enters Centre Square from the south, having turned 90° from its easterly course at the corner of Chestnut and Broad streets. Four other pipes pierce the foundation: two "Air Drains" to conduct air to the fireboxes of the boilers; the "Hot Water Drain" to carry away the surplus of the hot water cistern; and the "Principal Water Pipe" to connect the reservoirs to the distributing chest.

The heavy timbers necessary for the support of the steam engine lie within the circular foundation wall. In the upper half on the right side are two parallel timbers keyed into the wall which support the axle of the flywheel. In the lower half are six parallel timbers which support the cylinder (largest circle), and the condenser and air pump (smaller circles).

The notes penciled on the drawing were probably made much later by Frederick Graff, who was Latrobe's assistant during construction of the waterworks and from 1805 to 1847 the chief engineer of the waterworks. Within the circular foundation wall he outlined the wooden reservoirs located in the dome of the engine house, and the double-acting pump at the end of the brick tunnel. He placed the line of the three cylinders of the pump perpendicular to the axis of the conduit rather than parallel, as Latrobe and Charles Stoudinger did in other drawings. There is no clear evidence to indicate which placement was actually made.[1] Graff's marginal notes (inverted) show the pump with the air vessel and the valve in the rising main, which Latrobe had planned for Centre Square in 1801.[2]

1. Other drawings in the Graff Collection (Franklin Institute, Philadelphia, Pennsylvania) also show the pump cylinders perpendicular to the longitudinal axis of the brick conduit. All of the BHL and Stoudinger drawings show them parallel to the axis.
2. BHL, "First Report of Benjamin Henry Latrobe, to the American Philosophical Society, held at Philadelphia; in answer to the enquiry of the Society of Rotterdam, 'Whether any, and what improvements have been made in the construction of Steam-Engines in America?'" *Transactions of the American Philosophical Society* 6 (1809): 97–98, and plate 2, figure 9.

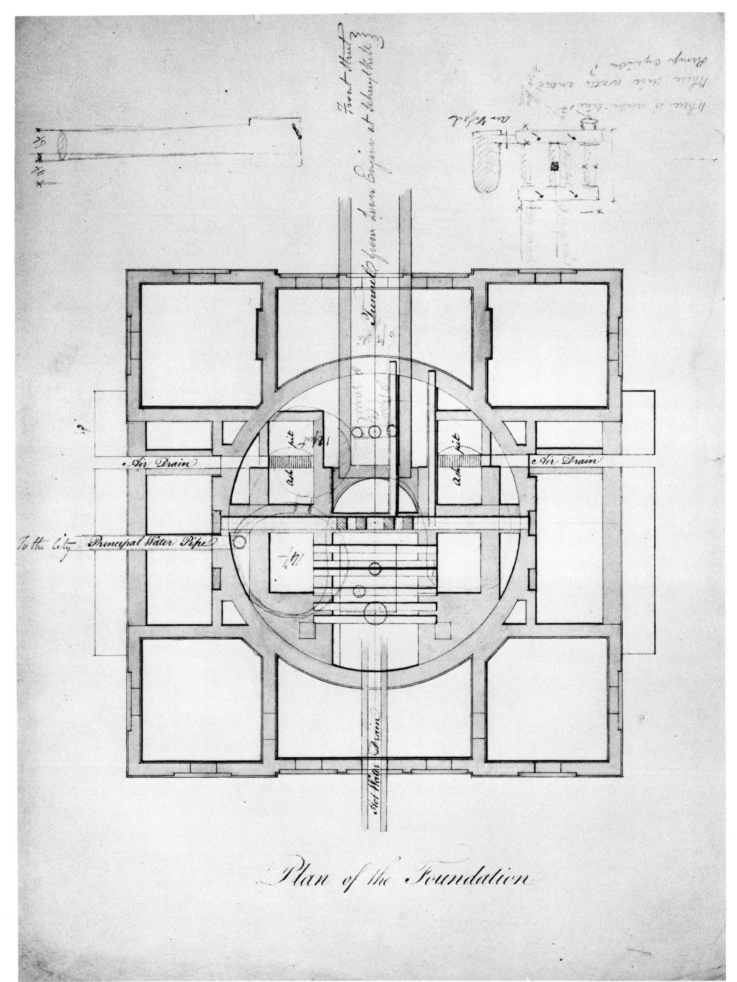

40 Plan of the foundation of the Centre Square Engine House

size: 11⅛" × 15" (28.2 cm × 38.2 cm)
media: pencil, pen and ink, watercolor
date: c. 1799
location: Frederick Graff Collection, Franklin Institute, Philadelphia, Pennsylvania

PHILADELPHIA WATERWORKS

41 "Second, or Center Engine house, No. II; Plan of the Ground Story"

In this plan of the ground story of the Centre Square Engine House Latrobe shows a number of details differently or more clearly than in his earlier sketch (no. 39) of March 1799. The pump now has the three cylinders of a double-acting pump (f and f^*). The 32-inch (81 cm) steam cylinder (e) rests on two heavy timber supports, and to its left are located the steam valves (directly to the left), condenser (bottom left), and air pump (to the left of the valves). The condenser and air pump are enclosed by the cold water cistern, which in its upper left corner adjoins the hot water cistern for receiving hot water (condensed steam) from the air pump. The hot water cistern extends to the left through the center of the engine house to the hot water pump (shown as a circle). Several other engine parts are referred to in the drawing's legend.

A notable feature of this plan is the provision for a second descending main (k) on the west side of the building to serve the area west of the Centre Square when the city had expanded in that direction.

Philadelphia Waterworks

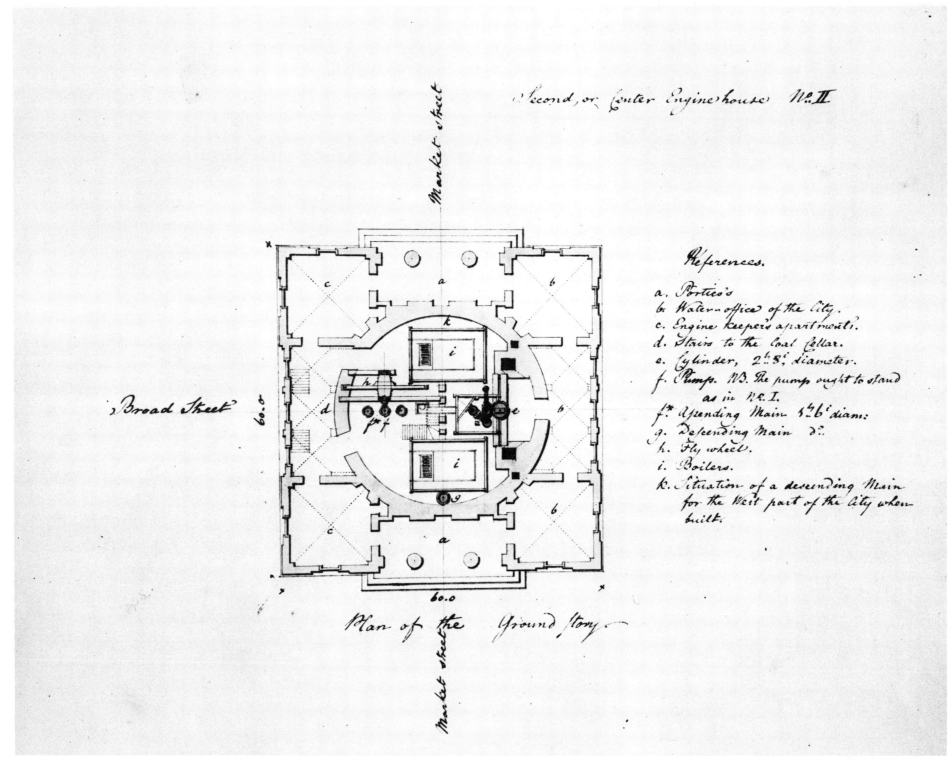

41 "Second, or Center Engine house, No. II; Plan of the Ground Story"

size: 13¼" × 11⅛" (33.7 cm × 28.3 cm)
media: pencil, pen and ink, watercolor
date: [1799]
location: The Historical Society of Pennsylvania, Philadelphia

PHILADELPHIA WATERWORKS

42 "Second, or Center Enge. He. No. V; Section from West to East, looking Northward"

The section of the Centre Square Engine House by Latrobe displays a number of features which are not as well shown elsewhere. The timber frame which dominates the inside of the building was a typical support for the working beam and the cylinder of a Boulton and Watt engine. The frame got its stability from horizontal timbers keyed into the walls of the building, and from two vertical and two inclined timbers, the former attached to the engine house floor with iron braces. At the top of the frame are the spring beams (seen in section) from which the working beam was hung, and the blockings. The working beam was probably regulated by pins extending from it horizontally so that they struck the blockings if the beam's stroke exceeded its normal amplitude. (The pins are not shown.)

This drawing also gives one of the few views of Latrobe's design for the reservoir in the dome of the Centre Square Engine House. In his sketch of March 1799 (no. 39) he drew two unequal sections of a reservoir which apparently encircled the inner rim of the dome. In this drawing the reservoir is still ring-shaped, but now the sections are equal. In both renderings the reservoir is stone; in this one it is definitely colored as marble. Latrobe's plan for the reservoir was not followed because the Watering Committee had reservations about the cost and leakiness of marble. A temporary reservoir installed in 1801 was made of timber, probably white pine plank. In 1807 or 1808 two larger reservoirs of cedar plank were installed to replace it.

Latrobe provided two pairs of openings (f, f, f, f) by which the water descending from the reservoirs was conveyed to the distributing chest. Only the east (right) openings were used; those on the west were for the possible future expansion of the city to that side of Centre Square.

REFERENCES

Farey, John. *A Treatise on the Steam Engine, Historical, Practical, and Descriptive.* London, 1827.
Poulson's American Daily Advertiser (Philadelphia). 30 October 1801.
Report of the Committee Appointed by the Common Council to Enquire into the State of the Water Works. Philadelphia, 1802.
Watering Committee. *Annual Reports.* Philadelphia, 1801, 1805, 1807.

Philadelphia Waterworks

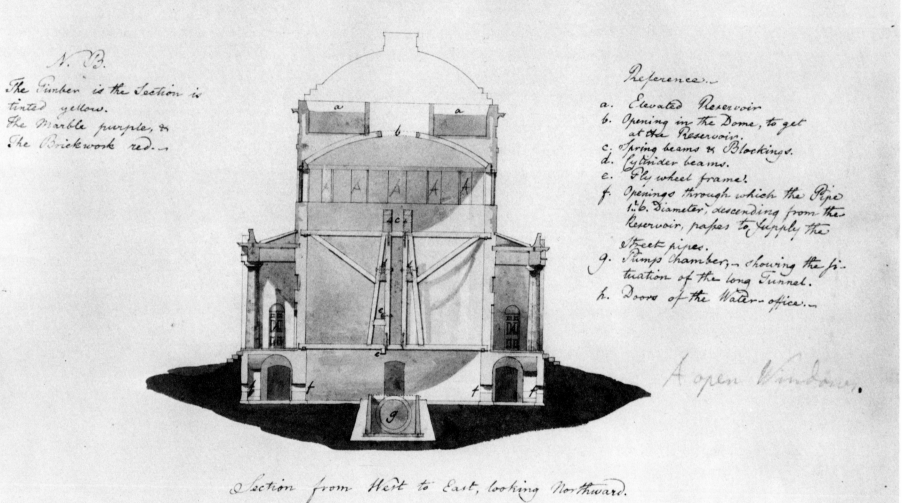

42 "Second, or Center Enge. He. No. V; Section from West to East, looking Northward"

size: 13¼" × 11⅛" (33.7 cm × 28.3 cm)
media: pencil, pen and ink, watercolor
date: [1799]
location: The Historical Society of Pennsylvania, Philadelphia

PHILADELPHIA WATERWORKS

Drawings of the Centre Square Engine

43 Plan of the Centre Square engine

44 Plans and sections of the wooden boiler

45 Section of the Centre Square engine

46 Section of the air pump and condenser

47 Section of the Centre Square engine

The most detailed drawings of the Centre Square engine and indeed of either engine of the waterworks are those which were probably drawn by Charles Stoudinger, draftsman of the Soho Works of northern New Jersey where the engines were built. The plan (no. 43) shows all the features in Latrobe's plan of the engine house (no. 41) but with a number of refinements. At the top of the plan are the two brick chimneys for the boilers, but without an indication of how they are to curve to the center for venting through the opening in the dome. In the upper middle of the plan is the 32-inch (81 cm) cylinder, resting on two heavy beams. The flanged top of the cylinder is clearly visible.

Directly below the cylinder are the valves; the superimposed circle is the entrance of the steam pipe from the boilers. To the left is the pipe taking steam to the bottom steam valve, the valve which admits steam into the cylinder on the piston's upward stroke. To the right of the valves is the exhaust pipe which connects the condenser to the exhaust valves at the top and bottom of the cylinder. The condenser is the larger circle directly beneath the exhaust pipe, and it is connected to the air pump which draws both air and condensed steam from the condenser. Both the condenser and air pump have flanged tops with boltholes for attaching the covers through which their piston rods go. Both the condenser and air pump are immersed in the cold water cistern which is shown as a box with the lower left corner reserved for part of the hot water cistern.

The hot water cistern has roughly the shape of a capital J with its crossing part being the corner of the cold water cistern just mentioned, its stem running toward the bottom of the plan, and the hot water pump located in the projection to the left at its bottom end. The hot water pump is indicated by a circle.

On either side of the two cisterns are the boilers, drawn here as wooden ones. The top of the left boiler is shown: it has a manhole in the center, and a fitting for the steam pipe in the upper right corner. The wooden casing, fire grate, and outline of the flue are shown for the right boiler. Originally the Centre Square engine was provided with only one wooden boiler, and that burned out in July 1801 after a few months of use. Undoubtedly the water level fell too low and the flue overheated and disintegrated. That boiler was replaced immediately with another wooden one which was, according to Latrobe, "by far the best" of those used in the Philadelphia Waterworks, whether wooden or cast iron.

Latrobe included two plans and two sections of that boiler (no. 44) in his report to the American Philosophical Society on steam engines in America. The main feature of the new boiler was the use of nine upright, cylindrical heaters through which water circulated to be raised to the steam point. In most boilers the heat from the fire was transmitted to the water by a circular or oval flue leading from the firebox to the chimney, but in this case there was a wide, flattened flue penetrated by the heaters leading from the firebox.

This new wooden boiler remained in use during the entire life of the Centre Square steam engine (until 1815), and after more than ten years' service waterworks superintendent Frederick Graff stated that "it far exceeds any boiler now in use at the water works . . . both in saving fuel and the ease with which the steam can be kept up."[1] In 1803 Latrobe reported that while the steam engine was operating at 12 strokes a minute for 16 hours a day, the boiler used "from 25 to 33 bushels of Virginia coals of the best sort," which was "a great saving" over other boilers.[2] Graff and Latrobe agreed that although the wooden casing saved fuel because it was a poor conductor of heat, it was also very difficult to keep steam-tight. There were leaks at boltholes and joints in the planking which were difficult to close, and the boiler had to be replanked occasionally. A cast-iron boiler was installed at Centre Square in 1804/05 to provide relief for the wooden boiler.

The remaining items in the plan of the Centre Square engine (no. 43) are the flywheel and the pump. On the left at the bottom of the plan is the wooden flywheel with its cast-iron rim. The flywheel was about 16 feet (4.9 meters) in diameter, and had a thick wooden hub hooped with iron. Bearing blocks bolted to heavy timbers held the axle on each side. Although

1. Frederick Graff, "Steam Engine Boiler," *Emporium of the Arts and Sciences*, new series, 3 (1814):275.
2. BHL, "First Report of Benjamin Henry Latrobe to the American Philosophical Society, held at Philadelphia; in answer to the enquiry of the Society of Rotterdam, 'Whether any, and what improvements have been made in the construction of Steam-Engines in America?'" *Transactions of the American Philosophical Society* 6 (1809):93–94.

the flywheel for the Schuylkill engine had the dual purpose of regulating the pumping and the motion of the rolling mill, the flywheel for the Centre Square engine had no function other than regulating the pumping. To the right of the flywheel are three circles which indicate the location of the double-acting pump, which was sunk into the oval pump well.

The two sections of the Centre Square engine are remarkably full of information. The upper right corner of the north-to-south section (no. 45) has a view of the parallel motion which connects the end of the beam to the piston rod of the cylinder and the piston rod of the air pump. The parallel motion permitted the rods of both pistons to work vertically rather than be pulled from side to side by the end of the working beam, which moved in an arc.

The cylinder and piston are shown in section, with the piston at the downward end of its stroke. The stuffing box at the top of the piston was usually filled with oiled hemp to provide a steam- and air-tight seal around the piston rod. The shallow groove in the rim of the piston was likewise filled with hemp to provide for a tight fit along the walls of the cylinder.

At the top and bottom of the cylinder are the openings which lead to the steam and exhaust valves. At the top of the valves is a section of the steam pipe which comes from the boiler. Directly below that is the steam valve, and below that the exhaust valve, for the top of the cylinder. The steam valve has a pipe running from it to the bottom steam valve, while the exhaust valve has a pipe connecting with the bottom exhaust valve and the condenser. The latter pipe has a portion removed to allow the bottom steam and exhaust valves to be visible. The valves were operated by the working gears tripped by the pump rod of the air pump; the working gears are not shown.

Directly below the valves and to the bottom left of the cylinder is the condenser, which has no moving parts. It and the air pump to its left are immersed in the cold water cistern, which was made of wood. This drawing of the air pump is one of the few representations of the double-acting air pump introduced by the Soho engineers, although Latrobe included a drawing of it (no. 46) in his 1803 report to the American Philosophical Society on steam engines in America. Describing it he said

> I will first remark, that by the air-pump of Bolton and Watt, the condenser is only once emptied, of its water of condensation and of the air produced, in every stroke. The superiority of our air-pump consists in its evacuating the condenser twice at every stroke, thereby creating a much better vacuum, and of course adding considerably to the power of our engine in proportion to the diameter of its cylinder without encreasing friction.[3]

In the section (no. 45) water from the air pump empties into the wooden hot water cistern to the left of the cold water cistern. At the left end of the hot water cistern is the hot water pump, which has its piston rod attached to the beam on the left of the axle. The hot water pump is rendered as a long, thin pipe with a small reservoir at the top. The reservoir has a circle drawn on it indicating the pipes which lead from the reservoir to each boiler.

The working beam at the top of the drawing is connected to the piston rod of the cylinder at the right end by the parallel motion, and to the crank rod of the flywheel and the pump rod at the left end. It rotates on a short axle resting in bearing blocks on the spring beams. The beam also operated the working gears, the air pump, and the hot water pump.

This drawing (no. 45) shows that the flywheel is turned by a crank attached to the connecting rod on the left end of the beam. The standard Boulton and Watt engines of the time used the sun-and-planet gearing—a Watt patent—to operate the flywheel, because the crank method had already been patented. The crank was just as effective as the sun-and-planet, however, and had fewer parts. There is no reason why Smallman or Latrobe should have preferred one type of linkage to the other, and both types were well known to steam engineers.

Both the crank arm and the flywheel were wood and iron composites. The arm was iron where it was attached to the beam and at the crank on the arm's lower end, but the rest was wood. The flywheel axle and arms were wood, but the rim was iron cast in four segments. In this drawing the rim of the flywheel appears to descend into the pump well, but in other drawings in this series the flywheel pit is shown to be separate from the well.

The east-to-west section of the Centre Square engine (no. 47) shows a few details not visible in other drawings. On the left there is a section of one of the boilers, with a deep ash pit below the grate in the firebox. The firebox is wide and rectangular and the flue to the upper left is also rectangular, indicating that the first Centre Square boiler differed in those respects from the second (see no. 44). In the upper left corner of the boiler are the water gauges, which penetrate to two different depths corresponding to the safe high and low water marks in the boiler. If the water level was above their bottom ends, steam pressure on the water kept the gauges full of water; if the water level

3. Ibid., p. 96.

was lower, only steam was emitted. To use the gauges, one opened the cocks at the top of the gauges; the water level was correct if the left gauge held water and the right gauge gave off steam.

To the upper right of the boiler are the hot water reservoir and its automatic feeding balance. A pipe leads from the reservoir of the hot water pump to the hot water reservoir of the boiler, and from the boiler reservoir a pipe leads down into the boiler. The pipe has a plug in it attached to a balance which has on one end a counterweight and on the other end a weight (probably a stone) which "floats" in the boiler water. A contemporary account describes the action of the feeding balance this way:

> ... when a heavy body is suspended in a fluid, it will lose as much of its own weight as the weight of that quantity of the fluid which it displaces. When the water in the boiler diminishes by the conversion of a part of it into steam, the upper surface of the stone will remain above the water, and consequently a less proportion of its weight will be supported by the fluid, because it ceases to bear up that part of the stone that is not immersed. By the action of this additional portion of its weight, the stone will overcome the balance-weight at the end of the lever ... so as to elevate that weight, and open the valve at the top of the pipe [i.e., the bottom of the hot water reservoir] ... to admit water into the boiler in sufficient quantity to replace that which is carried off by evaporation.[4]

At the top of the drawing is a cross section of the beam and the spring beams which run parallel to it, and to which the axle of the beam is attached. The linkage on the beam is attached to the air pump rod, which has iron portions but is wood in the center.

Below the beam and behind the air pump rod are the cylinder valves. At the top the steam pipe from the right boiler is fitted to the upper steam valve, but the pipe from the left boiler is not shown. To the left of the valves a pipe goes from the upper steam valve to the bottom steam valve to allow passage of steam for the up stroke. To the right an exhaust pipe connects the upper and lower exhaust valves to the condenser. The cylinder is just visible above and behind the exhaust pipe. At the bottom of the valves and the cylinder is the cold water cistern which contains the condenser and the air pump.

On the right side of the drawing is an exterior view of the right boiler. There are two openings to the firebox: one is shown with its door and the other without. Below the right boiler there is a large ash pit just like that for the left boiler.

The Centre Square engine was started on 27 January 1801, the date which is usually taken as the beginning of operation of the Philadelphia Waterworks. In 1803 Latrobe reported that the engine pumped water from the conduit to the reservoir 51 feet above in the dome, operating at the rate of 12 strokes per minute for 16 hours a day. But it had problems similar to those of the Schuylkill engine—leaky boilers, decaying wooden parts, and breakdowns—and was taken out of service on 7 September 1815, when the new steam pumping station at Fairmount was ready.

REFERENCES

Manuscript

Haverford, Pennsylvania. Haverford College Library. Quaker Collection. Thomas P. Cope Diaries, 1800–1, Acc. 975.

Printed

Farey, John. *A Treatise on the Steam Engine, Historical, Practical, and Descriptive*. London, 1827.
Graff, Frederick. "Steam Engine Boiler." *Emporium of the Arts and Sciences*. New series, 3 (1814): 275–76; with plate.
Latrobe, Benjamin Henry. "First Report of Benjamin Henry Latrobe, to the American Philosophical Society, held at Philadelphia; in answer to the enquiry of the Society of Rotterdam, 'Whether any, and what improvements have been made in the construction of Steam-Engines in America?'" *Transactions of the American Philosophical Society* 6 (1809): 89–98.
Pursell, Carroll W., Jr. *Early Stationary Steam Engines in America*. Washington: Smithsonian Institution Press, 1969.
Report of the Committee Appointed by the Common Council to Enquire into the State of the Water Works. Philadelphia, 1802.
Watering Committee. *Annual Reports*. Philadelphia, 1803–7.

---

4. John Farey, *A Treatise on the Steam Engine, Historical, Practical, and Descriptive* (London, 1827), pp. 453–54.

Philadelphia Waterworks

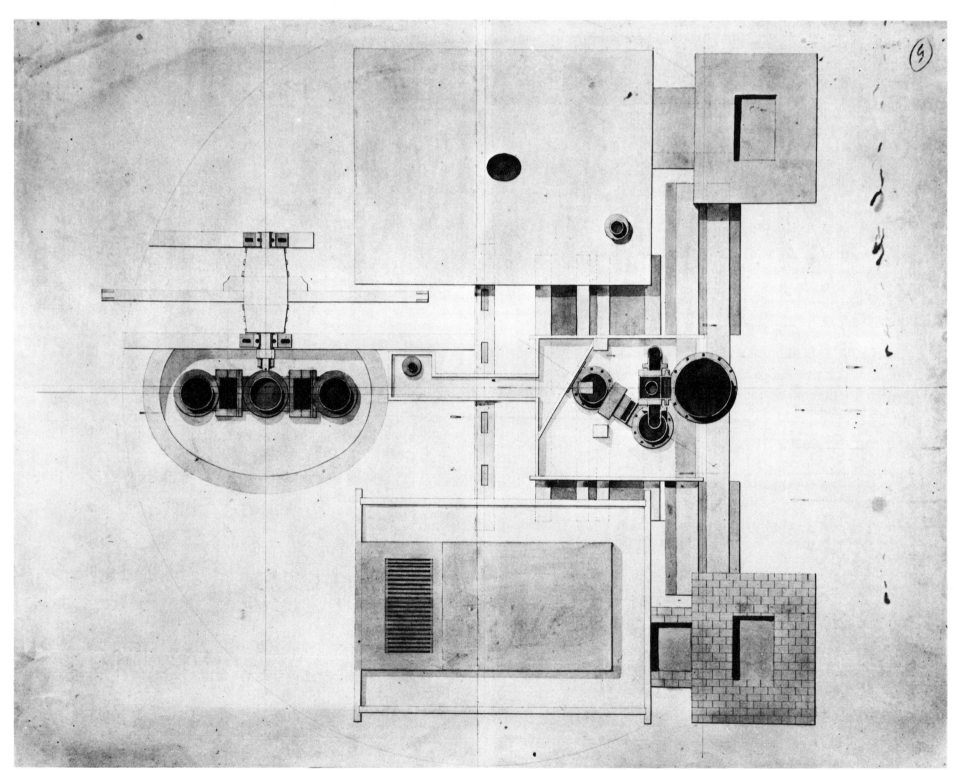

43 Plan of the Centre Square engine

size: 28⅛″ × 34⅜″ (71.4 cm × 87.4 cm)
media: pencil, pen and ink, watercolor
date: c. 1799
location: The New Jersey Historical
   Society, Newark

The Centre Square Engine

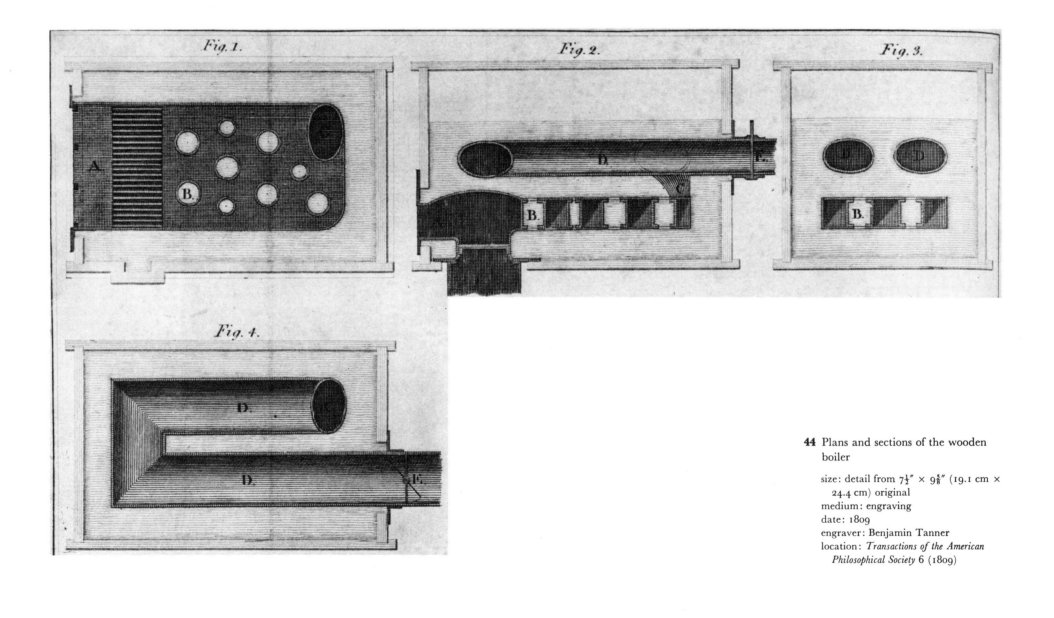

44 Plans and sections of the wooden boiler

size: detail from 7½″ × 9⅝″ (19.1 cm × 24.4 cm) original
medium: engraving
date: 1809
engraver: Benjamin Tanner
location: *Transactions of the American Philosophical Society* 6 (1809)

Philadelphia Waterworks

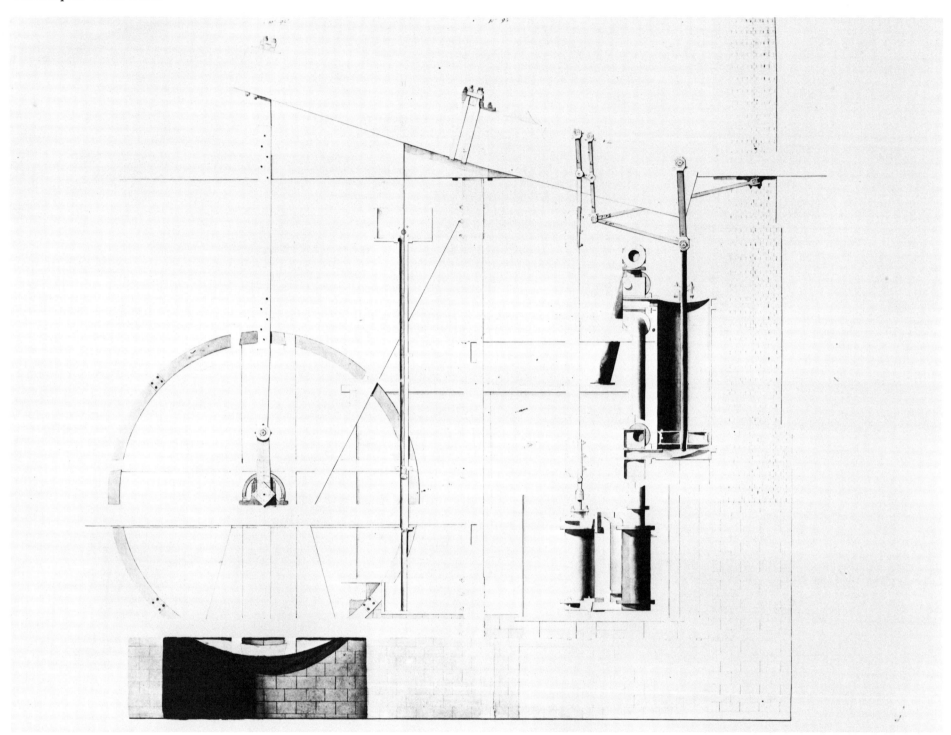

45 Section of the Centre Square engine

size: 28⅝″ × 38⅝″ (72.6 cm × 98.3 cm)
media: pencil, pen and ink, watercolor
date: c. 1799
location: The New Jersey Historical
 Society, Newark

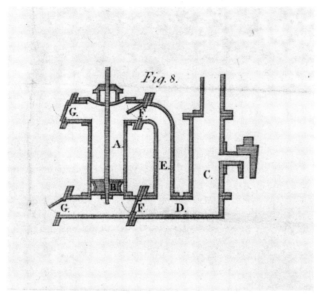

46 Section of the air pump and condenser

size: detail from 7½″ × 9⅝″ (19.1 cm × 24.4 cm) original
medium: engraving
date: 1809
engraver: Benjamin Tanner
location: *Transactions of the American Philosophical Society* 6 (1809)

Philadelphia Waterworks

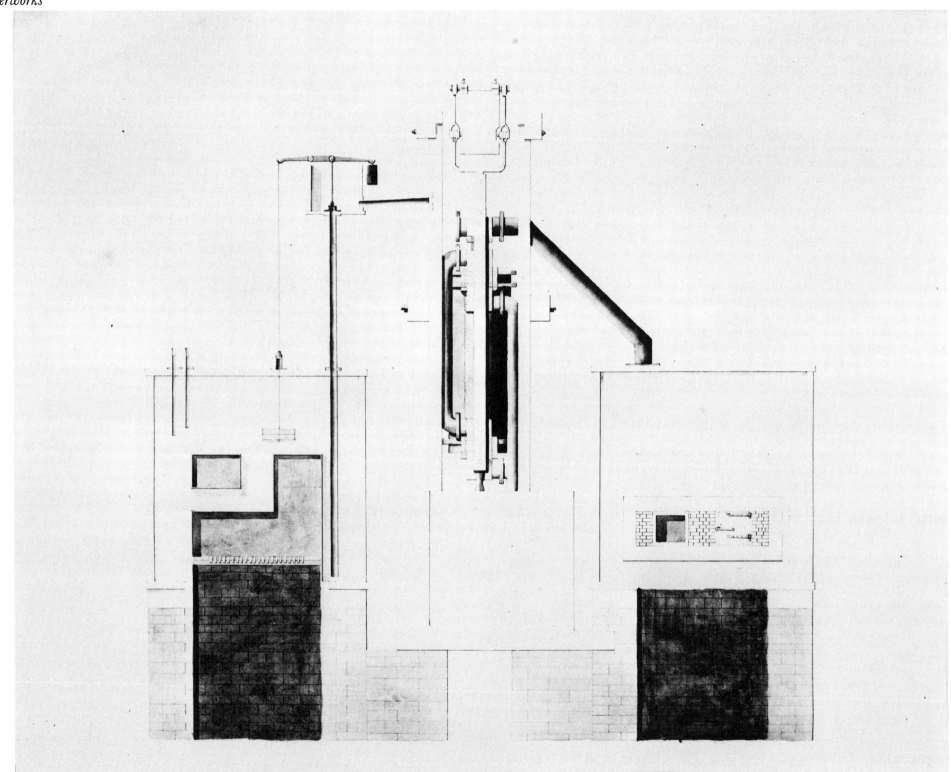

47 Section of the Centre Square engine

size: 27 5/16" × 35 5/16" (69.3 cm × 89.7 cm)
media: pencil, pen and ink, watercolor
date: c. 1799
location: The New Jersey Historical Society, Newark

PHILADELPHIA WATERWORKS

Drawings of the Flywheel and Pump at Centre Square

48 East-to-west section of the flywheel and pump

49 North-to-south section of the flywheel and pump

50 Section of the pump

The drawings (nos. 48 and 49 probably by Charles Stoudinger) of the pump and flywheel at Centre Square are excellent renderings which leave little to the imagination. The first section (no. 48) shows the mounting of the flywheel on the supporting timbers and bearing blocks, and the flywheel pit. The crank for the flywheel is attached to the right end of the flywheel axle. To the right of the crank rod is the 18-inch (45.7 cm) pipe which ascended 51 feet (15.6 meters) from the pump to the reservoir. In the outline behind the pipe are (from the top) the pump rod, the guide for the pump piston rod, the pump piston rod, and the piston.

All of these elements are revealed in the second section (no. 49). To the rear of the apparatus is the flywheel and on the left side is the pipe leading to the reservoir. In the center at the top is the pump rod which is worked by the beam. Attached to the pump rod is the piston rod guide which joins the pump rod and piston rod with a swivel joint. The guide worked up and down between two vertical posts so that the side to side motion of the pump rod was changed to a true vertical movement. Therefore the piston rod worked without any sideways pressure and its stuffing box retained a good seal.

The pump (there is a plan of it at the bottom of no. 49) is double-acting, meaning that it pumps water on both its upward and downward strokes. Latrobe's published diagram of the pump (no. 50) shows its operation somewhat better than Stoudinger's drawing. Included in the diagram is a conical valve (E) in the ascending pipe which was never installed, but which Latrobe believed necessary. It would have relieved the strain on the engine caused by the drop of the water column at the instant the pump piston reached the end of each stroke. Also included on the left side is an air vessel (H) which Latrobe believed would also relieve strain by keeping a constant air pressure on the water column. An air vessel was attached to the Centre Square pump in 1810.

Philadelphia Waterworks

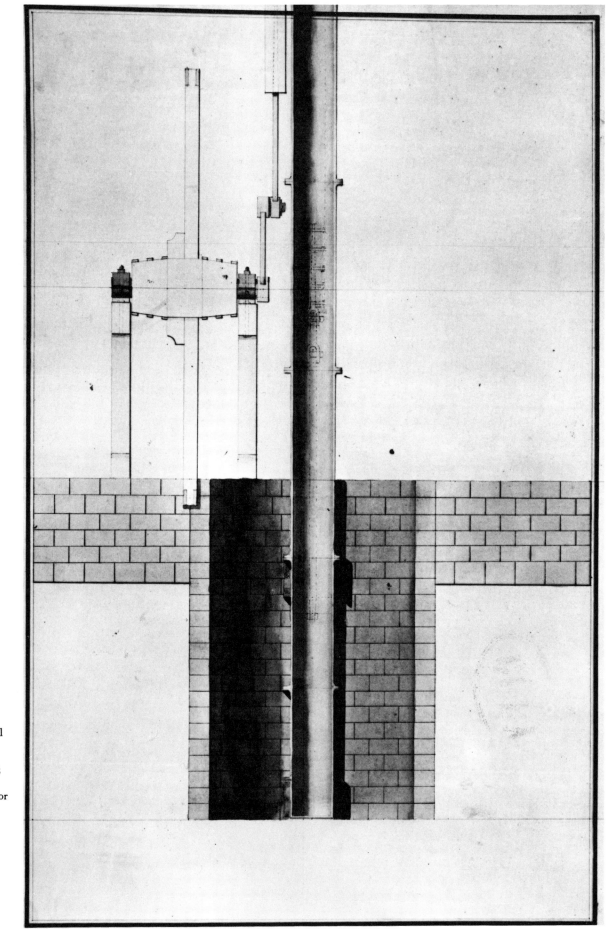

48 East-to-west section of the flywheel and pump

size: 27 3/16" × 18 3/16" (69.1 cm × 46.3 cm)
media: pencil, pen and ink, watercolor
date: c. 1799
location: The New Jersey Historical Society, Newark

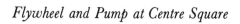

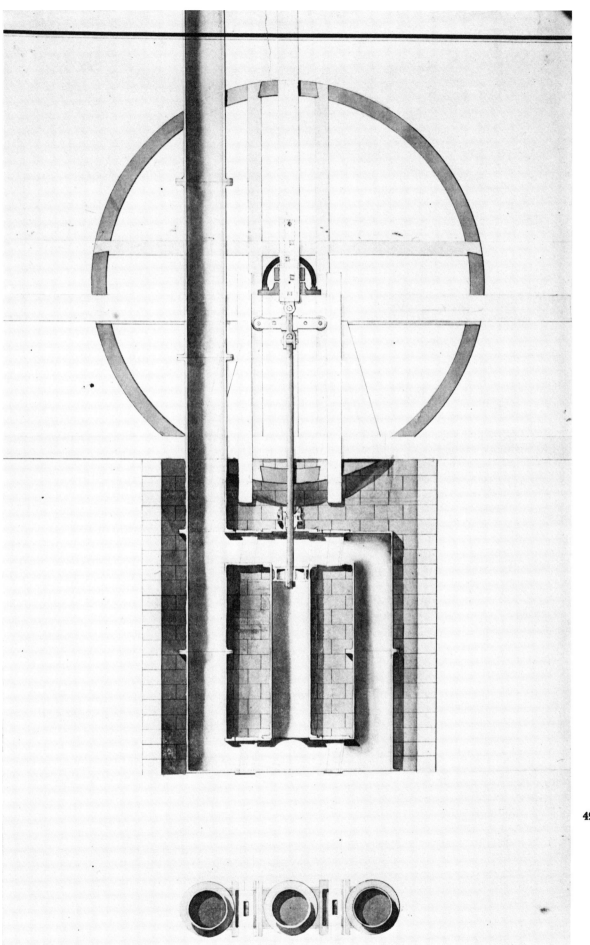

49 North-to-south section of the flywheel and pump

size: 27⅞″ × 19⅞″ (70.8 cm × 50.8 cm)
media: pencil, pen and ink, watercolor
date: c. 1799
location: The New Jersey Historical Society, Newark

Philadelphia Waterworks

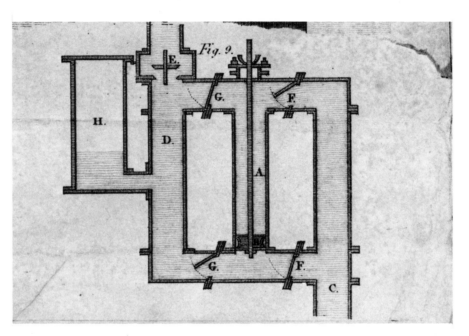

50 Section of the pump

size: detail from 7½″ × 9⅝″ (19.1 cm × 24.4 cm) original
medium: engraving
date: 1809
engraver: Benjamin Tanner
location: *Transactions of the American Philosophical Society* 6 (1809)

PHILADELPHIA WATERWORKS

Plans and Section of Centre Square Reservoirs

51 Plan of three reservoirs

52 Plan of the decking; section of a reservoir

In his earliest plans for the waterworks Latrobe provided for an elevated reservoir inside the Centre Square Engine House to give sufficient pressure for delivery of water through the pipes to the customers. Latrobe wanted to have a ring-shaped marble reservoir encircling the inside of the dome, but the Watering Committee decided to erect wooden reservoirs instead. They first put up a small tank originally meant to be a supply cistern for the Schuylkill engine boilers, and they did not replace it until 1807 or 1808 when two reservoirs of cedar plank were installed.

Frederick Graff, superintendent of the waterworks, designed the reservoirs. The plan (no. 51, with north to the left, and south to the right) shows three reservoirs within the dome, although the large oval one was not erected.[1] The pipe for water coming up from the pump ("ascending main") enters the smallest reservoir, and the pipe leading outside to the distributing chest connects to the middle-sized reservoir. The marginal "Bill for cedar plank for Reservoirs" is apparently an estimate made before the Watering Committee decided to erect only two reservoirs. Cedar was used because of its resistance to decay.

The plan of the decking in the second drawing (no. 52) has the same orientation as the first drawing. The decking is keyed into the boiler chimneys on the left and the inside wall of the dome on the right. To the right in the drawing is a section of the largest reservoir resting on the decking. In the upper left is a note about the dimensions of the Centre Square engine's wooden boiler (see no. 44).

---

1. "Upon a more minute examination of the inside of the Centre Square Engine house, it has been resolved to erect but two reservoirs, instead of three, the contracted space of the building will not admit a greater number, without intercepting the light, which is necessary in so complicated a machinery...." *Report of the Watering Committee to the Select and Common Councils, November 13th, 1807* (Philadelphia, 1807), p. 3.

REFERENCES

Latrobe, Benjamin Henry. *View of the Practicability and Means of Supplying the City of Philadelphia with Wholesome Water.* Philadelphia, 1799.

Poulson's American Daily Advertiser (Philadelphia). 30 October 1801.

Report of the Committee Appointed by the Common Council to Enquire into the State of the Water Works. Philadelphia, 1802.

Watering Committee. *Annual Reports.* Philadelphia, 1801, 1803, 1805, 1807, 1808.

Philadelphia Waterworks

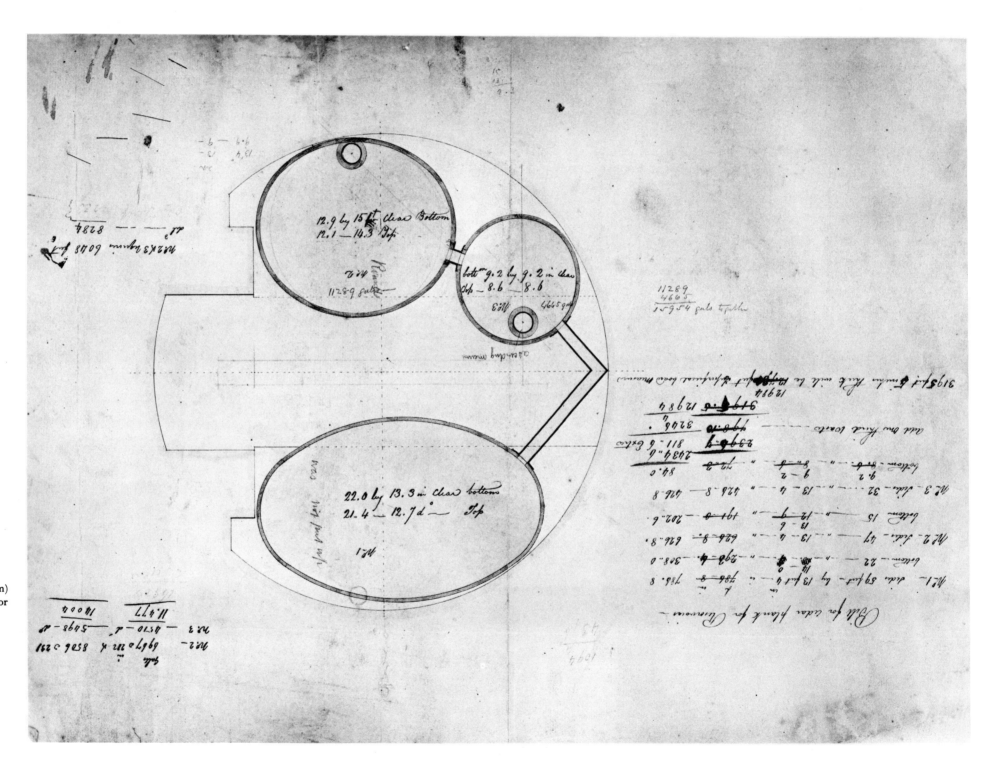

51 Plan of three reservoirs

size: 19⅛″ × 13½″ (48.5 cm × 34.3 cm)
media: pencil, pen and ink, watercolor
date: c. 1805–8
delineator: Frederick Graff
location: Frederick Graff Collection, Franklin Institute, Philadelphia, Pennsylvania

Centre Square Reservoirs

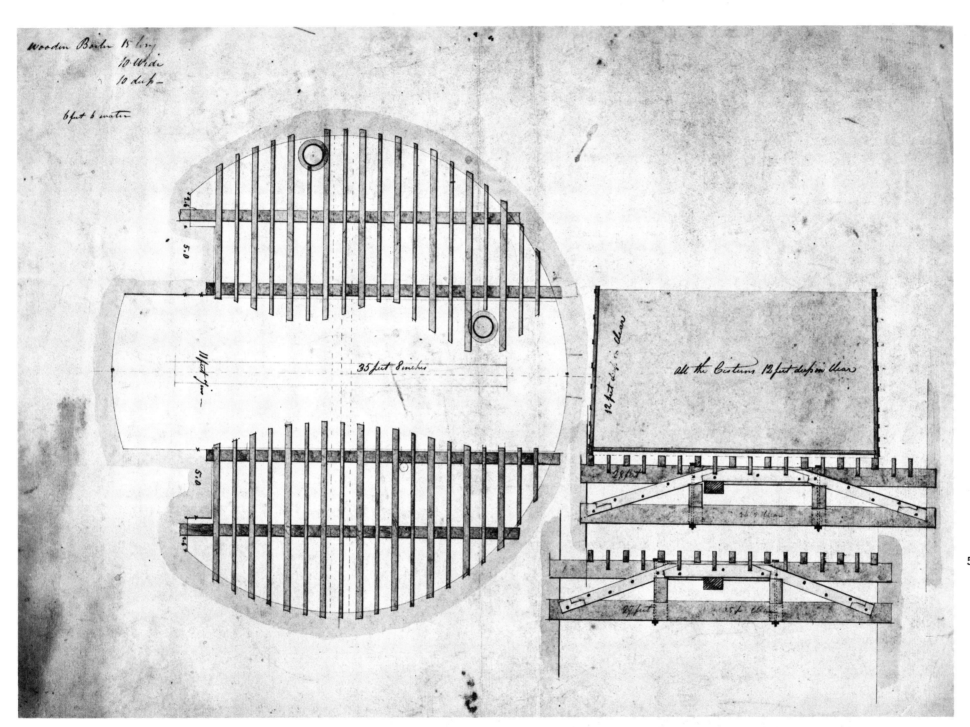

52 Plan of the decking; section of a reservoir

size: 19¼" × 13½" (49 cm × 34.3 cm)
media: pencil, pen and ink, watercolor
date: c. 1805–8
delineator: Frederick Graff
location: Frederick Graff Collection, Franklin Institute, Philadelphia, Pennsylvania

PHILADELPHIA WATERWORKS

Distribution Equipment of the Waterworks

53 Combined fireplug and hydrant, fireplug, and stopcock

54 Hydrant

From the reservoir in the dome of the Centre Square Engine House the Schuylkill water was conducted by an 18-inch (45.7 cm) pipe to a cast-iron distributing chest located underground on the east side of the building. The chest, which contained about 250 gallons (946 liters) of water, had several holes in its sides to which were attached water mains leading to the major streets of the city: Broad, Market, Arch, Race, and Chestnut. The chest had some spare holes to which future mains could be fixed. No drawing of the distributing chest exists, but it may have been similar to the receiver and distributor used by the London Bridge Waterworks in 1780.[1]

The pipes which carried the water through the streets were bored logs, at first white oak, but later yellow pine. Latrobe advertised in February 1799 for white oak logs "straight and free from shakes and knots" to be used for pipe, and he purchased many logs from Delaware raftsmen that spring. Local men were hired to bore the logs at their mills, as well as to put iron hoops on the ends of the logs to give joints extra strength. The first pipe was laid on 18 June 1799, and two and a half years later, in December 1801, almost 30,000 feet of pipe were in place. The main pipes were bored to 6 inches diameter, and branches were of $4\frac{1}{2}$, $3\frac{1}{2}$, and 3 inches.

By using logs for pipe the Philadelphia Waterworks followed the usual practice of London waterworks, which used wooden pipe by the seventeenth century and continued using it into the nineteenth. The Philadelphia Waterworks had the same problems with this kind of pipe as London waterworks did, including burst and rotted pipes and the impossibility of finding logs of sufficient diameter to carry large volumes of water. In Philadelphia as in London it was necessary to lay several pipes side by side (a "range") to deliver water to high-use areas.

Some iron pipe was tested by the Watering Committee in 1801, but it did not form an important segment of the waterworks distribution until 1820, when it was used to replace the six wooden pipes leading from the new reservoir at Fairmount to the distributing chest at Centre Square. By the 1830s it was clear that iron pipe was superior, and no more wooden pipe was laid.

At first most of the water was delivered to the citzens from public hydrants located at intervals along the water mains. Latrobe described them as "upright pipes rising three feet above the pavement, in each of them are 2 cocks, one of $\frac{3}{4}$ of an inch bore, for common use, the other of $2\frac{1}{2}$ bore to be used only to water the streets, and in cases of fire."[2] By December 1801 there were forty-two hydrants on the city's streets. The original combined hydrant and fireplug (top of no. 53, misdated 1803) was faulty, however, because it was easy to open the cocks and allow the water to run wastefully. In 1803 a new design was introduced which required anyone wanting water to pump it from a small cistern attached to the distributing pipes (no. 54), and in the same year cast-iron fireplugs were introduced for fire engines and street cleaning (lower right of no. 53). Thereafter it was more difficult to waste the "free" water.

Another appliance used to prevent waste was the stop or stopcock (left center of no. 54). Inserted into pipelines at regular intervals, stopcocks made it possible to cut off the flow of water to a leaky or broken pipe. Hydrants, fireplugs, stopcocks, and other devices were made as early as 1804 by smiths working at forges at Centre Square and the Schuylkill Engine House.

Citizens could, of course, have water piped directly into their residences and business establishments, and the subscribers to the water loans raised to finance the waterworks were given that privilege free. Other customers were charged an annual rate according to the expected amount of usage: the minimum was $5, but one bath house paid $50 for its water in 1811. Although there were few paying customers at first (only thirty-four near the end of 1801), each year larger numbers applied to the Watering Committee for permission to draw from the pipes.

These drawings illustrating the distribution appliances appear to provide reliable information. Frederic Graff (1817–90), the son of Latrobe's pupil Frederick Graff, had his father's drawings in his possession. He published several historical articles on the Philadelphia Waterworks and used his father's drawings as the basis for illustrations such as no. 53. Henry Latrobe's sketch (no. 54) was founded on personal observation, and may have been drawn from memory. (See description of no. 55, below.)

---

1. H. W. Dickinson, *Water Supply of Greater London* (London: Newcomen Society, 1954), pp. 29–30, fig. 13.

2. BHL, "The Waterworks," unidentified Philadelphia newspaper, 15 March 1801.

REFERENCES

Manuscript

Newark, New Jersey. New Jersey Historical Society. "Account of diverse articles belonging to the engines for the Corporation of Philadelphia," 1799–1800.

Haverford, Pennsylvania. Haverford College Library. Quaker Collection. Thomas P. Cope Diaries, 1800–1, Acc. 975.

Printed

Dickinson, H. W. *Water Supply of Greater London*. London: Newcomen Society, 1954.

Gazette (Philadelphia). 11 February 1799, 20 July 1799.

Report of the Committee Appointed by the Common Council to Enquire into the State of the Water Works. Philadelphia, 1802.

Singer, Charles et al. *A History of Technology*. 5 vols. Oxford: Clarendon Press, 1954–58. Vol. 4.

Watering Committee. *Annual Reports*. Philadelphia, 1799, 1803–7.

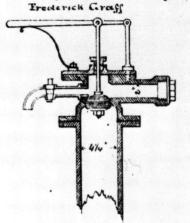

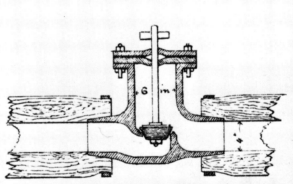

53 Combined fireplug and hydrant, fireplug, and stopcock

size: detail from 6⅜″ × 7⅞″ (16.2 cm × 20 cm) original
medium: photo-zincograph
date: 1876
delineator: Frederic Graff [Jr.]
location: *Journal of the Franklin Institute* 3d series, 72 (1876)

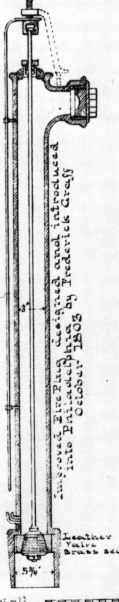

Philadelphia Waterworks

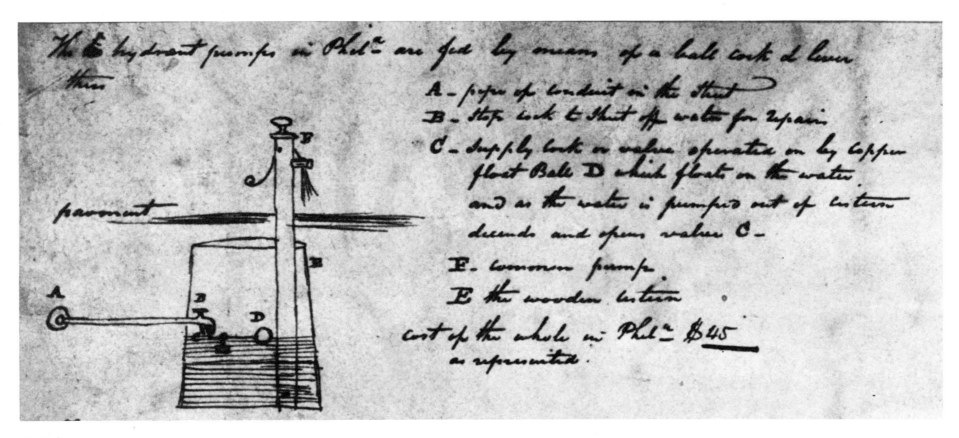

54 Hydrant
  size: detail from 14⅛″ × 20½″ (35.9 cm × 52.1 cm) original (no. 55)
  media: pencil, pen and ink, watercolor
  date: c. 1811
  delineator: Henry Sellon Boneval Latrobe
  location: Department of Streets, City of New Orleans, Louisiana

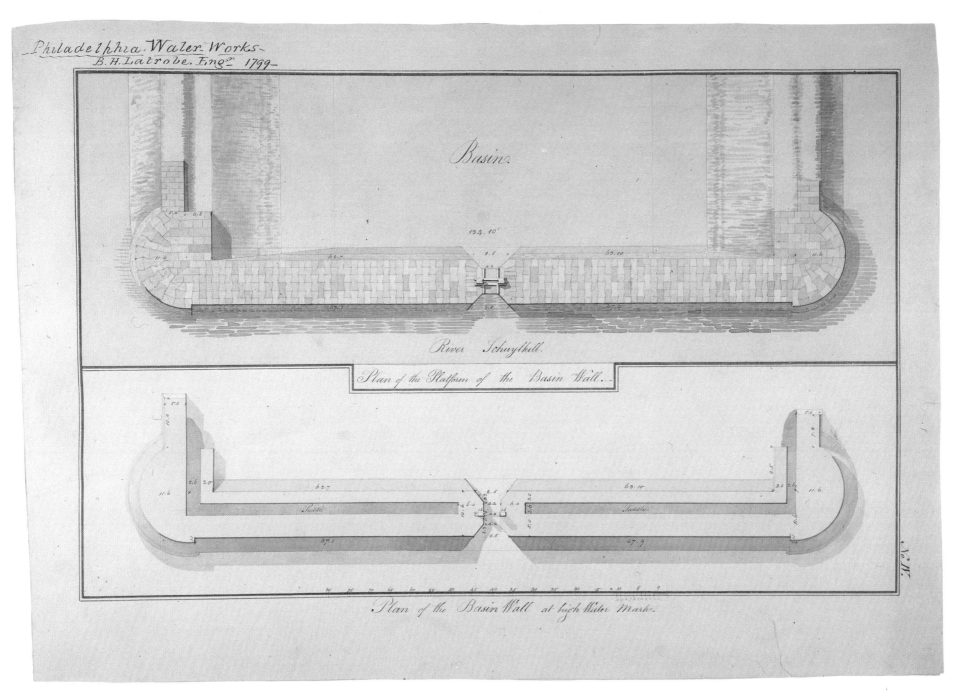

Plate 5. Philadelphia Waterworks—
"Section of the Basin; Plan of the Basin
Wall & of the Coffre-Dam," c. 1800.

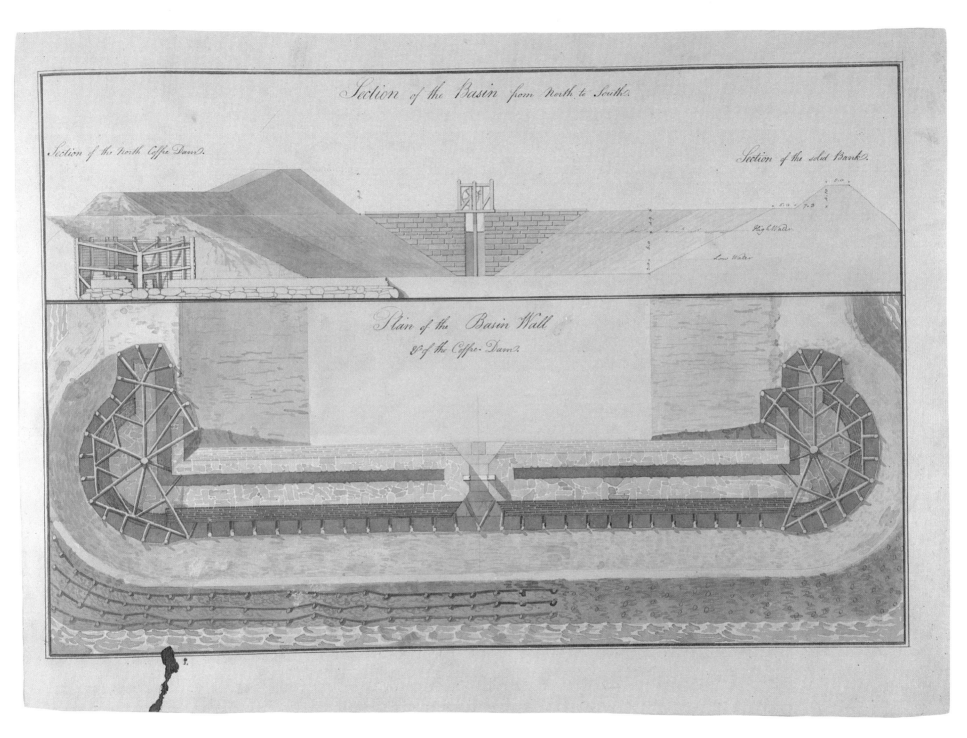

Plate 6. Philadelphia Waterworks—
"Plan of the Platform of the Basin Wall;
Plan of the Basin Wall at high Water
Mark," c. 1800.

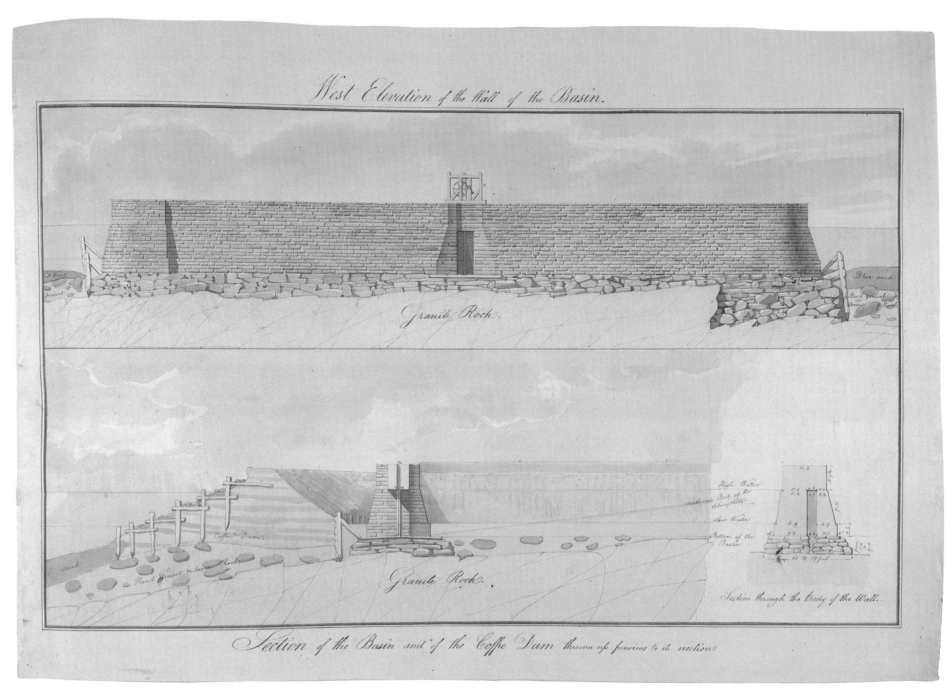

Plate 7. Philadelphia Waterworks—"West Elevation of the Wall of the Basin; Section of the Basin and of the Coffre Dam," c. 1800.

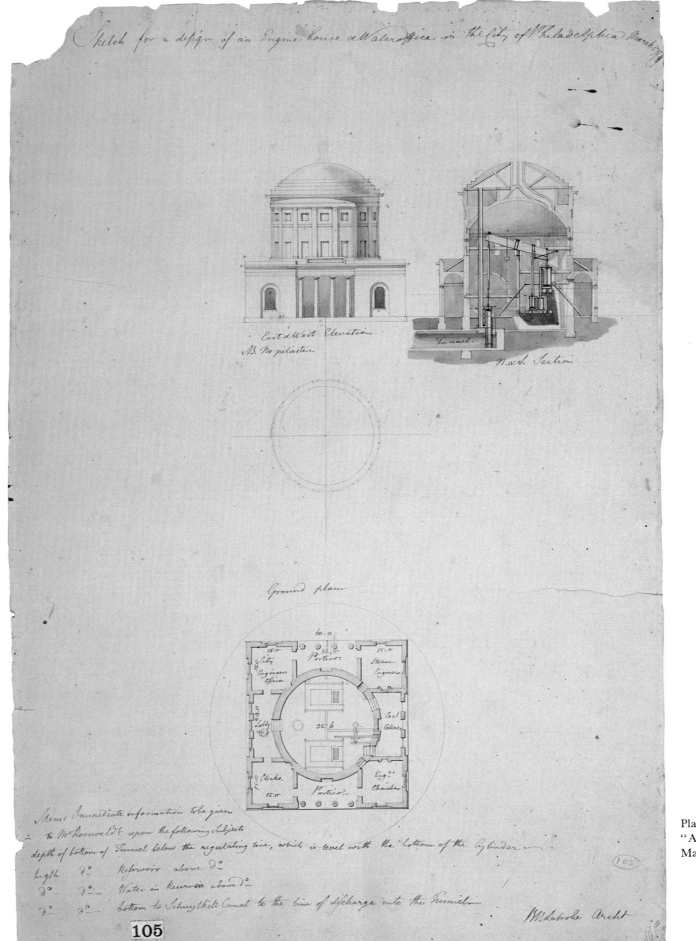

Plate 8. Philadelphia Waterworks—
"An Engine house and Water office,"
March 1799.

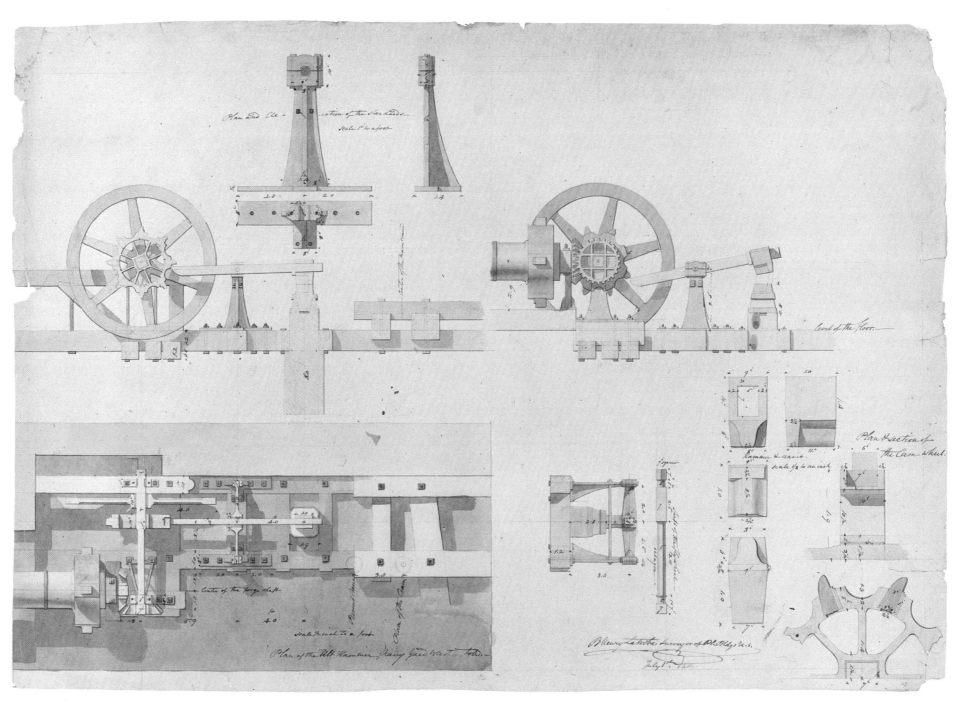

Plate 9. Washington Navy Yard Steam Engine—Forge hammer and gearing, 1 July 1811.

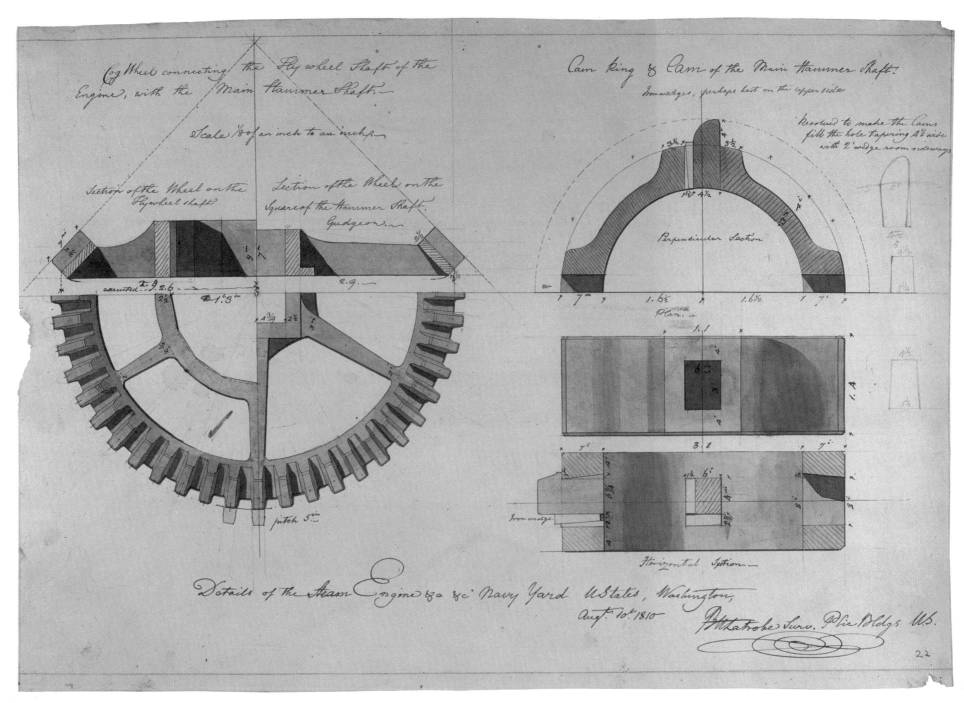

Plate 10. Washington Navy Yard Steam Engine—Cogwheels and cam ring, 10 August 1810.

NEW ORLEANS WATERWORKS

55 Map of New Orleans Waterworks with marginal notes and sketches

This map of the New Orleans Waterworks with marginal notes was probably drawn and annotated by Henry Latrobe during his visit to New Orleans in 1811. On the left-hand side of the drawing is a note stating that neither the size of the engine nor the height of the reservoirs was known, a situation existing only before the waterworks charter was granted in 1811: immediately afterward Benjamin Henry Latrobe placed preliminary orders for a steam engine and began to design an engine house. This drawing is the most detailed technical document relating to any stage of planning or execution of the New Orleans Waterworks, but because of its early date it leaves many questions unanswered.

The middle of the drawing contains a map showing the location of the "Engine" on the lower right at the river bank. Above the word "Engine" is a shaded polygon indicating the location of the steam engine, and to its left are six circles representing the reservoirs. The upper middle reservoir has a connection to a ten-inch iron pipe laid beneath the middle of the river-front street. Departing from that pipe is a network of wooden and iron pipes of differing diameters also laid in the middle of their respective streets. At many street corners there are dots representing hydrants and fireplugs. A fountain graces the central square.

Almost all of the marginal notes are concerned with the storage and distribution of the water once it is pumped up from the river. In the upper middle of the drawing is an elevation of the wooden reservoirs, with the pipe from the pump coming in at the lower right and the outlet to the pipes in the streets on the left. The building consists of a masonry platform supported by four transverse walls and a superstructure of timber sides and a roof. The reservoirs are "large timber casks" connected by pipes near the floor. The notes indicate that each reservoir would have a cutoff valve so that it could be repaired independently of the others, and that each should have a float "so as to give information to the Engineer whether the Reservoirs are full or near empty."

On the lower left are notes on the distributing pipes. The top sketch shows a four-way cast-iron connector for joining wooden pipes at street intersections. The bottom two sketches are of funnel mouth pipes: the upper, to join together 10-inch and 6-inch pipes in the street; the lower, to connect the pipe coming from the reservoir with the first pipe in the street. The sketch, however, labels the smaller end of the second funnel pipe as 12 inches (30.5 cm), while the map indicates that the first pipe is 10 inches (25 cm) in diameter.

In the upper left is a sketch of the public pumps used in Philadelphia.[1] The major elements of the pumps were a filler pipe connected to the street pipe, a wooden tub sunk below street level, and an upright pump for bringing water up from the tub. The key mechanism was a valve at the end of the filler pipe which opened as a copper float in the tub fell, thus keeping the tub constantly full. Clearly Henry intended to copy that apparatus for the New Orleans system. His notes in this part of the drawing also mention fireplugs, which if similar to those in Philadelphia would have been cast iron with screw fittings for hoses; and stopcocks, which were valves placed in the pipes at intervals to allow the isolation of leaks for repair.[2]

It seems likely that Henry made this drawing in order to give more detailed information to the New Orleans City Council or to local investors during the early months of the waterworks' promotion. Although his father had not yet made extensive plans, it was simple to draw upon the Philadelphia Waterworks for examples of successful equipment which could be easily understood, and which would undoubtedly be equally useful in New Orleans.

1. *Report of the Watering Committee to the Select and Common Councils, November 1, 1803* (Philadelphia, 1803), p. 4.
2. Ibid.; *Report of the Committee Appointed by the Common Council to Enquire into the State of the Water Works* (Philadelphia, 1802), p. 48. See drawing no. 53, above.

New Orleans Waterworks

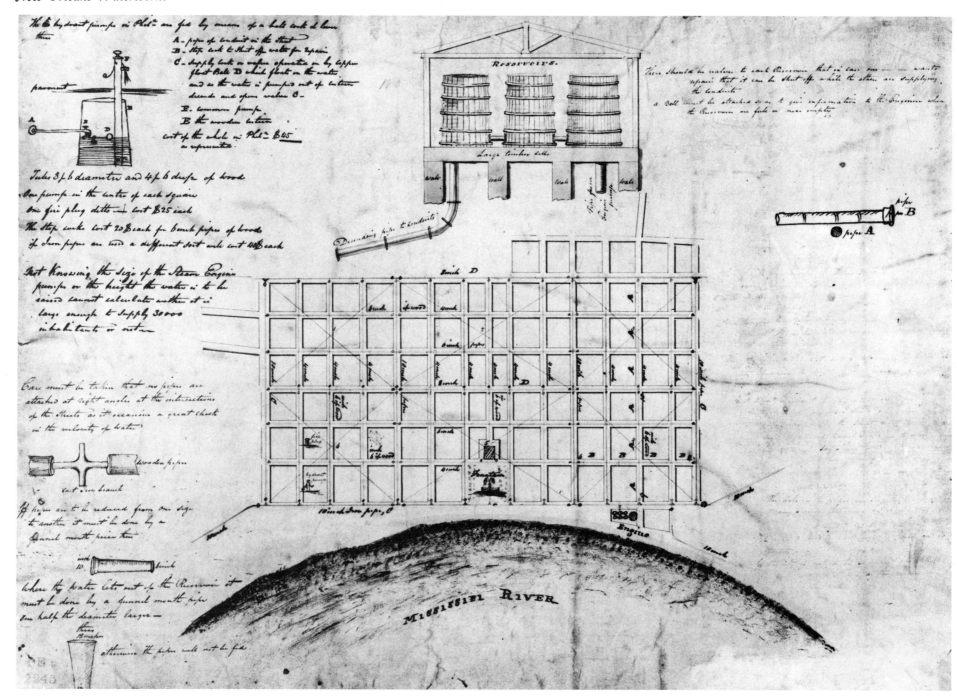

55 Map of New Orleans Waterworks
with marginal notes and sketches

size: 14⅛" × 20½" (35.9 cm × 52.1 cm)
media: pencil, pen and ink, watercolor
date: c. 1811
delineator: Henry Sellon Boneval
 Latrobe
location: Department of Streets, City
 of New Orleans, Louisiana

NEW ORLEANS WATERWORKS

56 "Details of a New mode of arranging the Side pipes & working Gear of a Steam Engine"

From about 1808 until near the end of the Pittsburgh steamboat episode in 1814, Latrobe frequently corresponded with Robert Fulton. Steam engines were often a topic of his letters. While he was setting up his machine shop in Washington in August 1811 to manufacture an engine for the New Orleans Waterworks he wrote to Fulton that ". . . I will send you drawings in detail of my works, and you will find perhaps something on which you may improve in the mode of my working gear and nozzles."[1] This drawing, which Latrobe dated only "1812," was probably a belated attempt to fulfill that promise, and certainly derived from his current work on the New Orleans engine. He may actually have manufactured an apparatus based on the plan in this drawing, since in the spring of 1812 he reported to his son Henry that "my working gear will answer admirably,—it is made and put together for trial."[2]

The drawing depicts Latrobe's scheme for operating the valves of a steam cylinder by inclined planes rather than by the ordinary levers. Figures 1 and 2 in the lower center of the drawing show the upper and lower steam valves, or "nozzles." Figure 1 is a plan of the lower valves, with the bonnets seated in them, and the steam and condensing pipes appearing as round openings at the top and bottom of the valves. To the left of the valves is an outline of the position of the cylinder. Figure 2 is a section of the upper valves, also from an overhead perspective. At the upper end of figure 2 Latrobe sketched a section of one of the bonnets of the upper valves (figure 3). Bonnets acted as guides and seals for the stems which operated the valves.

1. BHL to Fulton, 12 August 1811, Letterbooks.
2. BHL to Henry Latrobe, 17 May 1812, Letterbooks.

Figure 7, the broken sketch in the lower right corner of the drawing, shows the valves in section (appearing plano-convex) as well as the steam pipes. In a marginal note Latrobe indicated the location of the openings which would allow placement of the cores during the casting of the pipes.

Figures 5 (elevation) and 6 (section) are the most complete views of the working gear. The elevation depicts the valves and pipes as if the cylinder were behind them. The section indicates the connection of the working gear with the valve stems. The vertical rods pivot from side to side, turning the inclined planes (a, a). Thus the toothed racks (b, b) on the valve stems are raised or lowered, opening or closing the valves. Latrobe noted that the rods have stop rings to prevent excessive up and down motion.

Figures 8 and 9 are an elaboration of the means of giving the rods their side to side motion. On each rod there is a cam projecting horizontally in the direction of a nob attached to a horizontal slide. The slides are moved by a shaft located to the right in both figures; Latrobe noted that the shaft is "to be worked by any convenient libratory [i.e., side to side] motion of the Engine." The handles (figs. 6 and 8) would have been used for the manual operation of the engine, especially during "blowing thro'," or starting up.

This drawing is important because it shows that Latrobe had a detailed interest in steam engines, and was fascinated by innovations in them. The change of the working gear which he proposed did not have any significant result, but it may be regarded as an artifact of a technical discourse between two men intensely interested in steam engine development.

REFERENCES

Pursell, Carroll W., Jr. *Early Stationary Steam Engines in America*. Washington: Smithsonian Institution Press, 1969.

New Orleans Waterworks

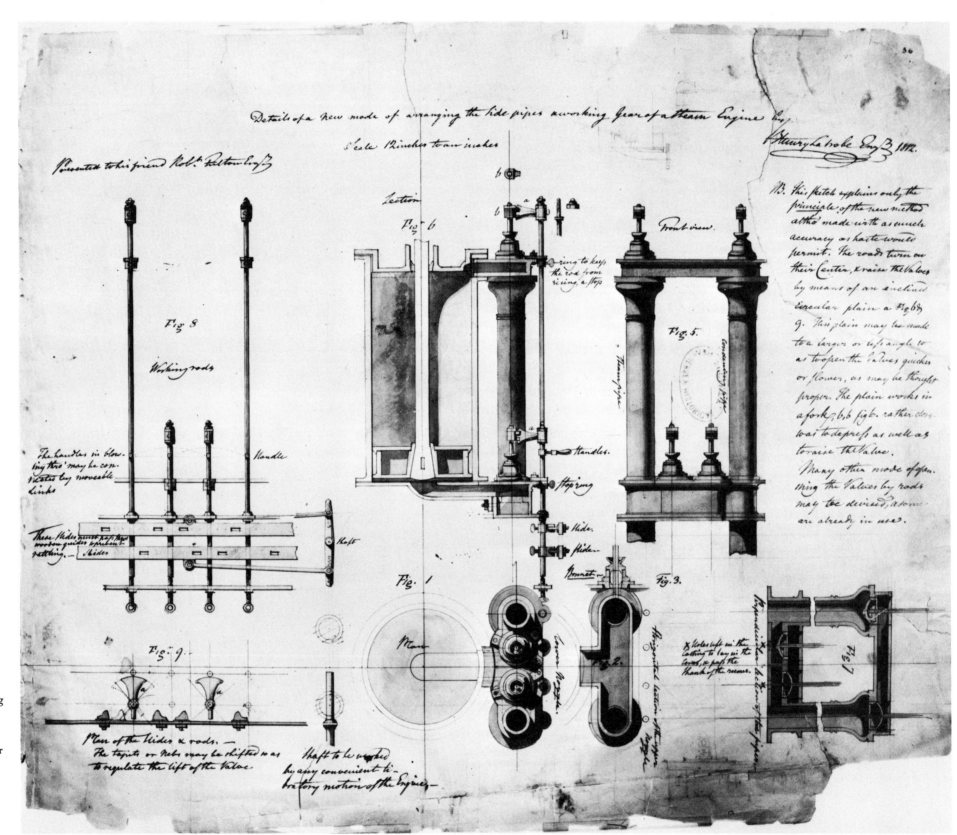

56 "Details of a New mode of arranging the Side pipes & working Gear of a Steam Engine"

size: 18¾" × 23" (47.6 cm × 58.4 cm)
media: pencil, pen and ink, watercolor
date: 1812
location: The New Jersey Historical Society, Newark

NEW ORLEANS WATERWORKS

57 "Design of a pier to cover the Suction pipe of the Pump for supplying Water to the city of New Orleans"

This is Latrobe's design for a pier to cover and protect the suction pipe carrying water from the Mississippi River to the New Orleans Waterworks engine house. In the lower right-hand corner of the drawing is a map with the location of the engine house, the suction pipe, and some of the pipes of distribution. At the right on the river bank is the pier (*Eperon*), with the suction pipe indicated by a pair of dotted lines. The pipe runs back to the engine house, shown as a rectangular building with a superimposed octagon. In the building is the engine, which draws water from the river, and behind it, indicated by a circle, is an elevated reservoir from which a pair of distribution pipes go into the city. On the lower left side of the engine house is a long rectangular structure housing the pipe-boring mill (*Moulin aux tuyaux*).

The other three parts of the drawing are views of the pier. In the lower left is a plan which indicates that the pier was tapered so that its base became wider from left to right. In the upper left of the drawing is a side elevation showing the pier's timber construction, and in the upper right is a bisected elevation of the end of the pier. The latter view indicates the width of the structure at the top and bottom (20 and 26 feet [6.1 and 7.9 meters], respectively) and shows the screen at the bottom, behind which the suction pipe opens. Latrobe also indicated the reach of high water (*Ligne des eaux hautes*) and low water (*Ligne des eaux basses*) so that it is clear that the suction pipe would always be under water.

From an early date in his planning for the New Orleans Waterworks Latrobe believed that the only means of obtaining a supply of water from the Mississippi was by drawing it with a suction pipe; but he also felt that it would be one of the most difficult aspects of construction. He anticipated that keeping the pipe airtight would be the major problem, but he also discovered that having cast-iron pipe made of the diameters he desired (16 and 18 inches [41 cm and 46 cm]) was nearly impossible. For some time he discussed with Henry Latrobe, his son, the possibility of making the suction pipe of wood, despite anticipated problems with air-tightness, but in 1816 he finally settled for eight-inch (20 cm) cast-iron pipe.

This drawing, made during the early weeks of Latrobe's first visit to New Orleans, indicates his rapid planning for completion of the waterworks.

REFERENCES

Manuscript

Baltimore, Maryland. Maryland Historical Society. Letterbooks of Benjamin Henry Latrobe. BHL to Fulton, 31 July 1811; BHL to Hughes, 18 August 1811, 21 December 1811; BHL to Henry Latrobe, 24 April 1812, 18 November 1812, 17 July 1814; BHL to McQueen, [12 July 1816]; BHL to Mark, 12 July 1816.
New Orleans, Louisiana. Tulane University Library. Benjamin Henry Latrobe Papers. Coulter to Raffignac, 14 December 1827.

New Orleans Waterworks

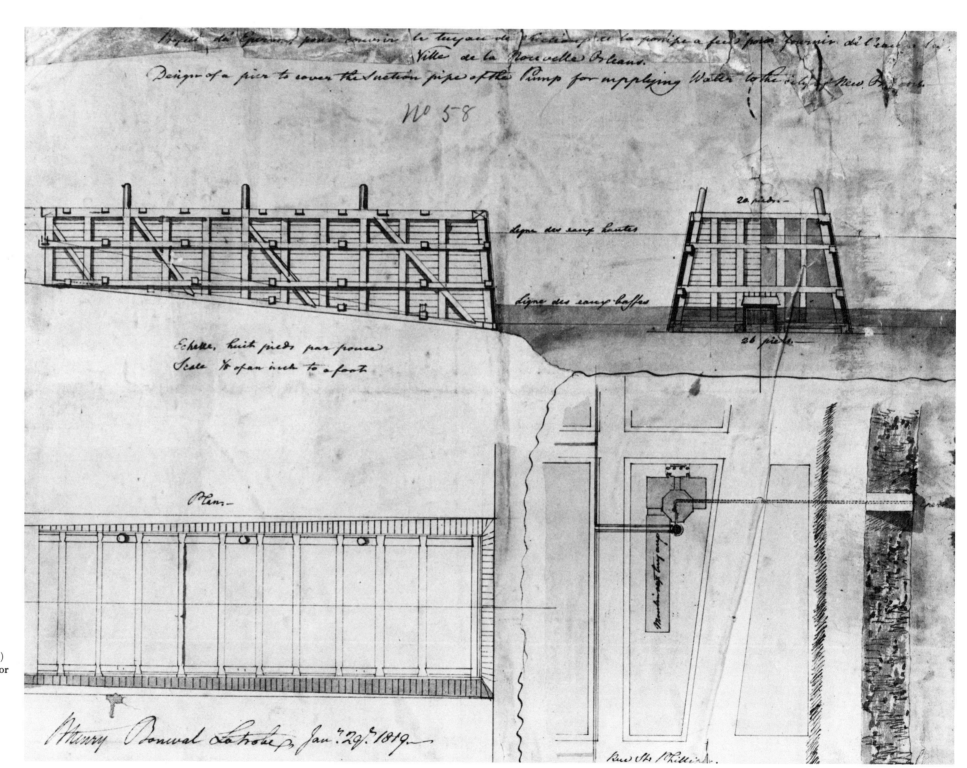

57 "Design of a pier to cover the Suction pipe of the Pump for supplying Water to the city of New Orleans"

size: 13½" × 18" (34.3 cm × 45.7 cm)
media: pencil, pen and ink, watercolor
date: 29 January 1819
location: Louisiana Division, New Orleans Public Library

MUD ISLAND BAR

58 Map of the Delaware River below Philadelphia

59 "Scheme Submitted to the Chamber of Commerce, June 6th, 1807, for blocking up the Passage between Hog & Maiden Islands in the Delaware, at the least expense"

On 25 May and 6 June 1807, Latrobe submitted reports to the Philadelphia Chamber of Commerce on shoaling in the Delaware River channel and accompanied them with these drawings. The Chamber of Commerce was concerned about the bar south of the city near Mud (or Fort) Island (no. 58) because it was hindering navigation, and asked Latrobe's opinion on how to prevent its increase or reduce it. He reported that shoaling in the rivers of the Atlantic coast, including the Delaware, resulted from the expanding cultivation of the hinterland and the subsequent carrying off of soil from the new fields by rain. He thought it unlikely that the accumulation of sediment would cease in the foreseeable future, and he noted that the islands in the vicinity of the bar (Hog, Maiden, and Mud islands) were the product of sediment brought into the Delaware by the Schuylkill River. Latrobe concluded: "... all that Art can effect, is to regulate the manner and place of its [the sediment's] Deposit; and the most powerful, and only competent instrument for this purpose, is the force of the water itself."[1]

Latrobe believed that the bar north of Mud Island could be eliminated by diverting the ebb tide over it and washing it away. He suggested doing it by blocking up the passage between Maiden and Hog islands with a series of wooden caissons 25 feet long and 15 feet wide having pointed pilings to hold them to the river bottom (no. 59). Between the caissons he planned for frames of five piles each, with the frames and caissons held together at the top by parallel timbers running across them. The caissons were "to be filled with Brush and Mud or Sand."

In the middle of the drawing Latrobe indicated how the caissons would connect the two islands, and measured the distance between a pair of caissons at 80 feet (24.4 meters). On the lower left there is a side view of a caisson showing pointed piles on the bottom. In the lower center is a plan of a caisson, connecting timbers, and a frame. In the lower right corner of the drawing is a side view of an "Intermediate Frame." Structures of this sort were common in European river improvements. Bernard Belidor showed wooden caissons filled with earth and stone being used for many purposes in his eighteenth-century work, *Architecture Hydraulique*, and they were already in use in America for bridge piers and docks.

Latrobe's plan was not acted on by the Chamber of Commerce, but it is an example of his careful consideration of the causes of engineering problems and his attempts to find effective long-term solutions.

REFERENCES

Manuscript

Baltimore, Maryland. Maryland Historical Society. Letterbooks of Benjamin Henry Latrobe. BHL to Fitzsimmons, 25 May 1807, 6 June 1807, 19 November 1807.

Printed

Belidor, Bernard. *Architecture Hydraulique*. 4 vols. Paris, 1737–53.

---

1. BHL to Fitzsimmons, 25 May 1807, Letterbooks.

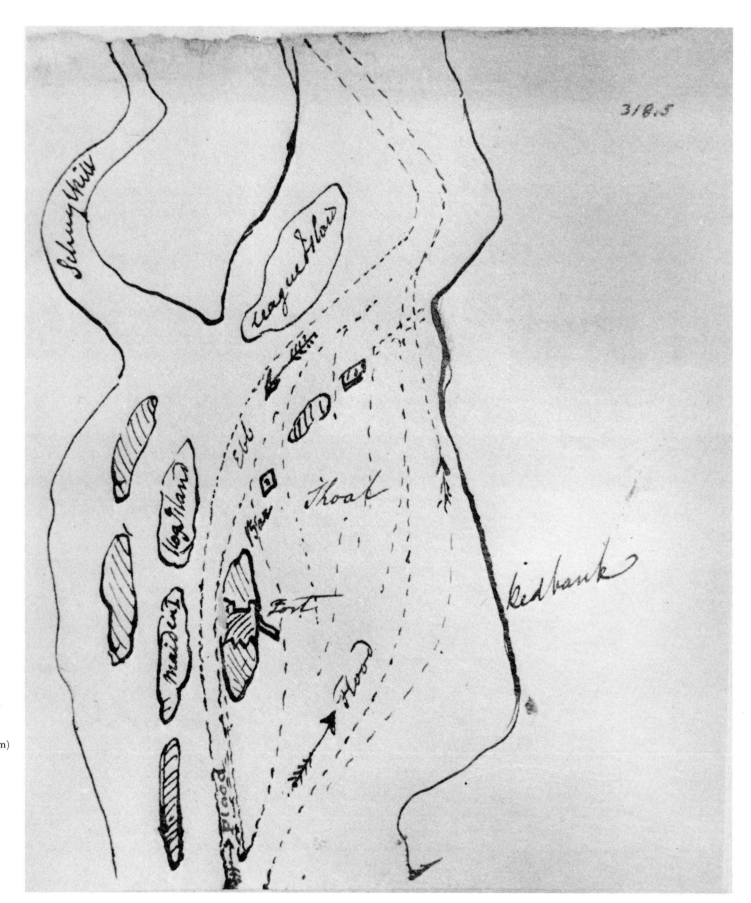

58 Map of the Delaware River below Philadelphia

size: 4 15/16″ × 5 15/16″ (12.5 cm × 15.1 cm)
medium: pen and ink
date: 25 May 1807
location: Letterbooks of Benjamin Henry Latrobe

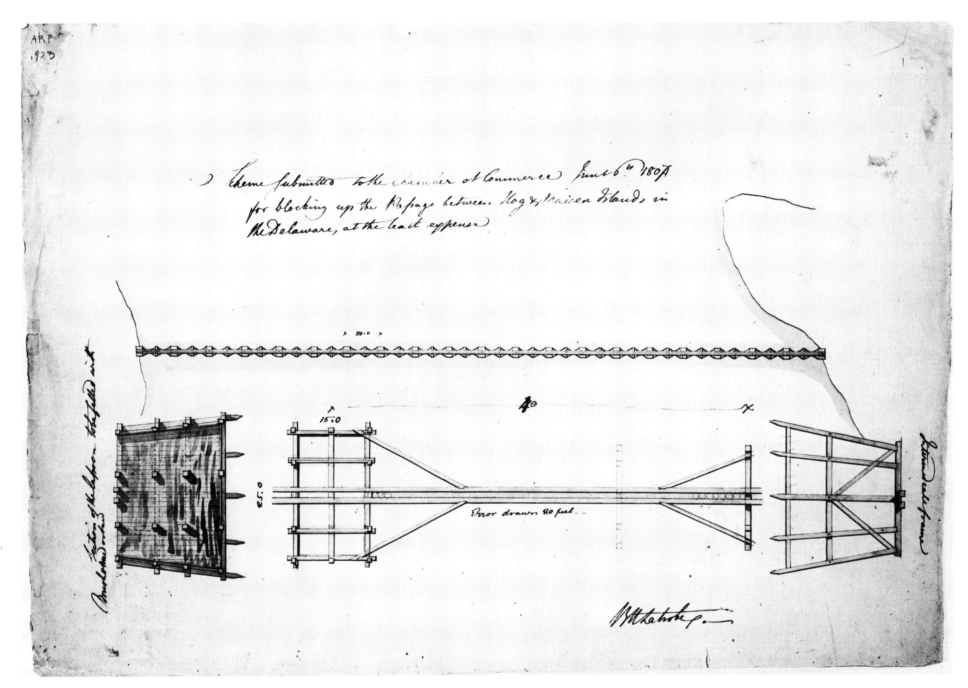

59 "Scheme Submitted to the Chamber of Commerce, June 6th, 1807, for blocking up the Passage between Hog & Maiden Islands in the Delaware, at the least expense"

size: 14 15/16" × 21 1/4" (38 cm × 54 cm)
media: pencil, pen and ink, watercolor
date: 6 June 1807
location: The Historical Society of Pennsylvania, Philadelphia

FASCINE WORKS

60 "I"—the Dagenham Gap
61 "II"—tools and materials
62 "III"—layout of boone
63 "IV"—construction of boone

Benjamin Henry Latrobe has left few clues concerning the engineering publications he acquired or consulted after his personal library was lost in transit to America. His essay and four drawings describing the German technique for constructing fascine river works, however, give us an idea of how closely he consulted European writings.

The occasion for Latrobe's essay was a serious flood at New Orleans in the spring of 1816 caused by a breach in the Mississippi levee. News of this flood and subsequent outbreaks of disease suspended shipment of steam engine parts from Baltimore to Latrobe's New Orleans Waterworks and made his steam engineer refuse to travel there. As a token of his personal interest in the problems of a city from which he had received the exclusive right to supply water to its citizens, but which he had not yet visited, Latrobe wrote for the city of New Orleans an "Essay on the Means of Preventing, Meeting and Repairing the Calamities Occasioned by Inundations." It suggested a solution to the problem of repairing breaches in the city's levees during a flood, and included four drawings.

The first illustrated the story of the Dagenham Gap in the Thames's banks east of London in 1720. Latrobe stated that its closure by Captain John Perry (1670–1732) with dovetailed piles was accomplished by luck rather than by sound engineering (no. 60). Latrobe then turned to the making of fascine works, a technique which had originated sometime in the medieval period in Baltic regions and which continued in use into the mid-nineteenth century. His consideration of the topic follows J. A. Eytelwein's *Praktische Anweisung zur Konstrukzion der Faschinenwerke* (1800) and at times is a translation of it. Latrobe's drawings are copies of the plates in the book. However, he freely reorganized much of Eytelwein's text. He described and illustrated how to build a fascine wing dam, which could be used to divert water away from a crevasse in the levee and make repair easier.

In the second drawing (no. 61) Latrobe drew a completed fascine (fig. 2). He suggested that it be made of straight willow or poplar branches an inch (2.5 cm) in diameter at their thickest part and about 10 feet (3 meters) long. A bundle of these were held together by withes, or ties, made from branches split into strands and woven to make a loop at one end (fig. 3). Slipped over a fascine the loop was tightened (fig. 4) and the ends of the withe were tucked into the bundle. Two tools useful in making fascines were a knife eight inches long for cutting and trimming the branches (fig. 7), and an iron gauge used to measure whether the thick end of a fascine was the correct diameter (fig. 9).

In this drawing Latrobe also included other tools and materials for the creation of fascine works. Stakes (fig. 8), and hammers to drive them (fig. 6), were necessary for holding the works to riverbanks and for holding the mass of fascines together. A rammer (fig. 5) packed earth into the completed work. The apparatus at the bottom of the drawing (fig. 10) was used to hold fascines and fascine bands while they were being tied with withes. Fascine bands were longer than fascines and were made up of overlapping branches, making the total length of a band 50 or 60 feet (15.2–18.3 meters). Bands held together large numbers of fascines in a river work.

In the next two drawings (nos. 62 and 63) Latrobe showed how to lay out and construct a fascine wingwork, or "boone" (from the German *Buhne*) along a shoreline. At first a straight line was laid out on the shore from stakes A and B which were at an angle of 60° to the direction of the river current (indicated by an arrow). A parallel line was laid out 12 feet downstream from stakes C and D to indicate the approximate width of the crown, or highest part of the boone. Given a depth of water of 36 feet at 50 feet from the shore, perpendiculars EF and GC were then laid out to equal 36 feet (11 meters). A line parallel to CD was then laid to the shore from G, making GH. This layout ensured that a boone with its ends at F and H, and its crown running along AB and CD, had sufficient depth to touch the river bottom. The curved line FKH enclosed with the shoreline an area which had to be excavated level with the waterline.

With a sufficient number of fascines, bands, and stakes on hand, the work was commenced. The fascines were laid into the water, thick end to shore and radiating from the crown, in layers slightly at an angle with one another. As each course of fascines was put in place a band was staked into the outer ends of the fascines and to the shore. A master workman superintended the placing of the fascines and determined when the work had reached sufficient size. The mass then extended well into the river, and as it was packed with earth and collected silt from the river it grew heavier and bent to the bottom. In Latrobe's words,

The work forms, as it were, a new shore of materials perfectly insoluble in water and bound together in every direction by bands and stakes, and which is loaded with earth so as to be immovable by the utmost violence of the stream. At last, when the Aquatic plants and shrubs, which are woven into its surface, take root, it becomes a texture inseparable by the most violent action of freshes, and in Northern countries is even perfectly indestructible by ice.

This essay is an instance of Latrobe's faith that the ancient civil engineering techniques of Europe could be transferred to the New World. There is, however, no evidence that anyone in New Orleans or Louisiana ever attempted to construct fascine works.

REFERENCES

Manuscript

Baltimore, Maryland. Maryland Historical Society. Letterbooks of Benjamin Henry Latrobe. BHL to the Mayor of New Orleans, 16 July 1816, 28 July 1816, 17 August 1816; BHL to Henry Latrobe, 16 August 1816.

Baltimore, Maryland. Maryland Historical Society. Mrs. Gamble Latrobe Collection. Benjamin Henry Latrobe, "Essay on the Means of Preventing, Meeting and Repairing the Calamities Occasioned by Inundations," draft, 10 June 1816.

New Orleans, Louisiana. New Orleans Public Library. Louisiana Division. Benjamin Henry Latrobe, "Essay on the Means of Preventing, Meeting and Repairing the Calamities Occasioned by Inundations," 10 June 1816.

Printed

Eytelwein, Johann Albert. *Praktische Anweisung zur Konstrukzion der Faschinenwerke und den dazu gedringen Anlagen an Flüssen und Strömen nebst einer Anleitung zur Beranschlagung dieser Baue.* Berlin, 1800.

Fascine Works

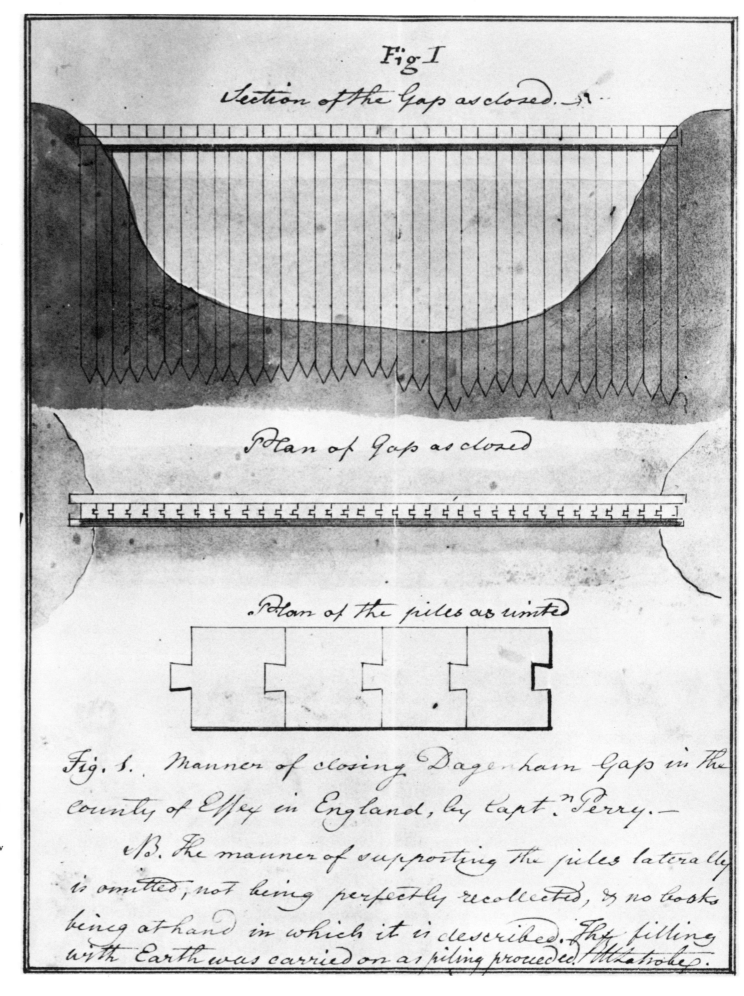

60 "I"—the Dagenham Gap

size: 8¼″ × 6¼″ (21 cm × 16 cm)
media: pen and ink, watercolor
date: June–July 1816
location: Louisiana Division, New Orleans Public Library

Fascine Works

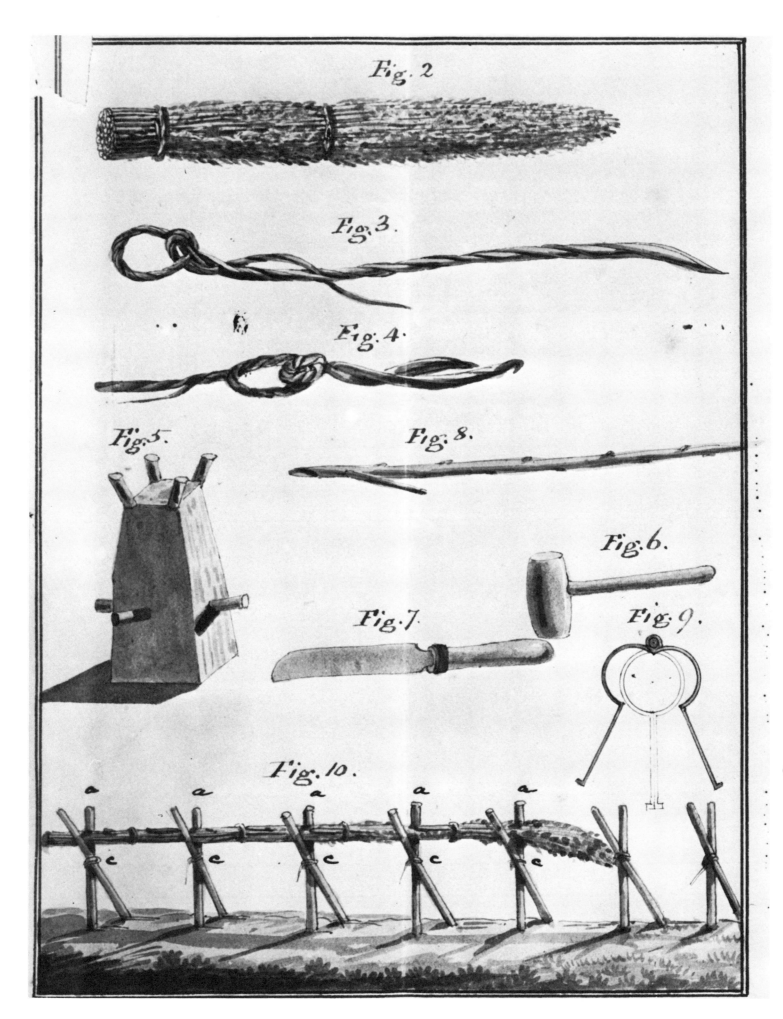

61 "II"—tools and materials

size: 8¼″ × 6¼″ (21 cm × 16 cm)
media: pen and ink, watercolor
date: June-July 1816
location: Louisiana Division, New Orleans Public Library

Fascine Works

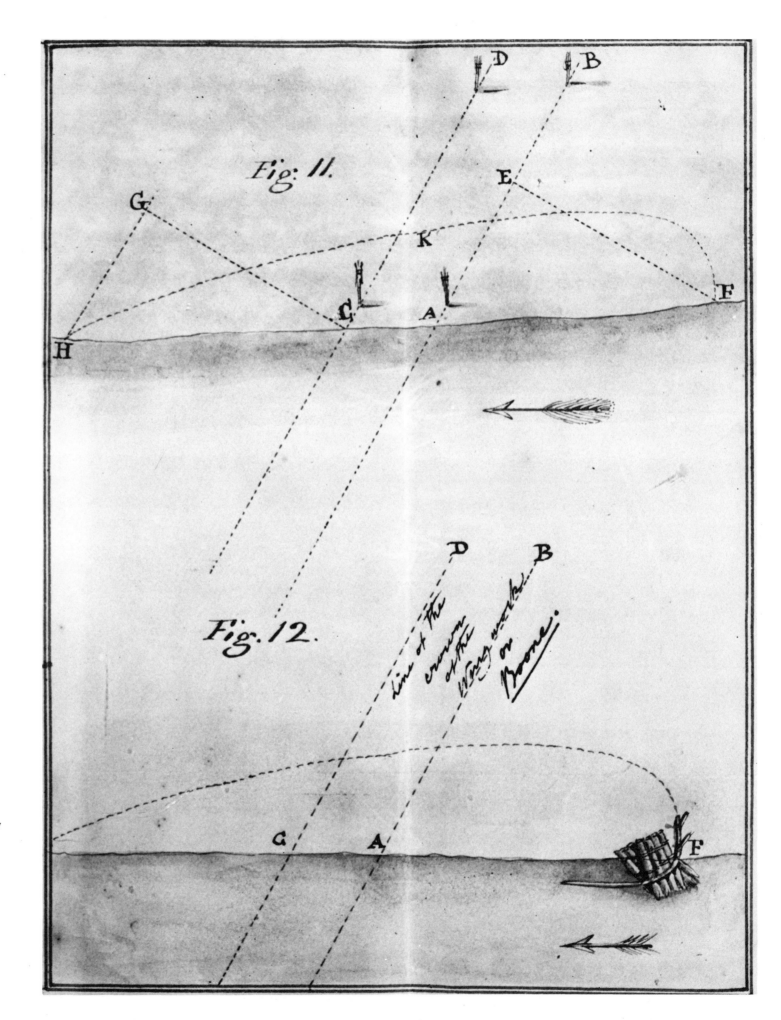

62 "III"—layout of boone

size: 7¾" × 6" (20 cm × 15.3 cm)
media: pen and ink, watercolor
date: June-July 1816
location: Louisiana Division, New Orleans Public Library

Fascine Works

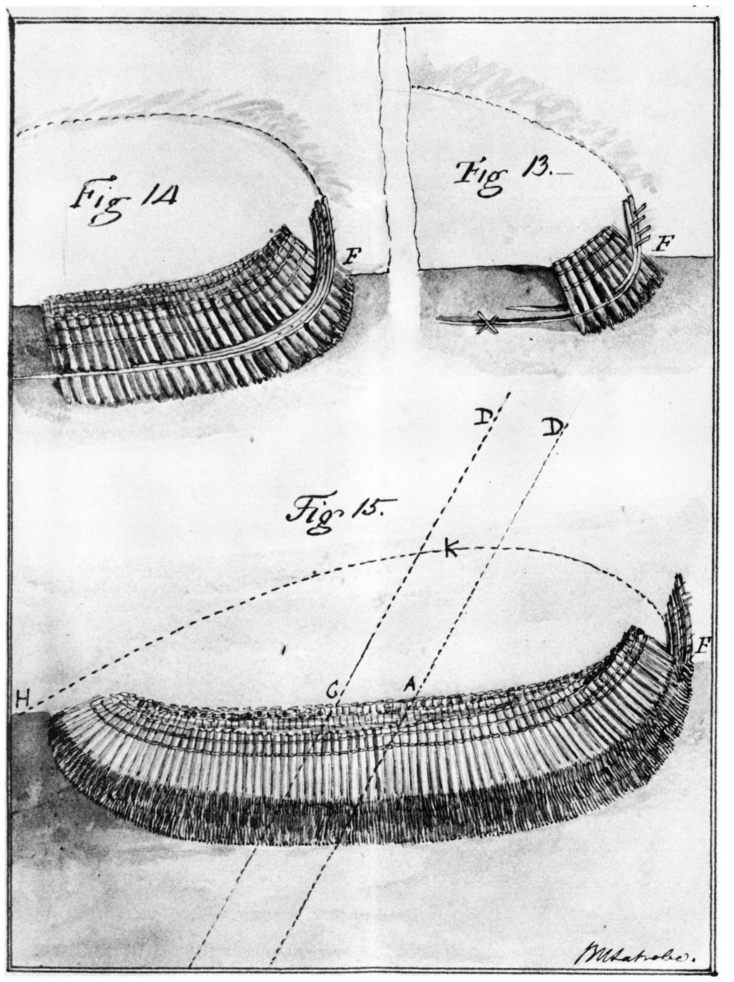

63 "IV"—construction of boone

size: 7¾" × 6" (20 cm × 15.3 cm)
media: pen and ink, watercolor
date: June-July 1816
location: Louisiana Division, New
Orleans Public Library

WASHINGTON NAVY YARD STEAM ENGINE

Plans for a Low-Pressure Steam Engine

64 Plan of a steam engine

65 Elevation of a steam engine

66 Elevation of a steam engine

67 Elevation of a beam, cranks, and flywheels

68 Section of the cylinder and plans of parts

69 "Nozzles" [steam valves]

These drawings were probably made by James Smallman in anticipation of a contract for building a steam engine for the Washington Navy Yard. They do not portray the engine as built, but rather a standard low-pressure rotative engine similar to others which Smallman constructed in his American career.

In the plan of the engine (no. 64) the boiler appears on the lower left as a large rectangular structure. Although in the plan the route of the flue is not fully shown, it does run from the grate above the firebox to the right side of the boiler, then counterclockwise (direction of arrows) to the chimney in the lower right corner. The lighter area around the grate and central part of the flue is the water which produces steam to operate the engine.

In the center of the drawing are the three cylinders of the engine, represented by circles. The lowermost, and largest, with a small square above it for the valves, is the steam cylinder. It stands on two timber supports running from left to right. To the left of the valves is the condenser which is immersed in a square wooden cistern along with the air pump. Above the air pump is the upper of a pair of boxed reservoirs which collect condensed steam from the air pump. The thin, tapered object with a superimposed circle at the top end is the cold water pump and filler trough which keep the cistern full of water.

The upper right-hand corner of the drawing is dominated by the two flywheels of 14 feet (4.3 meters) diameter. Between the flywheels are two sun-and-planet gears attached to the working beam crank rod (not shown in this drawing). On the right of the right-hand flywheel is a large gear which would, if meshed with a smaller gear, be able to drive machinery at a greater number of revolutions per minute than the flywheel shaft. The shaft of the left flywheel is extended to the left for connection with slower-moving machinery.

The first elevation (no. 65) exhibits the engine as if viewed with the boiler and steam cylinder to the front and flywheels to the rear. On the left is the boiler, showing the firebox door and an outline of the ash pit below it. Above that is the cast-iron boiler of a typically British type, the "waggon boiler," with a concave bottom and sides. The flue goes not only through the center of the boiler, but also around the sides. The pipe entering nearest the top of the boiler brings hot water to the boiler from the hot water pump. The sloping pipe below that pipe is the steam line going directly to the upper steam valve of the cylinder.

On the right side of the drawing at the bottom is the cold water cistern containing the condenser on the left side and the air pump on the right. Above the air pump in the drawing, and nearer to the observer, is the steam cylinder. It is made of cast iron and bound externally with three bands to prevent bursting. On the right side is the pipe which takes steam from the upper steam valve to the lower steam valve. At the top of the cylinder is the piston rod stuffing box (with the upper steam valve directly behind it), penetrated by the piston rod visible above it. At the top of the piston rod is a horizontal axle which has the piston rod guides at either end. The guides apparently move up and down in vertical slots to insure that the piston is not pulled from side to side by the semicircular motion of the beam end. Two rods connect the axle with the end of the beam.

The working beam is the rectangular block closest to the top of the drawing and just to the right of the chimney. It slants up and away from the viewer. The beam is pivoted in the middle on a horizontal axle held by a clamp extending around the beam's sides and top. The axle rests in two bearing blocks anchored to the spring beams parallel to and on either side of the beam.

The second elevation (no. 66) shows the engine as if being looked at from the right side of the plan (no. 64). To the left is the outline of the chimney and boiler. Next are the cylinder, its supports, and its connection with the working beam. It is notable that the engine lacks the parallel motion. The cylinder's vertical and horizontal supports are held together by long bolts. To the right of the cylinder are the upper and lower steam and exhaust valves, shown in section. The upper and lower exhaust valves are connected by the exhaust pipe, just as in the previous elevation (no. 65) the steam pipe is shown connecting the steam valves.

To the right of the cylinder supports is the condenser which, because it is immersed in the cold water cistern, condenses the steam which passes out of the cylinder into the exhaust valves and pipe. To the right of the condenser is a double air pump of the type which Smallman and Roosevelt first introduced for the Livingston-Roosevelt steamboat experiments and then at the Philadelphia Waterworks. It has valves at the top and bottom so that steam, air, and hot water are drawn out of the condenser on each stroke of the engine. The air pump contains a piston worked by a rod attached to the working beam. The rod has metal segments at each end, but the middle part is wooden and probably was planned to have pegs to trip the working gears for the steam and exhaust valves. To the right of the air pump is the hot water pump which has reservoirs for hot water coming from the air pump. The hot water pump is operated by a piston at the end of a rod attached to the beam, and has a small reservoir at the top of an ascending pipe. The previous elevation (no. 65) depicts the route the hot water follows to enter the boiler.

All of the moving equipment mentioned thus far is attached to the left side of the working beam. On the right side are the cold water pump and the sun-and-planet gear. The cold water pump required a well or some other source of cold water at the engine site. In the drawing the pump extends down out of the drawing into a well, but the pump piston is just visible. Water drawn up the pump flows left to the cold water cistern through a trough just behind the left side of the flywheel.

On the right side of this elevation is the flywheel and its connection with the beam. The flywheel crank rod is connected to the underside of the right end of the beam by a bearing block bolted through the beam. The crank rod has cast-iron connections to the beam and the planet gear, and a middle section of wood bolted to the cast-iron segments. Smallman notes near the upper end of the crank rod that it has a four-foot stroke.

The crank arm is attached to the axle of the planet gear which works around the sun gear on the flywheel axle. The flywheel is made of cast iron (or possibly cast-iron arms and a wooden rim) and is mounted on bearing blocks bolted to heavy horizontal timbers. The flywheel extends below the rest of the engine into a flywheel pit lined with brick.

The third elevation (no. 67) reveals a view of the flywheels and gearing similar to that in the plan (no. 64), and a view of the beam similar to that in the first elevation (no. 65), although from the opposite end. The composite nature of the crank rod is clearly shown here, with bolts joining the wood and iron segments of the rod. For no obvious reason this drawing shows the flywheels' shafts extending for an indefinite length in each direction.

The first detailed drawing (no. 68) has a section of the cylinder and plans of some of its parts. At the top of the drawing is the piston rod broken so that its connection with the working beam is not included. The piston passes through the stuffing box, which is packed with oiled hemp to insure a good seal, and then to the piston, shown in a raised position. The piston itself is composed of two pieces: the bottom, comprising the bulk of the piston, is shown in plan to the left of the cylinder, but the top is visible only in section. The two pieces are bolted together tightly enough to hold firmly against the cylinder wall the oiled hemp which fills the groove in the piston's edge.

The cylinder is 5 feet 2 inches (1.57 meters) long and is composed of three major pieces. The cylinder proper is a hollow tube of cast iron hooped with iron bands to give it strength. On the right side of the drawing is a plan of the cylinder top with boltholes for connection to the cylinder and a hole in the center for the piston rod. Below the cylinder is a plan of the cylinder bottom which reveals the location of the opening for admitting steam and exhausting condensed steam in the appropriate cycles of the engine. A similar opening for the upper valves is at the top right-hand corner of the piston.

The second detailed drawing (no. 69) depicts the nozzles, or steam valves, of which there are identical sets located at the top and bottom of the cylinder. On the left side of the drawing is a vertical section of a set of valves, the top being the steam valve and the bottom the exhaust valve. The valves proper are the slightly beveled discs above and below the centerline of the section. Each opens upward when a toothed sector moves a rack attached to the valve. When the upper valve opens it admits steam to the cylinder on the left; and when the lower valve opens steam is exhausted. The circular outlines behind the valves are the steam and exhaust pipes (above and below, respectively).

On the right-hand side of the drawing is a section of the valves, drawn as if the cylinder were behind them. At the top of the upper valve the steam pipes open to either side: from one side steam comes to the valve directly from the boiler and it passes through to a steam pipe leading to the lower steam valve. Intruding into the steam pipes is the spindle valve which is the axle of the sector. The spindle is rotated by the working gears of the engine. The lower valve is much the same as the upper, except that it is connected to an exhaust pipe which carries exhausted steam to the lower exhaust valve, and thence to the condenser.

Although these drawings represent early plans and not the actual form of

Washington Navy Yard Steam Engine

the steam engine erected at the Washington Navy Yard under Latrobe's direction, they are one of the most detailed series of drawings made in the earliest years of American steam engine technology.

REFERENCES

Farey, John. *A Treatise on the Steam Engine, Historical, Practical, and Descriptive.* London, 1827.

Latrobe, Benjamin Henry, "First Report of Benjamin Henry Latrobe, to the American Philosophical Society, held at Philadelphia; in answer to the enquiry of the Society of Rotterdam, 'Whether any, and what improvements have been made in the construction of Steam-Engines in America?'" *Transactions of the American Philosophical Society* 6 (1809): 89–98.

Pursell, Carroll W., Jr. *Early Stationary Steam Engines in America.* Washington: Smithsonian Institution Press, 1969.

Low-Pressure Steam Engine

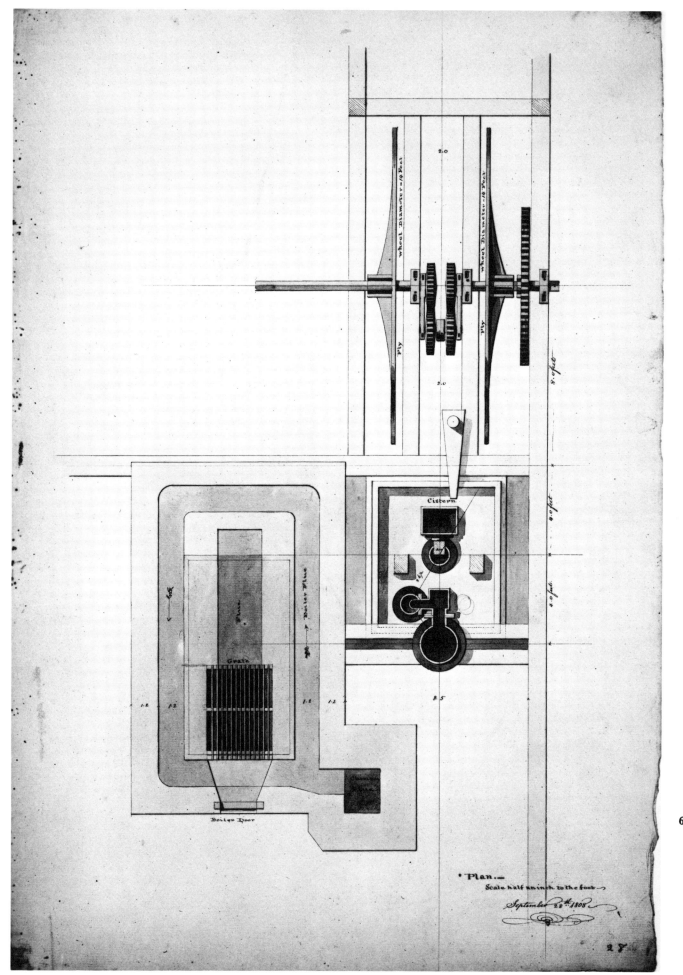

64 Plan of a steam engine

size: 21⅞″ × 14⅞″ (55.5 cm × 37.8 cm)
media: pencil, pen and ink, watercolor
date: 28 September 1808
delineator: James Smallman
location: Library of Congress,
  Washington, D.C.

Washington Navy Yard Steam Engine

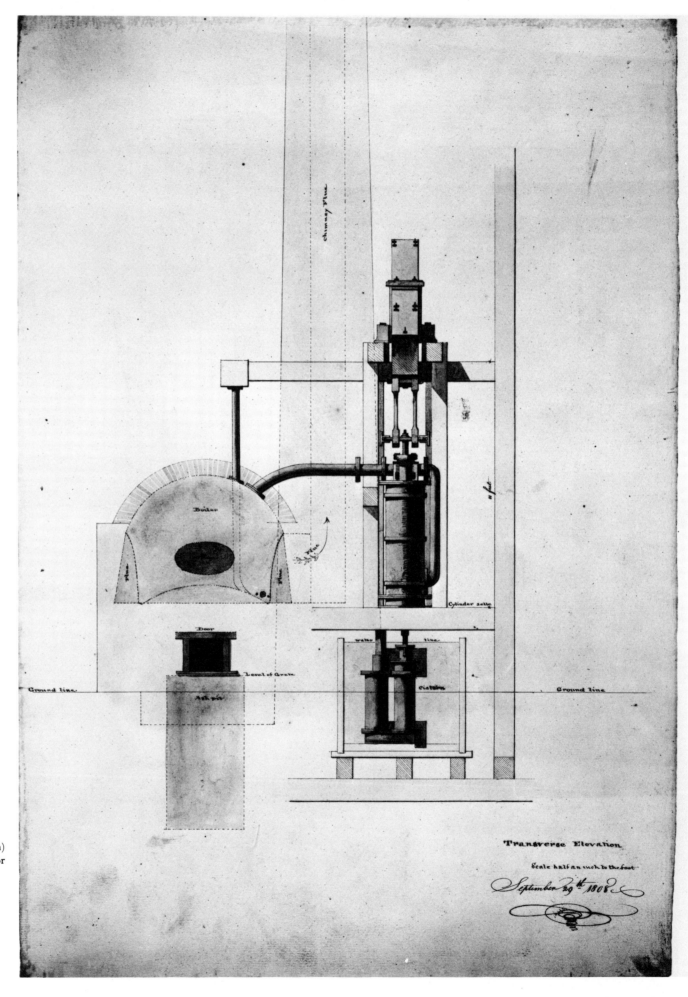

65 Elevation of a steam engine

size: 21⅞″ × 14⅞″ (55.5 cm × 37.8 cm)
media: pencil, pen and ink, watercolor
date: 29 September 1808
delineator: James Smallman
location: Library of Congress, Washington, D.C.

Low-Pressure Steam Engine

66 Elevation of a steam engine

size: $14\frac{3}{16}'' \times 21\frac{7}{8}''$ (37.6 cm × 55.5 cm)
media: pencil, pen and ink, watercolor
date: 29 September 1808
delineator: James Smallman
location: Library of Congress, Washington, D.C.

Washington Navy Yard Steam-Engine

67 Elevation of a beam, cranks, and flywheels

size: $21\frac{13}{16}'' \times 14\frac{13}{16}''$ (55.4 cm × 37.7 cm)
media: pencil, pen and ink, watercolor
date: 30 September 1808
delineator: James Smallman
location: Library of Congress, Washington, D.C.

Low-Pressure Steam Engine

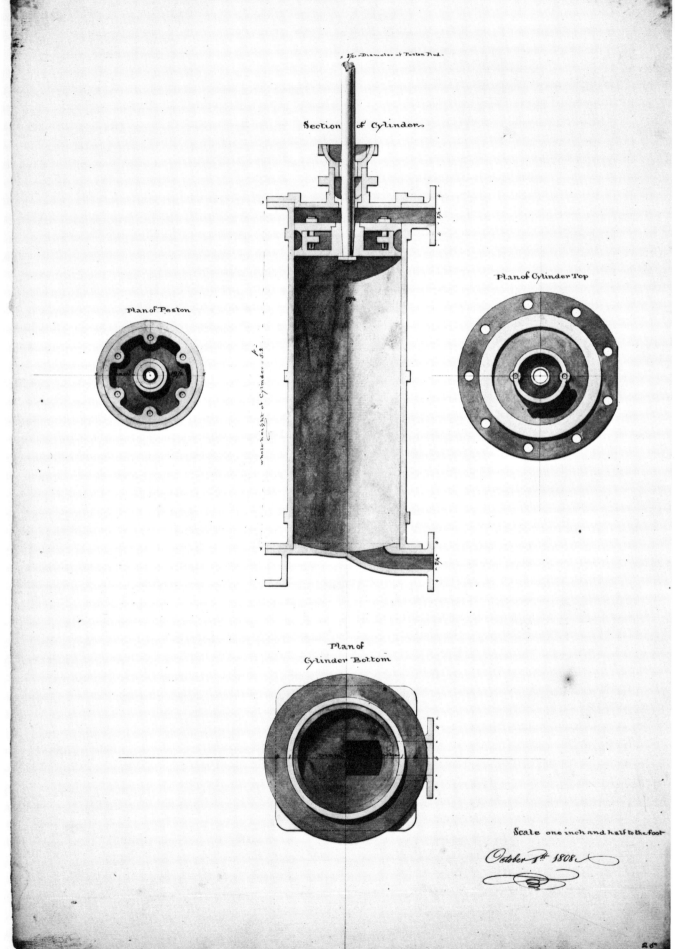

68 Section of the cylinder and plans of parts

size: 21⅞" × 14¹⁵⁄₁₆" (55.6 cm × 37.9 cm)
media: pencil, pen and ink, watercolor
date: 1 October 1808
delineator: James Smallman
location: Library of Congress, Washington, D.C.

Washington Navy Yard Steam Engine

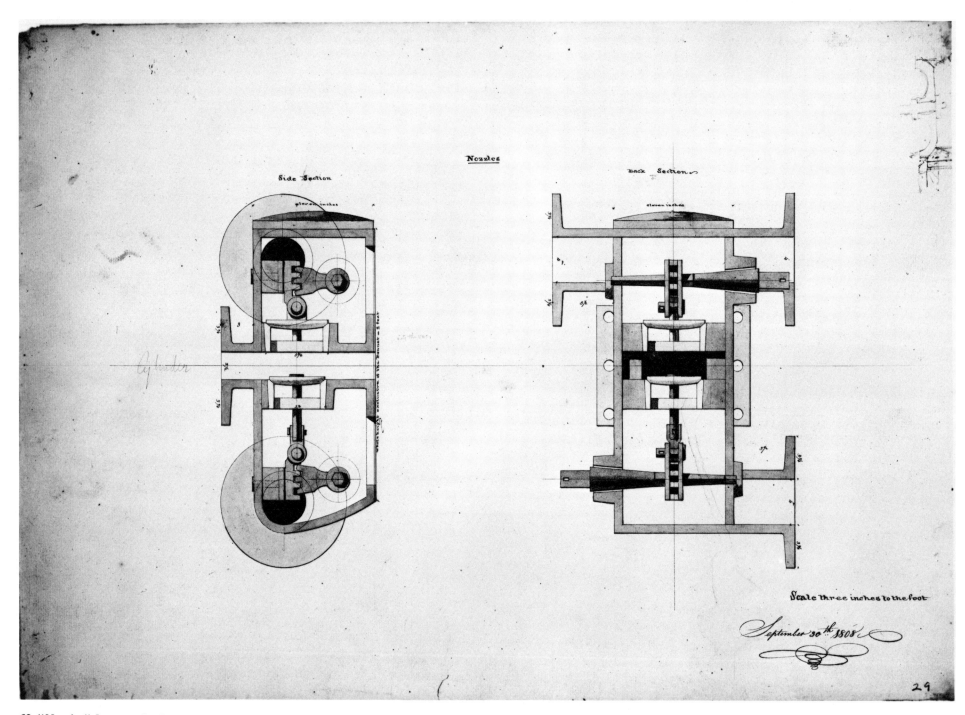

69 "Nozzles" [steam valves]
size: 14 15/16" × 21 7/8" (37.9 cm × 55.5 cm)
media: pencil, pen and ink, watercolor
date: 30 September 1808
delineator: James Smallman
location: Library of Congress, Washington, D.C.

WASHINGTON NAVY YARD STEAM ENGINE

70 Plan of a steam engine

71 Elevation of a steam engine

72 Elevation of a steam engine

73 Elevation of the beam, cranks, and flywheels

74 "Plan of the Timber for the Engine [at the] Navy Yard, Washington"

75 "Side Elevation of the Timber for the Engine [at the] Navy Yard, Washington"

76 "End Elevation of the Timber for the Engine [at the] Navy Yard, Washington"

77 Elevations of the working gears

78 Details of the flywheel and cam connection

In August 1809, after discussions and negotiations extending over some months, Latrobe notified James Smallman of Philadelphia that he was awarded a contract for making a steam engine for the Washington Navy Yard. During the winter of 1809/10 Smallman manufactured the engine in Philadelphia, meanwhile making these drawings so that Latrobe could prepare the navy yard site for the engine. They depict an engine similar to that shown in Smallman's earlier drawings of a standard low-pressure engine, but there are significant variations.

The plan (no. 70) includes the major engine elements in the same relationships as in the earlier drawings (nos. 64–69) except that the boiler appears on the opposite side of the cylinder. The flywheels are planned for a 16-foot (4.9 meter) diameter instead of a 14-foot (4.3 meter) diameter, and they are much thicker than before. The connection of the crank arm and the flywheel's shaft is not a sun-and-planet gear, but a simple crank. Both connections were used extensively in Britain and they worked equally well, so it is not surprising to find that Smallman chose to use the mechanically simpler device.[1] Both flywheel shafts are considerably extended, the one to the top of the drawing ending in a large cam ring (see no. 80).

The first elevation (no. 71) portrays little that is different from Smallman's earlier design (no. 65) with the exceptions of the shift of the boiler to the opposite side and the addition of the cam ring.

The second elevation (no. 72), viewed as if from the left side of the plan, includes a wooden beam, cast-iron cylinder, condenser, double ("American") air pump, and hot and cold water pumps. The pump rods are quite similar to those in the earlier design, including the composite wood and iron rods for the air pump and crank. The flywheel, however, follows the pattern of the flywheel of the Centre Square engine of the Philadelphia Waterworks. Both flywheels have cross-frames rather than spokes, the latter construction being the traditional one for flywheels. In addition, portions of the flywheels are made of wood instead of cast iron, and the additional weight necessary for the regulating function is made up by placing lead weights on the rim.[2] The drawings of the Centre Square flywheel included in the Philadelphia Waterworks section of this volume (nos. 45, 49) should be compared with this view.

The third elevation (no. 73) shows the engine as if the viewer were facing the flywheels, with the cylinder and boiler to the rear. The double crank connecting the crank rod and the flywheels' axle is located between the two flywheels. Both the flywheels' shafts are drawn broken, indicating that they were to be extended to some unknown length.

The three drawings of the timber framing for the engine (nos. 74, 75, 76) contain several interesting features. Two of them (nos. 74, 75) show the well for the cold water pump located to the side of the cistern. Although Latrobe attempted to dig a well in that planned location, he found that it had a very difficult "quicksand." Rather than cope with that substance he laid a drain to the engine from the basin of the navy yard, and that arrangement was satisfactory.

Most of the timbers are marked to be a foot or more wide, undoubtedly so thick to withstand the constant vibration and strain caused by the engine. Moreover, the trussing (angled) timbers are held to the horizontal supports by iron braces for added solidity (nos. 75, 76). On the second drawing (no. 75) a note is added giving the size of the working beam ("Main Lever Beam") as 20 feet 7 inches (6.27 meters) long, with a cross section of 2 feet by 16 inches (61 cm x 41 cm). The addition of such a note on the drawing is a clear indication that the working beam was also to be made of wood.

The working gears and valves (no. 77) are shown on the left side of the

---

1. John Farey, *A Treatise on the Steam Engine, Historical, Practical, and Descriptive* (London, 1827), pp. 409–26.

2. *Report of the Watering Committee to the Select and Common Councils, November 13th, 1807* (Philadelphia, 1807), p. 4.

Washington Navy Yard Steam Engine

drawing with the steam cylinder to their rear, and on the other side of the drawing with the cylinder outlined to their right. The working gears are pivoted on two parallel axles which are bolted to fixed vertical timbers. Shown directly in front of the valves on the left side of the drawing, and furthest to the left on the right side, is the wooden segment of the air pump rod. It operates the working gears through the movement of the two oblong blocks bolted to one side.

The final drawing of this series (no. 78) depicts the connector between one of the flywheel shafts and the cam ring. On the left in the middle of the drawing is the flywheel gudgeon, in the upper middle is the hollow wooden connector where the gudgeon will fit, and in the upper right in elevation and section is the gudgeon of the cam ring. Basically the connector is only an extension of the flywheel shaft, but it allows new connections to be made by removing only the gudgeon at the cam ring end.

REFERENCES

Manuscript

Baltimore, Maryland. Maryland Historical Society. Letterbooks of Benjamin Henry Latrobe. BHL to Smallman, 4 August 1809, 3 February 1810, 5 April 1810, 28 April 1810; BHL to Hamilton, 28 March 1810; BHL, "Facts respecting the Engine drain in the Navy Yard," [23 January 1812].

Printed

Farey, John. *A Treatise on the Steam Engine, Historical, Practical, and Descriptive*. London, 1827.
Pursell, Carroll W., Jr. *Early Stationary Steam Engines in America*. Washington: Smithsonian Institution Press, 1969.

Washington Navy Yard Steam Engine

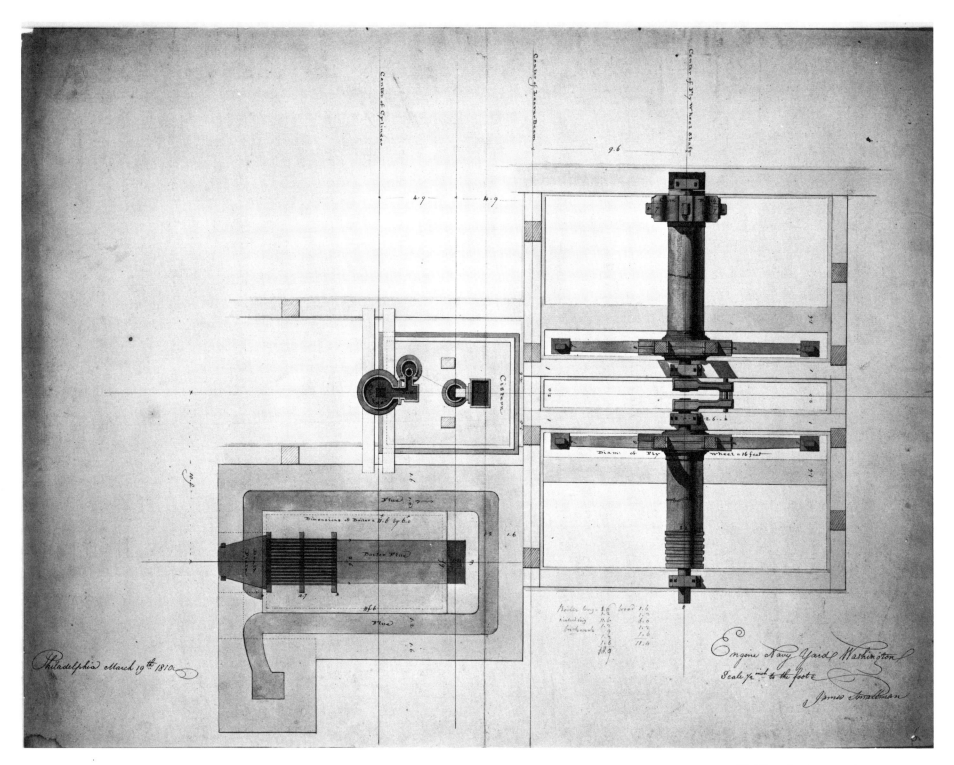

70 Plan of a steam engine

size: $21\frac{7}{8}'' \times 30\frac{1}{16}''$ (55.6 cm × 76.3 cm)
media: pencil, pen and ink, watercolor
date: 19 March 1810
delineator: James Smallman
location: Library of Congress, Washington, D.C.

Washington Navy Yard Steam Engine

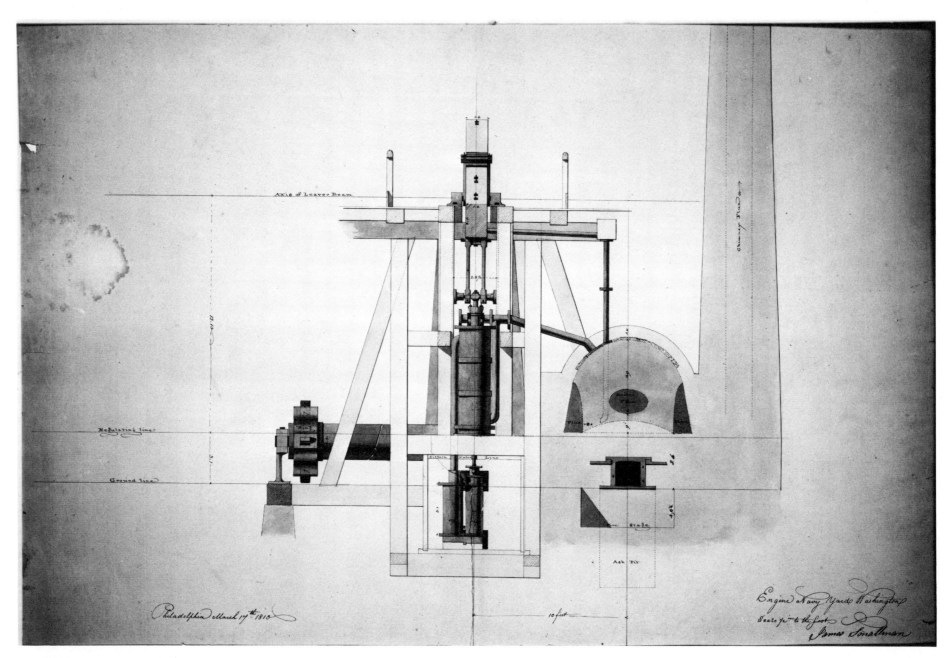

71 Elevation of a steam engine

size: $21\frac{11}{16}'' \times 30\frac{3}{16}''$ (55.1 cm × 76.7 cm)
media: pencil, pen and ink, watercolor
date: 17 March 1810
delineator: James Smallman
location: Library of Congress, Washington, D.C.

Washington Navy Yard Steam Engine

72 Elevation of a steam engine

size: 21 7/16" × 28 1/16" (54.5 cm × 71.3 cm)
media: pencil, pen and ink, watercolor
date: 22 March 1810
delineator: James Smallman
location: Library of Congress, Washington, D.C.

Washington Navy Yard Steam Engine

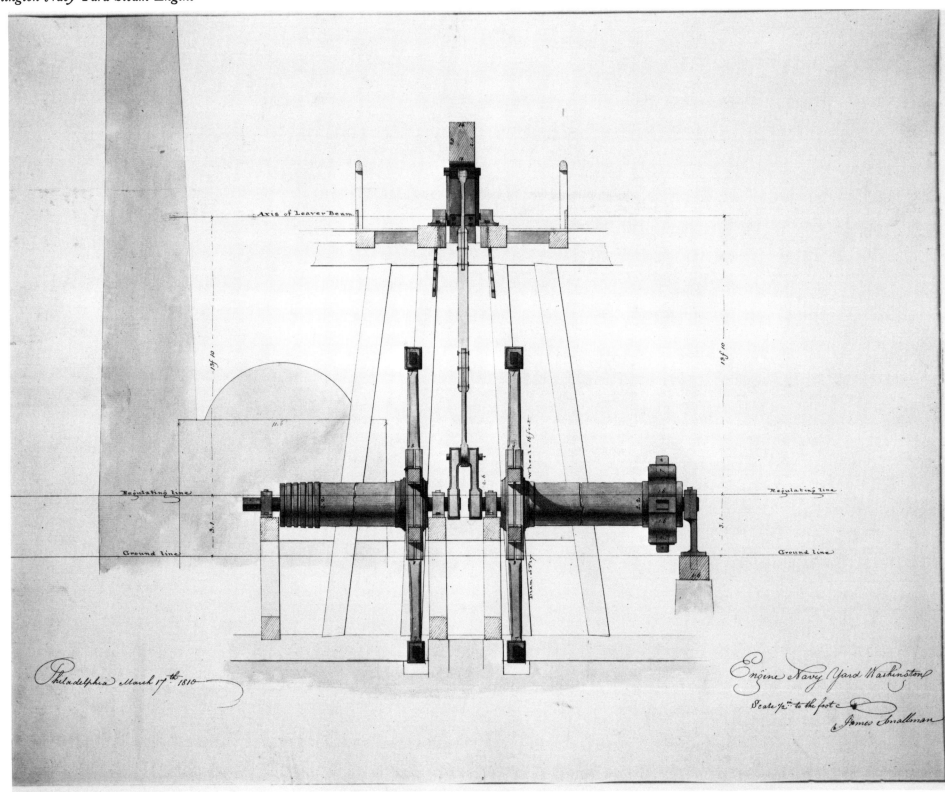

73 Elevation of the beam, cranks, and flywheels

size: 21 1/16" × 27 3/4" (55.7 cm × 70.4 cm)
media: pencil, pen and ink, watercolor
date: 17 March 1810
delineator: James Smallman
location: Library of Congress, Washington, D.C.

Washington Navy Yard Steam Engine

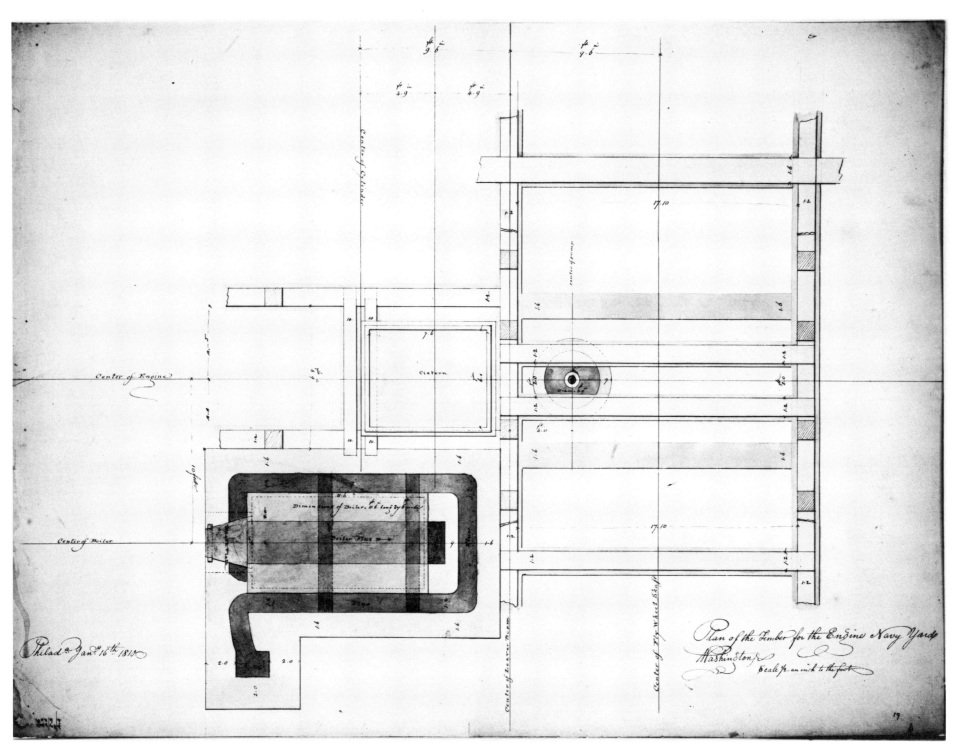

74 "Plan of the Timber for the Engine [at the] Navy Yard, Washington"

size: 22" × 29 11/16" (55.8 cm × 75.4 cm)
media: pencil, pen and ink, watercolor
date: 16 January 1810
delineator: James Smallman
location: Library of Congress, Washington, D.C.

Washington Navy Yard Steam Engine

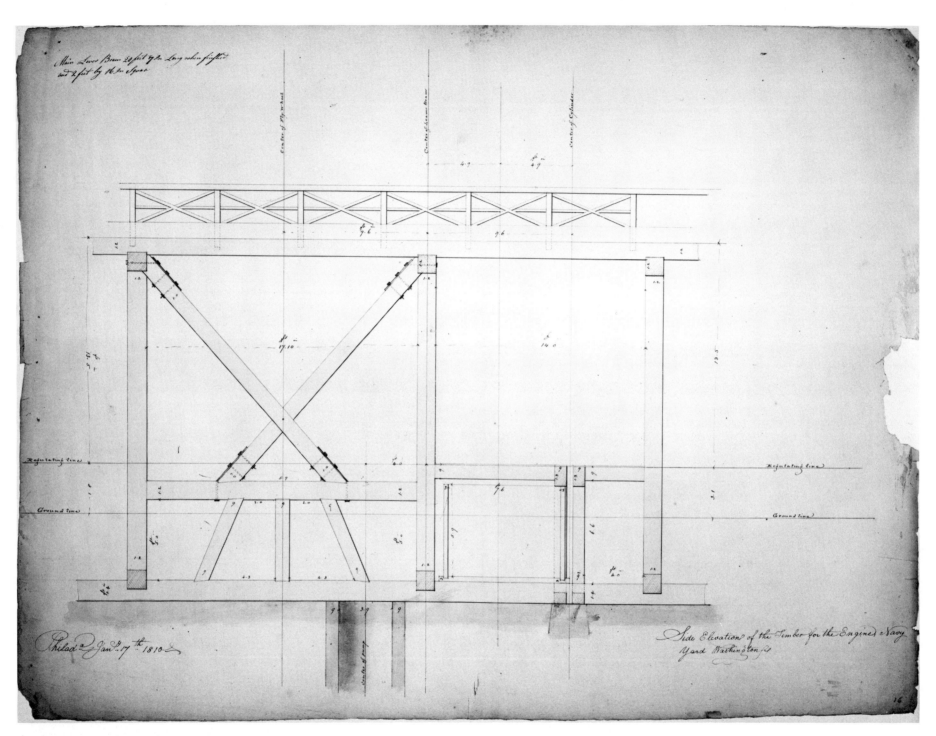

75 "Side Elevation of the Timber for the Engine [at the] Navy Yard, Washington"

size: 21⅞" × 29 13/16" (55.6 cm × 75.7 cm)
media: pencil, pen and ink, watercolor
date: 17 January 1810
delineator: James Smallman
location: Library of Congress, Washington, D.C.

Washington Navy Yard Steam Engine

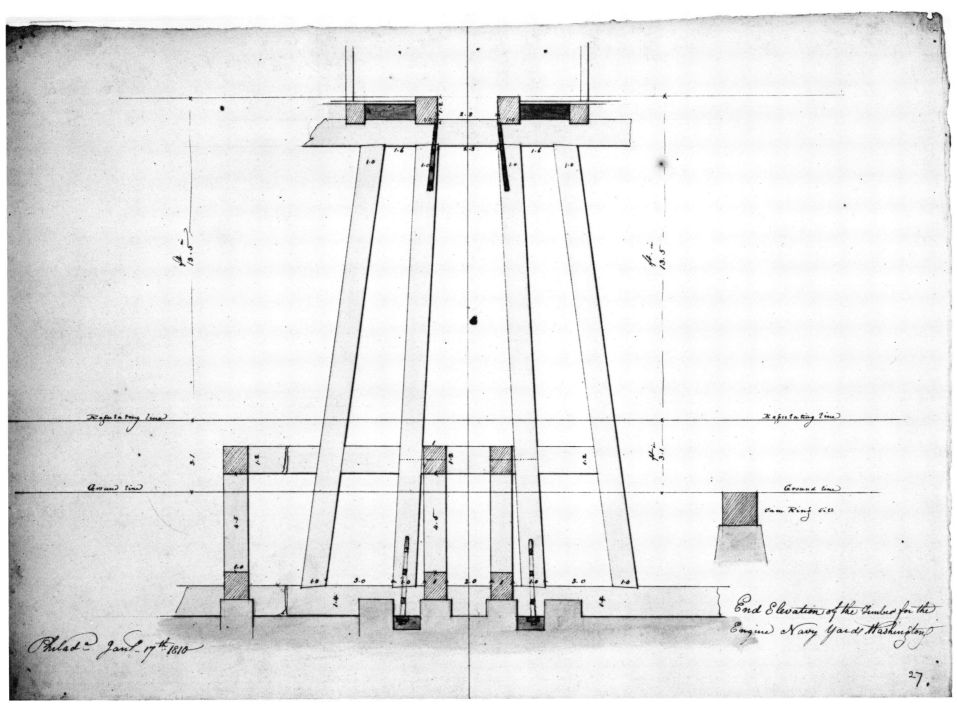

76 "End Elevation of the Timber for the Engine [at the] Navy Yard, Washington"

size: $14\frac{3}{4}'' \times 21\frac{11}{16}''$ (37.5 cm × 55.7 cm)
media: pencil, pen and ink, watercolor
date: 17 January 1810
delineator: James Smallman
location: Library of Congress, Washington, D.C.

Washington Navy Yard Steam Engine

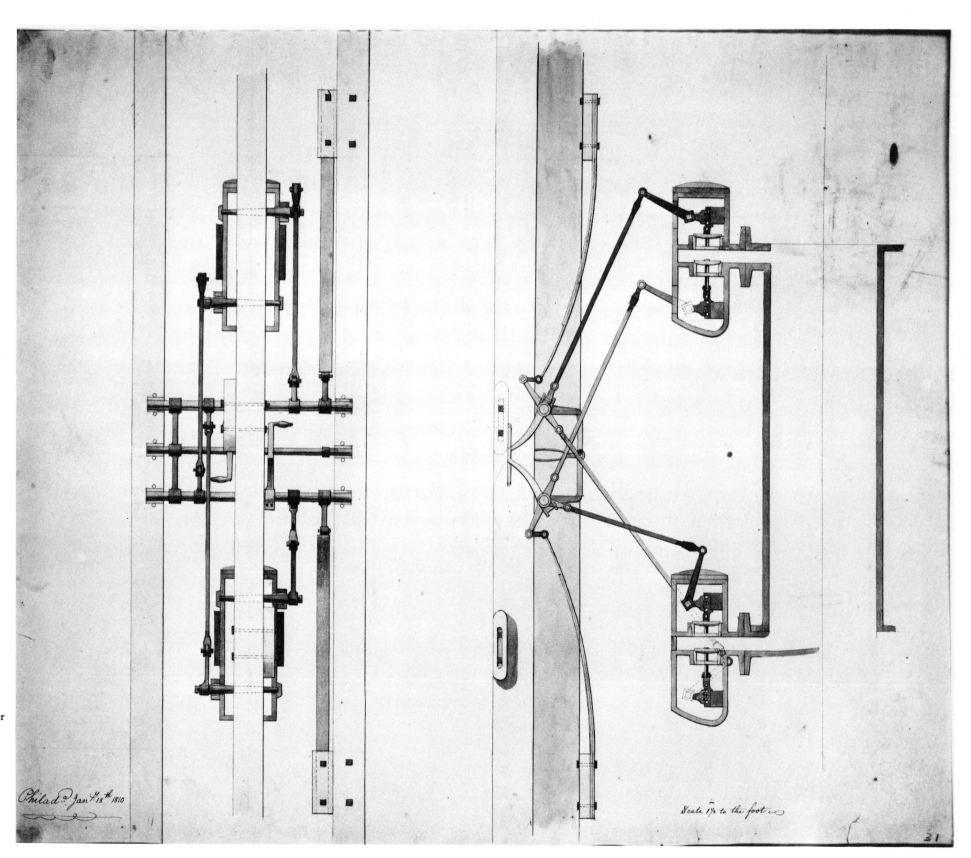

77 Elevations of the working gears

size: 19 15/16" × 23 13/16" (50.6 cm × 60.5 cm)
media: pencil, pen and ink, watercolor
date: 18 January 1810
delineator: James Smallman
location: Library of Congress, Washington, D.C.

Washington Navy Yard Steam Engine

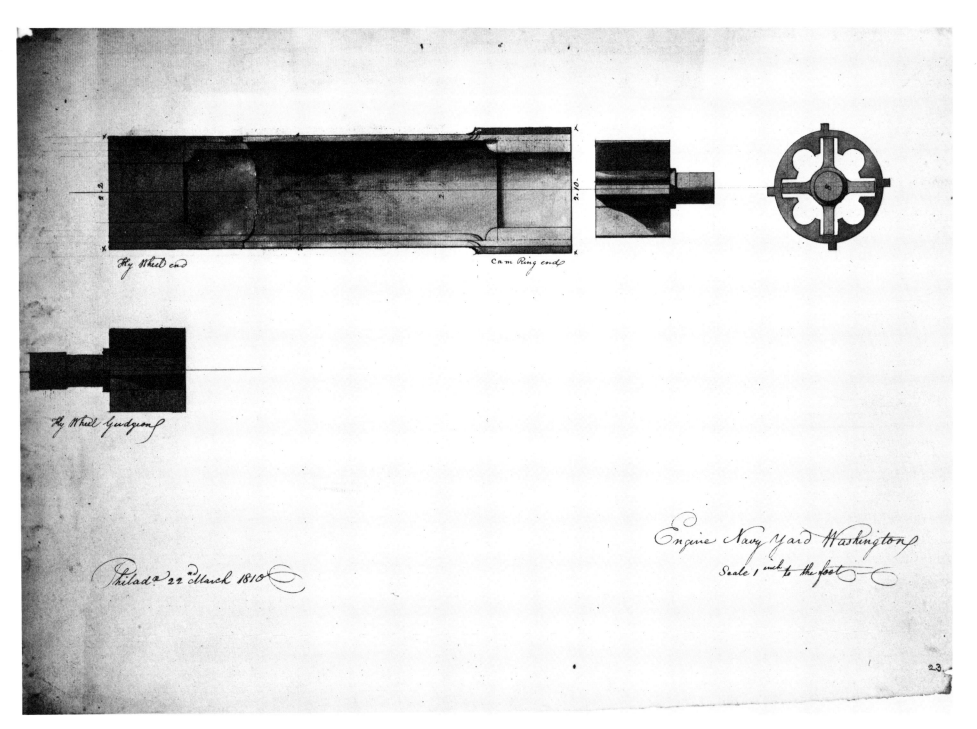

78 Details of the flywheel and cam connection

size: 15 1/16" × 21 13/16" (38.3 cm × 55.3 cm)
media: pencil, pen and ink, watercolor
date: 22 March 1810
delineator: James Smallman
location: Library of Congress, Washington, D.C.

WASHINGTON NAVY YARD STEAM ENGINE

Drawings of the Forge

79 Plan, elevations, and details of the forge hammer and gearing (See also colorplate 9.)

80 Elevations and section of the cogwheels and cam ring (See also colorplate 10.)

Although the erection of the Washington Navy Yard steam engine in 1810 was James Smallman's responsibility, Latrobe directed the installation of the manufacturing machinery, including the forge. Two surviving drawings reflect his role. On the lower left of the first drawing (no. 79) is a beautifully colored plan of the forge hammer. Coming in at the lower left and terminating in a large beveled cogwheel is the shaft of one of the engine's flywheels. That cogwheel meshes with another beveled cogwheel on the end of a slender shaft extending toward the top of the plan. Near the end opposite the cogwheel is a flywheel, and below that is the cam ring which operates the forge hammer. The hammer helve extends from left to right in the center of the plan, with its axle near its middle and the hammerhead and anvil at its right end. Beneath these structures the plan shows a network of heavy supporting timbers.

In the upper right corner of the full drawing is an elevation of the same machinery, but in the upper left corner of the drawing is a slightly different elevation. The cogwheels are lacking, but there is a full view of the cam ring and its flywheel. Undoubtedly the cam flywheel regulated the motion of the cam so that it did not slow down drastically when engaging the hammer helve or speed up excessively when not engaged. To the right of the cam ring are the helve, axle, hammer and anvil in section with the hammer at rest. To the right of the anvil is a platform marked "center of the main crane." Although there is no other evidence for the existence of a crane at the forge site, it would have been useful for handling the masses of heated scrap metal which were brought to the forge for reworking.[1]

The other sketches on the drawing are details of various elements shown in the plan and elevations. In the upper center are a plan and elevations of the standards which held the bearings for the axle of the hammer helve. In the lower right, the sketch nearest to the forge plan is also an elevation of the standards, and it shows the hammer helve axle to be tapered from the middle toward its bearing blocks.

The sketches in the lower right include the shaft of the cam and flywheel, the cam ring in plan and section, and views of the hammer and anvil. Two elevations of the hammerhead are side by side and a plan and elevation of the anvil are below them. Both the hammer and anvil are cross-faced, giving less striking area but allowing easier working of the forgings, according to Latrobe.[2]

The second drawing (no. 80) contains details of the cogwheels and the cam ring. On the left Latrobe depicts half sections of both the cogwheel on the engine flywheel shaft and the cogwheel on the cam ring shaft. Below the sections he provides quarter-elevations of each cogwheel, both of which are made of cast iron. On the right side of the drawing are sections and an elevation of the cam ring. It is interesting to note that the cam ring was to be cast with the location of the cams hollow. The cams themselves were wood, but made so that they did not fill completely the hollow area. An iron wedge was inserted next to them to hold them fast. This sort of construction was also used for cogwheels in this era, especially when wear was heavy and replacement of the cogs was expected.

These two drawings indicate Latrobe's active involvement in the erection and superintendence of the navy yard engine: one drawing (no. 80) was made while the forge was being installed, and the other (no. 79) after it had been in operation several months.

1. BHL to Hamilton, 3 December 1810, Letterbooks.
2. BHL to Johnson, 4 January 1813, Letterbooks.

The Forge

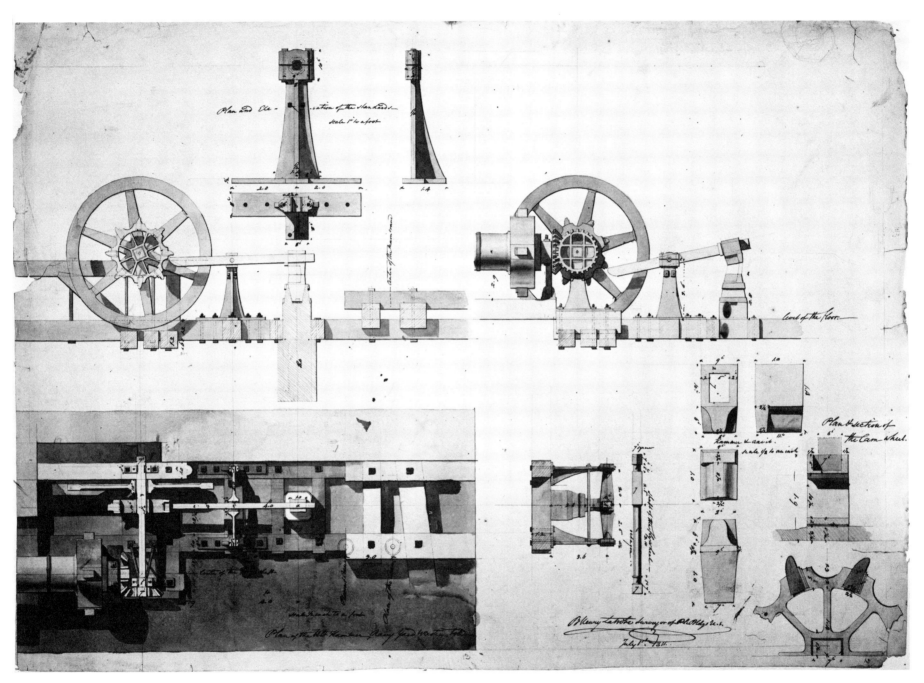

79 Plan, elevations, and details of the forge hammer and gearing (See also colorplate 9.)

size: 19¼" × 27⅝" (48.9 cm × 70.2 cm)
media: pencil, pen and ink, watercolor
date: 1 July 1811
location: Library of Congress, Washington, D.C.

Washington Navy Yard Steam Engine

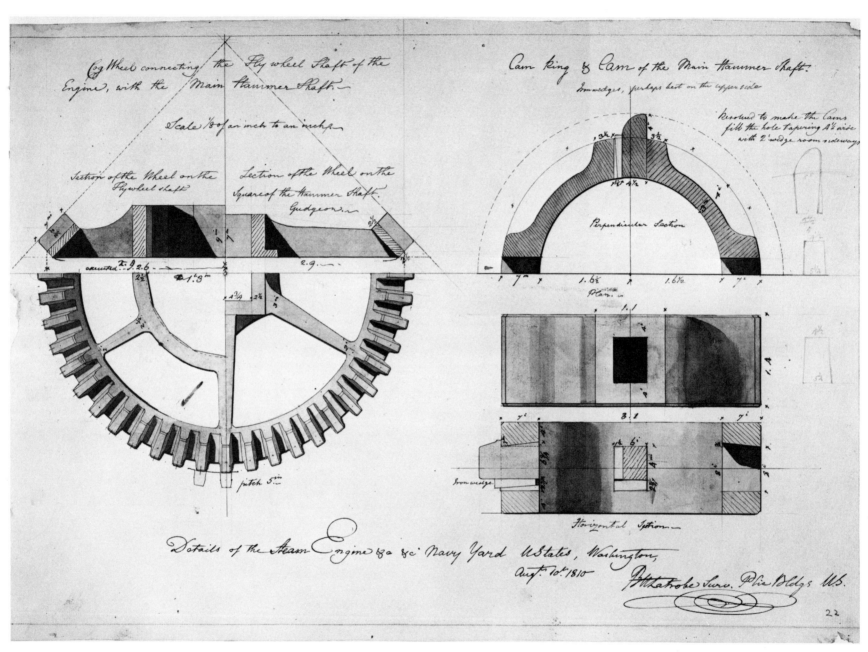

80 Elevations and section of the cogwheels and cam ring (See also colorplate 10.)

size: 13⅜″ × 19⅜″ (34.0 cm × 49.3 cm)
media: pencil, pen and ink, watercolor
date: 10 August 1810
location: Library of Congress, Washington, D.C.

MACHINERY

Block Mill at Alexandria, Virginia

81 Machinery for a block mill

82 "Hints for a block Mill"

In November of his first year in America Latrobe traveled to Alexandria, Virginia, where he saw an "experimental" pulley block mill. Alexandria was an active seaport near the head of tidal navigation on the Potomac River and had a good market for ships' supplies, including pulleys. Pulleys were an essential part of nautical rigging and were installed and replaced in large numbers. For centuries they had been made by hand from wood. Not until the British navy set up a series of pulley block machines at Portsmouth in the first decade of the nineteenth century was the handcraft element of pulley manufacture significantly reduced.

What Latrobe saw at Alexandria was an early American attempt to partially mechanize pulley manufacture. The American mill had two sets of saws performing two consecutive operations. The first set cut lumber of the size known as *scantling* into small lengths. This machinery (no. 81) is shown in figure 4 (looking at the circular saw edges) and figure 5 (as if viewed from the right end of the saw axle). Two important parts of this machine were the swing for holding the scantling, shown in figure 5, and the stop, shown as a blunt rod to the right of the larger saw blades in figure 4. When set correctly the stop was the same distance from the right blade as the two blades were from each other. Thus, when the scantling was set in the swing, adjusted so that its end was against the stop, and pushed against the saws, two blocks of equal dimensions were cut off.

As shown in figure 2 (no. 82), each block was then centered on the four parallel circular saws (also shown in figure 3) and pressed against them until it met the saws' axle. Next it was turned around and its other side cut in the same manner. Thus the saws outlined the spaces for two sheaves, making a double pulley. Not shown but mentioned by Latrobe in his notes were the augers which bored holes through the blocks at the top and bottom of the saw cuts, and which may have bored the axle hole through the center of the block as well.

From this point on the block had to be finished by hand. The remainder of the wood occupying the space for the sheaves had to be removed, the sheaves (the grooved wheels which carried the ropes) made and inserted, and an axle pushed through the block and sheaves. The entire block also had to be shaped into the characteristic oval form of a pulley; Latrobe's sketches suggest that the shaping may have been done before the sheave spaces were sawn. In all, the machinery Latrobe described and sketched eliminated some measurement and the initial rough cutting of the pulley block, but there remained a large amount of handwork to complete it.

Latrobe's interest in this block mill is an example of his recognition that mechanization could regularize and increase production of pulley blocks. From an early date the Washington Navy Yard (of which he was engineer) had a small block mill, and in 1812 Latrobe urged the installation there of a more complicated and sophisticated mill to manufacture standardized pulley parts. His interest in this field suggests that he knew it was one of the rapidly developing areas of engineering.

REFERENCES

K. R. Gilbert, *The Portsmouth Blockmaking Machinery: A Pioneering Enterprise in Mass Production*. London: Her Majesty's Stationery Office, 1965.

Machinery

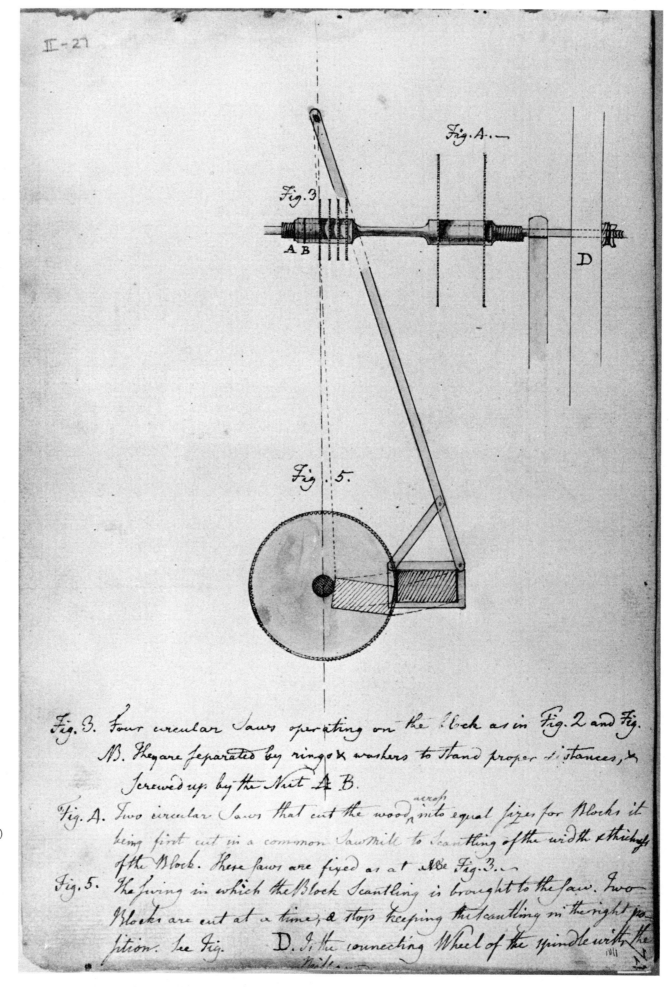

81 Machinery for a block mill

size: 6⅛" × 9" (15.5 cm × 22.8 cm)
media: pen and ink, watercolor
date: [23 November 1796]
location: Latrobe Sketchbook II

Block Mill at Alexandria, Virginia

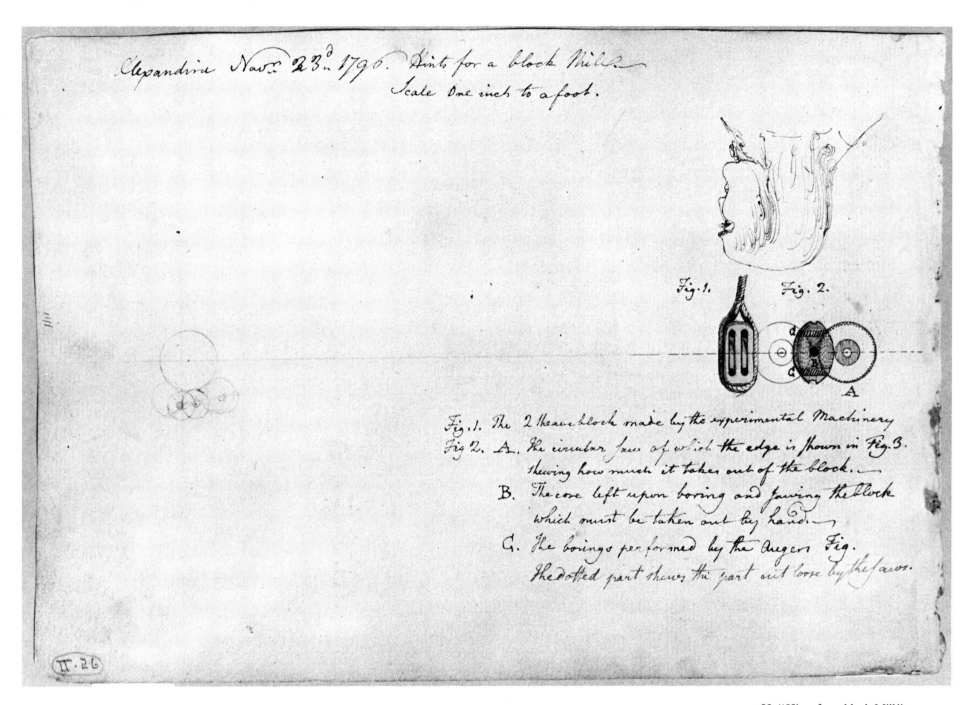

82 "Hints for a block Mill"
size: 9⅛" × 6 5/16" (23.1 cm × 16.0 cm)
media: pen and ink, watercolor
date: 23 November 1796
location: Latrobe Sketchbook II

MACHINERY

83 Hoisting machine

James Mease's edition of *The Domestic Encyclopaedia* includes an entry which depicts one of the hand-cranked hoisting machines that Latrobe used to raise stone for the construction of the Bank of Pennsylvania and the Centre Square Engine House. Each machine had a white oak frame 10 feet (3 meters) long and 5 feet (1.5 meters) wide which supported iron gear wheels and a crank. The largest pair of wheels was 4 feet (1.2 meters) in diameter and had a windlass between them. The crank had the four smaller gear wheels on its axle. Two wheels 8 inches (20 cm) in diameter meshed with the windlass gears when the crank was set in the nearer bearing blocks, and two of 12 inches (30 cm) meshed when set in the further blocks.

When in use the frame was securely weighted and staked to the ground, with a rope fixed to the windlass. The rope went through a pulley fixed above the spot to which the stone had to be raised, then down to the stone itself. Several men turned the crank to draw up the stone. According to *The Domestic Encyclopaedia*,

> In that part of the roof of the Pennsylvania Bank, which is of marble, there are several blocks of from 5 to $7\frac{1}{2}$ tons [4.5 to 6.8 metric tons] weight, the heaviest of which were hoisted by two of these machines, and *eight* men, in the short space of fifty-five minutes. And the columns of the Centre Square Engine House, which weighed about 9 tons [8.2 metric tons] each were set by the same force. The same kind of machines were employed in lowering weighty stone used in the piers and abutments of the Schuylkill Permanent Bridge, as well as in unloading the shallops employed in transporting large stones from the different quarries.[1]

Apparently these devices became standard equipment in Philadelphia for heavy construction. Although they were probably well-known machines in England, and contained no single innovative element, Latrobe presumably introduced them to America. He later used the same machines at the Capitol.[2]

Adam Traquair, who made the drawing on which the engraving is based, served in Latrobe's office during the latter stages of the construction of the bank and engine house, and worked as a draftsman for the Schuylkill Permanent Bridge.

1. A. F. M. Willich, *The Domestic Encyclopaedia; or, A Dictionary of Facts, and Useful Knowledge*, ed. James Mease. 5 vols. (Philadelphia, 1803–4), 5:368–69.
2. BHL to Lenthall, 14 May 1803, Lenthall Papers, LC. Note the similarity of BHL's plan and the plan described as "Mr. Smeaton's design for a Crane, for the Wool Quay, Custom House, London, 1789." Abraham Rees, ed., *The Cyclopaedia*, 1st American ed., 40 vols. (Philadelphia, 1805–24), plates, s.v. Mechanics, plate XIX.

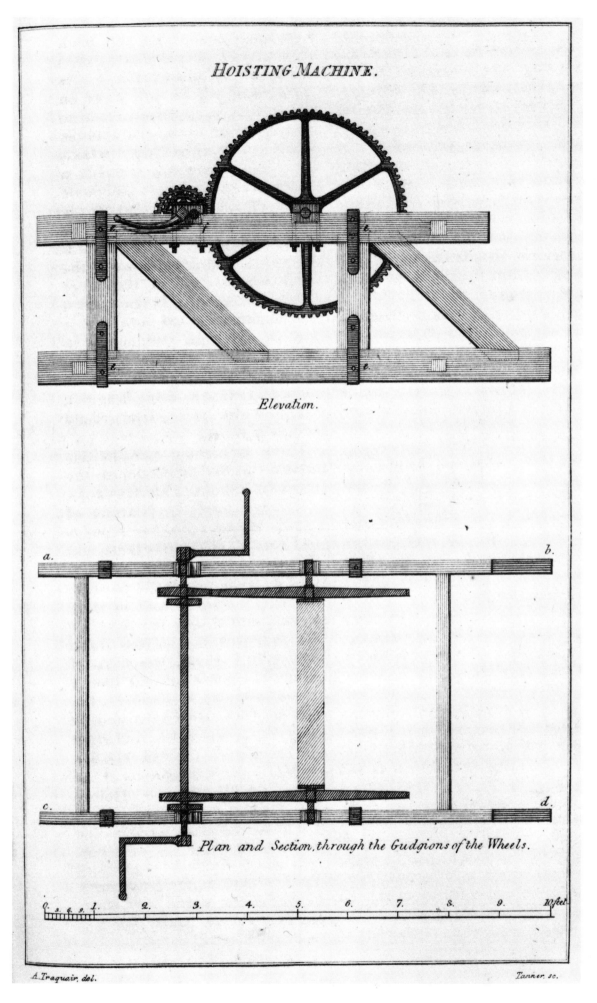

83 Hoisting machine

size: 4 1/16" × 6 3/4" (10.3 cm × 17.1 cm)
medium: engraving
date: 1804
delineator: Adam Traquair
engraver: Benjamin Tanner
location: A. F. M. Willich, *The Domestic Encyclopaedia; or, a Dictionary of Facts, and Useful Knowledge*. Edited by James Mease. Philadelphia, 1803–4. Vol. 5: facing p. 368.

MACHINERY

84 "Sketch of the landing place of the Columns of the House of Repr."

The original stone columns of the House of Representatives in the Capitol were ruined in the fire started by raiding British troops in August 1814. When Latrobe resumed his position as architect of the Capitol in 1815 one of his greatest problems was creating the colonnade for the new chamber. In 1815 he discovered a deposit of beautiful puddingstone in Montgomery County, Maryland, on the Potomac River. Latrobe anticipated that the quarried stone could be delivered by boats traveling down the Potomac, passing through the locks around the Great and Little falls above Washington, and entering the city by the Washington Canal. That would require land carriage of less than a mile from the landing to the Capitol.

This drawing shows Latrobe's plan for receiving the columns from the boats at a wharf on the Washington Canal. On the left side is the plan of the deck of the wharf: an open lattice of heavy timbers making a square 25 feet on a side. At the lower left and lower right corners of the deck are the bases of the cranes for lifting the columns. On the right side of the drawing is an elevation of the platform, showing its upward slope to the land side of the wharf. There is also an apparently incomplete drawing of a crane lifting a column from a boat.

As depicted, there is nothing to keep the crane in an upright position. Presumably several radiating guy ropes would have been fastened to the top to hold it, with the additional value that the angle of the crane (and thus the location of the stone) could be changed by adjusting their length. But this would have been a slow method, and the nature of the crane's operation is unclear.

By late June 1817 several sections of the House of Representatives columns had arrived at the Capitol. They probably came by the intended route, since a city newspaper reported on 1 July 1817 that "marble, stone &c. are now landed at the foot of the Capitol."[1] Shortly thereafter, however, Latrobe wrote to Jefferson that "the locks of the Potomac lower Canal having fallen in, beyond the power of Art to restore them, we suffer difficulty in getting down our Columbian Marble, but a great effort will be made to bring them thence by Land."[2] Latrobe's successor, Charles Bulfinch, completed the task of assembling the columns.

1. *National Intelligencer* (Washington), 1 July 1817.
2. BHL to Jefferson, 24 July 1817, Letterbooks.

84 "Sketch of the landing place of the Columns of the House of Repr."

size: 12¾" × 19¾" (32.4 cm × 50.2 cm)
media: pencil, pen and ink, watercolor
date: 3 May 1817
delineators: William Small and Benjamin Henry Latrobe
location: Library of Congress, Washington, D.C.

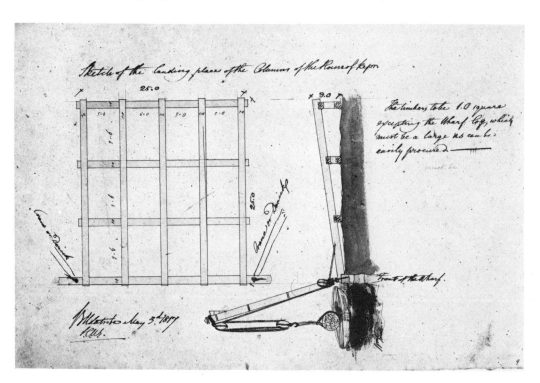

APPENDIX

Additional Drawings by Benjamin Henry Latrobe

Thirty-two drawings signed by or attributable to Latrobe are listed here but not published in this volume. All but two may be found in Thomas E. Jeffrey, ed., *The Papers of Benjamin Henry Latrobe: The Microfiche Edition* (Clifton, N.J.: James T. White & Co., 1976). Most of them are maps too large for publication or drawings with features similar to those in other renderings. Others will be published and described in Charles E. Brownell, ed., *The Architectural Drawings of Benjamin Henry Latrobe* (New Haven: Yale University Press, forthcoming).

INDEX

Potomac Canal Extension	no. 85
Chesapeake and Delaware Canal	nos. 86–92
Washington Canal	nos. 93–96
Harford Run Bridge	nos. 97–98
Franks Island Lighthouse	no. 99
Philadelphia Waterworks	nos. 100–11
Potomac Wharves	no. 112
Jones' Falls Improvement	no. 113
Water Rice Machine	no. 114
Fort Nelson	nos. 115–16

85 "No. III, of Plans, and Sections, of the Proposed continuation of the Canal at the Little Falls of the Potomak to the Navy Yard in the City of Washington"
size: $22'' \times 32''$ (55.8 cm × 81.3 cm)
media: pencil, pen and ink, watercolor
date: December 1802
delineator: Benjamin Henry Latrobe
location: Library of Congress, Washington, D.C.
microfiche: 267/B1

This shows the canal passing through the western portion of Georgetown, D.C. It is one of a series of four maps; details of the others are published as nos. 5–8.

86 "Section of the Northern Course of the Canal from the Tide in the Elk River at Frenchtown to the forked [oak] in Mr. Rudulph's Swamp"
size: $20\frac{5}{16}'' \times 30\frac{13}{16}''$ (51.6 cm × 78.4 cm)
media: pencil, pen and ink, watercolor
date: 3–10 August 1803
delineator: Benjamin Henry Latrobe
location: Library of Congress, Washington, D.C.
microfiche: 288/A1

The result of an early survey for a possible western route and terminus of the Chesapeake and Delaware Canal. It forms a series with the following two maps (nos. 87 and 88), each containing considerable detail.

87 "No. II. Plot of the Ground from the forked Oak to the Summit Maple at Oliver Howell, & of the Road to the Head of the Tide in Elk River"
size: $25\frac{5}{8}'' \times 38\frac{7}{16}''$ (65.1 cm × 97.6 cm)
media: pencil, pen and ink, watercolor
date: August 1803
delineator: Benjamin Henry Latrobe
location: Pennsylvania Historical and Museum Commission, Harrisburg
microfiche: 288/A3

88 "No. III. Map of the Ground between the Summit at the Maple opposite to Oliver Howell, and at the Old School House, & Aitkentown: shewing two practicable lines for the Canal"
size: $25\frac{11}{16}'' \times 38\frac{1}{8}''$ (65.2 cm × 96.9 cm)
media: pencil, pen and ink, watercolor
date: August 1803
delineator: Benjamin Henry Latrobe

Appendix: Additional Drawings

location: Pennsylvania Historical and Museum Commission, Harrisburg
microfiche: 288/B1

89 Survey taken near Bear, Delaware
size: $25\frac{5}{8}'' \times 38\frac{1}{2}''$ (65.1 cm × 97.8 cm)
media: pencil, pen and ink, watercolor
date: late 1803 or early 1804
delineator: possibly Benjamin Henry Latrobe
location: Pennsylvania Historical and Museum Commission, Harrisburg
microfiche: 289/A3

This entry and the next two (nos. 90 and 91) are apparently unfinished survey maps of a possible eastern route and terminus for the Chesapeake and Delaware Canal.

90 Possible canal terminus at New Castle, Delaware
size: $26'' \times 38\frac{3}{8}''$ (66.0 cm × 97.5 cm)
media: pencil, pen and ink
date: late 1803 or early 1804
delineator: possibly Benjamin Henry Latrobe
location: Pennsylvania Historical and Museum Commission, Harrisburg
microfiche: 288/B3

91 Possible canal terminus at New Castle, Delaware
size: $25\frac{11}{16}'' \times 38\frac{1}{4}''$ (65.3 cm × 97.2 cm)
media: pencil, pen and ink
date: late 1803 or early 1804
delineator: possibly Benjamin Henry Latrobe
location: Pennsylvania Historical and Museum Commission, Harrisburg
microfiche: 289/A1

92 Survey taken at Newark, Delaware
size: $25\frac{13}{16}'' \times 38\frac{9}{16}''$ (65.6 cm × 98.0 cm)
media: pencil, pen and ink, watercolor
date: [1804]
delineator: possibly Benjamin Henry Latrobe
location: Pennsylvania Historical and Museum Commission, Harrisburg
microfiche: 289/B1

Records elevations above tide level along several roads. BHL made this survey to determine the feasibility of using White Clay Creek as a source for the Chesapeake and Delaware Canal feeder.

93 "Plan of the Washington Canal No. I"
size: $21'' \times 31\frac{3}{4}''$ (53.3 cm × 80.6 cm)
media: pencil, pen and ink, watercolor
date: 5 February 1804
delineator: Benjamin Henry Latrobe
location: Library of Congress, Washington, D.C.
microfiche: 280/A1

Includes an outline of the mouth of the Tiber Creek and design for making it a basin for the Washington Canal.

This and the next drawing (no. 94) accompanied BHL's report to the first Washington Canal Company. Noteworthy on this drawing are a plan and elevation of a masonry lock and bridge combination. A third drawing in the series is missing.

94 "Washington Canal No. II"
size: $21\frac{1}{2}'' \times 31\frac{3}{4}''$ (54.8 cm × 80.6 cm)
media: pencil, pen and ink, watercolor
date: 5 February 1804
delineator: Benjamin Henry Latrobe
location: Library of Congress, Washington, D.C.
microfiche: 280/A3

95 "Map, exhibiting the property of the U.S. in the vicinity of the Capitol"
size: $23'' \times 18\frac{1}{2}''$ (58.4 cm × 47 cm)
media: pencil, pen and ink, watercolor
date: 3 December 1815
delineator: Benjamin Henry Latrobe
location: Library of Congress, Washington, D.C.
microfiche: 266/A1

Shows part of the Washington Canal and auxiliary improvements, but it is unknown whether the improvements were planned or extant.

96 "Plan of the West end of the public Appropriation in the city of Washington, called the Mall, as proposed to be arranged for the Site of the University"
size: $12\frac{1}{2}'' \times 18\frac{1}{2}''$ (31.7 cm × 47 cm)

Appendix: Additional Drawings

media: pencil, pen and ink, watercolor
date: January 1816
delineator: Benjamin Henry Latrobe
location: Library of Congress, Washington, D.C.
microfiche: 266/A3

97 "Design for a bridge proposed to be built over Hartford [Harford] run in Wilkes Street"
size: $17'' \times 20''$ (43.2 cm × 50.8 cm)
media: pencil, pen and ink, watercolor
date: 14 March 1818
delineator: Benjamin Henry Latrobe
location: The Peale Museum, Baltimore, Maryland
microfiche: 287/B3

Plans, sections, and an elevation of a bridge planned by BHL in conjunction with his Jones' Falls improvement scheme at Baltimore, Maryland. This drawing is very intricate, employing an elaborate version of the split section technique.

98 "Design of a Tunnel proposed to be built in Wilkes, for the passage of Hartf[ord (Harford) Run]"
size: $25'' \times 19\frac{1}{2}''$ (63.5 cm × 49.5 cm)
media: pencil, pen and ink, watercolor
date: 16 March 1818
delineator: Benjamin Henry Latrobe
location: The Peale Museum, Baltimore, Maryland
microfiche: 287/B5

This drawing concentrates on the stream passage under the bridge shown in no. 97. A few dimensions and details vary, however.

99 United States Lighthouse at Frank's Island, Louisiana
size: $22\frac{1}{2}'' \times 38''$ (57.1 cm × 96.5 cm)
media: pencil, pen and ink, watercolor
date: 7 November 1816; annotated 26 September 1817
delineators: Henry S. B. Latrobe and Benjamin Henry Latrobe
location: RG 26, National Archives, Washington, D.C.
microfiche: 281/B1

100 Plan of the Schuylkill Engine House
size: $23\frac{1}{8}'' \times 27\frac{5}{8}''$ (58.6 cm × 70.1 cm)
date: c. 1799
media: pencil, pen and ink, watercolor
delineator: possibly Benjamin Henry Latrobe
location: Maryland Historical Society, Baltimore
microfiche: 299/A1

Includes flywheel and pump well.

101 "Ground Plan of the Schuylkill Engine House. Philadelphia"
size: $14\frac{7}{8}'' \times 21\frac{5}{16}''$ (37.9 cm × 54.2 cm)
date: c. 1799
media: pencil, pen and ink, watercolor
delineator: Benjamin Henry Latrobe
location: The New Jersey Historical Society, Newark
microfiche: 296/A1

Has dimensions, but is otherwise almost identical to no. 28.

102 "Section of the Schuylkill Engine house, from South to North looking to the West"
size: $14\frac{3}{4}'' \times 21\frac{7}{16}''$ (37.6 cm × 54.4 cm)
date: c. 1799
media: pencil, pen and ink, watercolor
delineator: Benjamin Henry Latrobe
location: The New Jersey Historical Society, Newark
microfiche: 296/B9

Compare with no. 30, which is probably by Charles Stoudinger.

103 "Section of the Schuylkill Engine house from West to East, looking to the North"
size: $14\frac{5}{8}'' \times 21\frac{3}{8}''$ (37.1 cm × 54.3 cm)
date: c. 1799
media: pencil, pen and ink, watercolor
delineator: Benjamin Henry Latrobe
location: The New Jersey Historical Society, Newark
microfiche: 296/B11

Compare with no. 33, which is probably by Charles Stoudinger.

Appendix: Additional Drawings

104 Side elevation of the Schuylkill Engine House
size: 15⅛″ × 21⅜″ (38.4 cm × 54.4 cm)
media: pencil, pen and ink, watercolor
date: c. 1799
delineator: Benjamin Henry Latrobe
location: The New Jersey Historical Society, Newark
microfiche: 296/A7

Sparsely dimensioned.

105 "Schuylkill Enge. House, No. III. North and South Elevation as along Chesnut Street"
size: 13¼″ × 11⅛″ (33.1 cm × 27.8 cm)
media: pencil, pen and ink, watercolor
date: c. 1799
delineator: Benjamin Henry Latrobe
location: The Historical Society of Pennsylvania, Philadelphia
microfiche: 295/A1

Similar to no. 104, but without dimensions.

106 "Schuylkill Enge. He. No. IV; West Elevation, looking to the Schuylkill"
size: 13¼″ × 11⅛″ (33.1 cm × 27.8 cm)
media: pencil, pen and ink, watercolor
date: c. 1799
delineator: Benjamin Henry Latrobe
location: The Historical Society of Pennsylvania, Philadelphia
microfiche: 295/A3

Has some annotation.

107 East elevation of the Schuylkill Engine House
size: 15″ × 21⅜″ (38.1 cm × 54.3 cm)
media: pencil, pen and ink, watercolor
date: c. 1799
delineator: Benjamin Henry Latrobe
location: The New Jersey Historical Society, Newark
microfiche: 296/A9

Has annotation and dimensions, which are lacking in the similar view below (no. 108).

108 "Schuylkill Engine House No. V; East Elevation, along Front Street, Schuylkill"
size: 13¼″ × 11⅛″ (33.1 cm × 27.8 cm)
media: pencil, pen and ink, watercolor
date: c. 1799
delineator: Benjamin Henry Latrobe
location: The Historical Society of Pennsylvania, Philadelphia
microfiche: 295/A5

109 Plan of the foundation of the Centre Square Engine House
size: 15⅛″ × 21 5/16″ (38.3 cm × 54.1 cm)
media: pencil, pen and ink, watercolor
date: c. 1799
delineator: possibly Benjamin Henry Latrobe
location: Maryland Historical Society, Baltimore
microfiche: 299/A5

Similar to no. 40.

110 "Plan of the Ground Floor [of the Centre Square Engine House]"
size: 10⅜″ × 15″ (26.3 cm × 38.2 cm)
media: pencil, pen and ink, watercolor
date: c. 1799
delineator: Benjamin Henry Latrobe
location: Frederick Graff Collection, Franklin Institute, Philadelphia, Pennsylvania. (This drawing will be included in the supplement to *The Microfiche Edition of The Papers of Benjamin Henry Latrobe*.)

BHL's hand identifies the major elements in the plan, but Frederick Graff later made pencil annotations. This drawing appears to have been in a set with no. 40. Compare it to no. 41 and the plan on no. 39.

111 Section of the Centre Square Engine House
size: 27⅜″ × 23″ (69.5 cm × 58.4 cm)
media: pencil, pen and ink, watercolor
date: c. 1799–1800
delineator: Benjamin Henry Latrobe
location: Maryland Historical Society, Baltimore
microfiche: 299/B3

Details similar to no. 42, but with dimensions.

Appendix: Additional Drawings

112 Plan of wharves on the Eastern Branch of the Potomac River
size: 23″ × 17½″ (58.4 cm × 44.5 cm)
media: pencil, pen and ink
date: 3 March 1817
delineator: Benjamin Henry Latrobe
location: Library of Congress, Washington, D.C.
microfiche: 266/B1

The context of this commission is unknown.

113 Map and sections of the proposed canal for improvement of Jones' Falls
size: 24⅞″ × 30¾″ (63.2 cm × 78.1 cm)
media: pencil, pen and ink, watercolor
date: post 1 May 1818
delineators: Joseph Jeffers and Benjamin Henry Latrobe
location: Maryland Historical Society, Baltimore
microfiche: 287/B1

This map is on very fragile paper and has broken into fragments which were reassembled for the microfiche entry. Fig. 18, p. 47, is based on this map.

114 "An Inside View of a Water Rice Machine as used in South Carolina"
size: 10⅝″ × 6⅝″ (27.1 cm × 16.7 cm)
medium: engraving
date: [1802]
engraver: J. Akin
delineator: Benjamin Henry Latrobe
location: John Drayton. *A View of South-Carolina, As Respects Her Natural and Civil Concerns.* Charleston, 1802. (This engraving will be included in the supplement to *The Microfiche Edition of The Papers of Benjamin Henry Latrobe.*)

John Drayton made a drawing of a water-powered mill for hulling rice to illustrate a portion of his book, *A View of South-Carolina*. His drawing, which survives in the book manuscript at the Charleston Library Society, is incorrect in its perspective and awkward in composition. The book's engraver, J. Akin of Philadelphia, apparently requested BHL to make a corrected version, which became the basis of the published engraving.

115 "Plan of Fort Nelson, in Virginia, shewing the exact state of the Works in July 1798"
size: 21 7/16″ × 16⅜″ (54.5 cm × 41.5 cm)
media: pencil, pen and ink, watercolor
date: July 1798
delineator: Benjamin Henry Latrobe
location: RG 77, National Archives, Washington, D.C.
microfiche: 309/B1

Drawing of a fort protecting Norfolk, Virginia. Deteriorated and unfinished parts of the works are included.

116 "Design for compleating Fort Nelson, Virginia"
size: 23⅞″ × 17½″ (60.6 cm × 44.4 cm)
media: pencil, pen and ink, watercolor
date: c. July 1798
delineator: Benjamin Henry Latrobe
location: RG 77, National Archives, Washington, D.C.
microfiche: 309/B3

BHL's proposal for a substantial reconstruction of the fort shown in no. 115.

BIBLIOGRAPHY

This bibliography is in two parts, the first prepared by the editor, Darwin H. Stapleton, and the second by the author of the Susquehanna map essay, Stephen F. Lintner.

The first part is a list of printed works utilized in the preparation of the introductory essay and the annotation of the engineering drawings. Only those works which were fundamental resources of facts and ideas are included, and many peripheral works, even if they are cited in the footnotes, are not listed. Manuscript collections do not form part of the bibliography, since the bulk of the manuscripts concerned is in the collection of *The Papers of Benjamin Henry Latrobe*, Maryland Historical Society, Baltimore, Maryland. They are available in a microfiche edition. For the few other manuscript collections consulted, the reader will find complete citations in the footnotes.

The second part of the bibliography is a list of sources pertaining to the Susquehanna map essay. This includes a wide range of materials which bear on all phases of the map's history and significance. The reader will find items which will suggest further study in many areas, and which will enlarge his understanding of subjects more briefly examined in the essay.

I.

THE ENGINEERING PRACTICE OF BENJAMIN HENRY LATROBE

Ashton, Thomas Southcliffe. *Iron and Steel in the Industrial Revolution*. 2d ed. Manchester: Manchester University Press, 1951.

Bacon-Foster, Corra. *Early Chapters in the Development of the Patowmac Route to the West*. Washington: Columbia Historical Society, 1912.

Banks, Joseph et al., eds. *Reports of the late John Smeaton, F.R.S., Made on Various Occasions, in the Course of his Employment as a Civil Engineer*. 3 vols. London, 1812.

Bathe, Greville, and Bathe, Dorothy. *Oliver Evans: A Chronicle of Early American Engineering*. Philadelphia: Historical Society of Pennsylvania, 1935.

Bishop, J. Leander. *A History of American Manufactures from 1608 to 1860*. 2 vols. Philadelphia, 1864.

Blake, Nelson Manfred. *Water for the Cities*. Maxwell School Series, III. Syracuse, N.Y.: Syracuse University Press, 1956.

Boucher, Cyril T. G. *John Rennie, 1761–1821: The Life and Work of a Great Engineer*. Manchester: Manchester University Press, 1963.

Brown, Alexander Crosby. *The Dismal Swamp Canal*. Chesapeake, Va.: Norfolk County Historical Society, 1970.

Brown, A. F. J. *Essex at Work, 1700–1815*. Chelmsford: Essex County Council, 1969.

Bryan, Wilhelmus Bogart. *A History of the National Capitol*. 2 vols. New York: Macmillan Company, 1914–16.

Burton, Anthony. *The Canal Builders*. London: Eyre Methuen, 1972.

Caldwell, J. A. *History of Belmont and Jefferson Counties, Ohio*. Wheeling, W.Va., 1880.

Calhoun, Daniel H. *The American Civil Engineer: Origins and Conflict*. Cambridge, Mass.: M.I.T. Press, 1960.

Carter, Edward C., II, with Darwin H. Stapleton and Lee W. Formwalt. "Benjamin Henry Latrobe and Public Works: Professionalism, Private Interest, and Public Policy in the Age of Jefferson." *Essays in Public Works History* 3 (December 1976): 1–29.

Chesapeake and Delaware Canal Company. *First General Report of the President and Directors. June 4, 1804*. [Philadelphia, 1804].

Chesapeake and Delaware Canal Company. *Second General Report of the President and Directors. June 3, 1805*. [Philadelphia, 1805].

Cochran, Thomas, ed. *The New American State Papers, 1789–1860. Transportation*. 7 vols. Wilmington, Del.: Scholarly Resources, 1972.

Cole, Arthur Harrison. *The American Wool Manufacture*. 2 vols. Cambridge, Mass.: Harvard University Press, 1926.

Condie, Thomas, and Folwell, Richard. *History of the Pestilence, Commonly Called Yellow Fever, Which Almost Desolated Philadelphia, in the Months of August, September & October 1798*. Philadelphia, [1799?].

Condit, Carl. *American Building*. Chicago: University of Chicago Press, 1968.

[Cramer, Zadok]. *The Navigator: Containing Directions for Navigating the Monongahela, Allegheny, Ohio and Mississippi Rivers*. Pittsburgh, 1814.

Crone, G. R. *Maps and Their Makers: An Introduction to the History of Cartography*. London: Hutchinson's University Library, 1953.

[Davis, John]. "Autobiography of John Davis, 1770–1864." *Maryland Historical Magazine* 30 (1935): 11–39.

Dickinson, H. W. *Robert Fulton: Engineer and Artist, His Life and Works*. London: John Lane, 1913.

———. *Water Supply of Greater London*. London: Newcomen Society, 1954.

Duffy, John. "Nineteenth Century Public Health in New York and New Orleans: A Comparison." *Louisiana History* 15 (1974): 325–37.

———, ed. *The Rudolph Matas History of Medicine in Louisiana*. 2 vols. Baton Rouge, La.: Louisiana State University Press, 1958.

Ellsworth, Henry L. *A Digest of Patents Issued by the United States, from 1790 to January 1, 1839*. Washington, 1840.

Eytelwein, Johann Albert. *Praktische Anweisung zur Konstrukzion der Faschinenwerke und den dazu gedringen Anlagen an Flüssen und Strömen nebst einer Anleitung zur Beranschlagung dieser Baue.* Berlin, 1800.

Facts and Observations Respecting the Chesapeake and Delaware Canal, and its Extension into Pennsylvania. n.p., [1805].

Farey, John. *A Treatise on the Steam Engine, Historical, Practical, and Descriptive.* London, 1827.

Ferguson, Eugene S. *Bibliography of the History of Technology.* Cambridge, Mass.: Society for the History of Technology and M.I.T. Press, 1968.

———. "Mr. Jefferson's Dry Docks." *The American Neptune* 11 (1951): 108–14.

———, ed. *Early Engineering Reminiscences (1815–1840) of George Escol Sellers.* United States National Museum Bulletin 238. Washington: Smithsonian Institution, 1965.

Formwalt, Lee W. "Benjamin Henry Latrobe and the Development of Internal Improvements in the New Republic, 1796–1820." Ph.D. dissertation, The Catholic University of America, 1977.

Gallagher, H. M. Pierce. *Robert Mills: Architect of the Washington Monument, 1781–1855.* New York: Columbia University Press, 1935.

Gilchrist, Agnes A. *William Strickland.* 2d ed. New York: Da Capo Press, 1969.

Gillespie, W. M. *A Manual of the Principles and Practice of Road-Making.* 9th ed. New York, 1868.

Gilpin, Joshua. *A Memoir on the Rise, Progress, and Present State of the Chesapeake and Delaware Canal, Accompanied with Original Documents and Maps.* Wilmington, Del., 1821.

Graff, Frederick. "Steam Engine Boiler." *Emporium of the Arts and Sciences.* New series, 3 (1814): 275–76, plate.

Graff, Frederic. "Notice of the Earliest Steam Engines used in the United States." *Journal of the Franklin Institute* 3d series, 25 (1853): 269–71.

Gray, Ralph D. *The National Waterway: A History of the Chesapeake and Delaware Canal, 1769–1965.* Urbana: University of Illinois Press, 1967.

Hadfield, Charles. *British Canals.* 4th ed. New York: Augustus M. Kelley, 1969.

———. *The Canals of South and South East England.* New York: Augustus M. Kelley, 1969.

Haller, Mabel. *Early Moravian Education in Pennsylvania.* Nazareth, Pa.: Moravian Historical Society, 1953.

Hamlin, Talbot. *Benjamin Henry Latrobe.* New York: Oxford University Press, 1955.

Harris, Robert. *Canals and Their Architecture.* New York: Frederich A. Praeger, 1969.

Heine, Cornelius W. "The Washington City Canal." *Records of the Columbia Historical Society* 53 (1953–54): 1–27.

Hibben, Henry B. "History of the Washington Navy Yard." U.S. Congress, Senate, Executive Document 22, 51st Congress, 1st Session, 1889–90.

Hindle, Brooke. *Technology in Early America.* Chapel Hill: University of North Carolina Press, 1966.

"History of the Steam Engine in America, The." *Journal of the Franklin Institute* 3d series, 72 (1876): 253–68.

Hunter, Louis C. *Steamboats on the Western Rivers: An Economic and Technological History.* Cambridge, Mass.: Harvard University Press, 1949.

Hunter, Robert F. "Turnpike Construction in Antebellum Virginia." *Technology and Culture* 4 (1963): 177–200.

Hunter, W. H. "The Pathfinders of Jefferson County." *Ohio Archaeological and Historical Publications* 6 (1898): 95–313.

[Jefferson, Thomas]. *Message from the President of the United States, Transmitting Plans and Estimates of a Dry Dock, for the Preservation of our Ships of War.* Washington, 1802.

Jeremy, David. "Innovation in American Textile Technology during the Early 19th Century." *Technology and Culture* 14 (1973): 40–76.

Johnston, George. *History of Cecil County, Maryland.* Elkton, Md., 1881. Reprint. Baltimore: Regional Publishing Co., 1967.

Jordan, Philip D. *The National Road.* Indianapolis: Bobbs-Merrill Co., 1948.

Journal of the House of Representatives of the United States, at the First Session of the Eleventh Congress. Washington, 1826.

Kirby, Richard Shelton. "William Weston and his Contribution to Early American Engineering." *Transactions of the Newcomen Society* 16 (1935–36): 111–27.

Kirby, Richard Shelton, and Laurson, Philip Gustave. *The Early Years of Modern Civil Engineering.* New Haven: Yale University Press, 1932.

Kite, Elizabeth, ed. *L'Enfant and Washington.* Baltimore: Johns Hopkins Press, 1929.

Langton, Edward. *History of the Moravian Church.* London: George Allen & Unwin, 1956.

Latrobe, Benjamin Henry. *American Copper Mines.* n.p., [1800].

———. *Answer to the Joint Committee of the Select and Common Council of Philadelphia, on the subject of a plan for supplying the city with water, &c.* [Philadelphia, 1799].

———. "First Report of Benjamin Henry Latrobe, to the American Philosophical Society, held at Philadelphia; in answer to the enquiry of the Society of Rotterdam, 'Whether any, and what improvements have been made in the construction of Steam-Engines in America?'" *Transactions of the American Philosophical Society* 6 (1809): 89–98.

———. *Opinion on a Project for Removing the Obstructions to a Ship Navigation to Georgetown, Col.* Washington, 1812.

———. *Proposals for the Establishment of a Company for the Supply of New Orleans with Water.* [Washington, 1812].

———. *Remarks on the Address of the Committee of the Delaware and Schuylkill Canal Company.* Philadelphia, 1799.

———. "To the Editor of the Emporium [on the construction of the National Road]." *Emporium of the Arts and Sciences* new series, 3 (1814): 284–97.

———. *View of the Practicability and Means of Supplying the City of Philadelphia with Wholesome Water.* Philadelphia, 1799.

———. *The Virginia Journals of Benjamin Henry Latrobe, 1795–1798.* Edited by Edward C. Carter II et al. 2 vols. New Haven: Yale University Press, 1977.

Latrobe, Benjamin Henry, Jr., Trimble, Isaac Ridgeway, and Tegmeyer, John H. *Report of the Board of Engineers upon Changing the Course of Jones' Falls, With a View to Prevent Inundations, to the Mayor and City of Baltimore.* Baltimore, 1868.

Latrobe, J. H. B. *The First Steamboat Voyage on the Western Waters.* Maryland Historical Society Fund Publication No. 6. Baltimore, 1871.

Letters to the Honorable Albert Gallatin, Secretary of the Treasury of the United States: And other Papers Relative to the Chesapeake and Delaware Canal. Philadelphia, [1808].

McAdam, John Loudon. *Remarks on the Present System of Road Making.* 9th ed. London, 1827.

M'Clure, David. *Report on the Survey of a Section of the River Delaware, From One Mile Below Chester, to Richmond, Above Philadephia.* Philadelphia, 1820.

McFadden, William H. "A Brief History and Review of the Water Supply of Philadelphia." In *Annual Report of the Chief Engineer of the Water Department.* Philadelphia, 1876.

McKee, Harley J. *Introduction to Early American Masonry.* Washington: National Trust for Historic Preservation, 1973.

McMurtrie, Henry. *Sketches of Louisville and its Environs.* Louisville, 1819.

Bibliography

Marestier, Jean Baptiste. *Memoir on Steamboats of the United States of America.* Translated by Sidney Worthington. Mystic, Conn.: The Marine Historical Association, Inc., 1957.

Matthews, William. *Hydraulia: An Historical and Descriptive Account of the Water Works of London.* London, 1835.

Mease, James. *The Picture of Philadelphia.* Philadelphia, 1811.

Norton, Paul F. "Jefferson's Plans for Mothballing the Frigates." *U.S. Naval Institute Proceedings* 82 (1956): 736–41.

Nurnberger, Ralph D. "The Great Baltimore Deluge of 1817." *Maryland Historical Magazine* 69 (1974): 405–8.

[Peters, Richard]. *A Statistical Account of the Schuylkill Permanent Bridge, Communicated to the Philadelphia Society of Agriculture, 1806.* Philadelphia, 1807.

Peterson, Charles E. "Iron in Early American Roofs." *Smithsonian Journal of History* 3 (1968–69): 41–76.

Philadelphia. City Councils. *Joint Committee to Whom was Referred the Memorial and Remonstrance of Nicholas J. Roosevelt. Report.* Philadelphia, [1801 or 1802].

Philadelphia. City Councils. *Report of the Committee Appointed by the Common Council to Enquire into the State of the Water Works.* Philadelphia, 1802.

Philadelphia. City Councils. Watering Committee. *Report of the Committee for the Introduction of Wholesome Water into the City of Philadelphia.* Philadelphia, 1801.

Philadelphia. City Councils. Watering Committee. *Report of the Joint Committee, Appointed by the Select and Common Councils for the purpose of Superintending and Directing the Water Works.* Philadelphia, 1802.

Philadelphia. City Councils. Watering Committee. *Report of the Joint Committee of the Select and Common Councils, Appointed to Receive Information on the Subject of Watering the City.* Philadelphia, 1799.

Philadelphia. City Councils. Watering Committee. *Reports of the Watering Committee to the Select and Common Councils.* Annual. Philadelphia, 1803–19.

Philadelphia. City Councils. Watering Committee. *Report to the Select and Common Councils, on the Progress and State of the Water Works, on the 24th of November, 1799.* Philadelphia, 1799.

Proposals for Establishing a Company, for the purpose of employing the surplus power of the Steam-Engine Erected near the river Schuylkill: Under the Title and Firm of the Philadelphia Rolling Company. [Philadelphia, 1801]. Only known copy in Hist. Soc. Pa.

Pursell, Carroll W., Jr. *Early Stationary Steam Engines in America: A Study in the Migration of a Technology.* Washington: Smithsonian Institution Press, 1969.

Redlich, Fritz. *Essays in American Economic History: Eric Bollmann and Studies in Banking.* Ann Arbor, Mich.: Edwards Brothers, Inc., 1944.

———. "The Philadelphia Water Works in Relation to the Industrial Revolution in the United States." *Pennsylvania Magazine of History and Biography* 69 (1945): 243–56.

Rees, Abraham, ed. *The Cyclopaedia.* 40 vols. Philadelphia, 1805–24.

Richeson, A. W. *English Land Measuring to 1800: Instruments and Practices.* Cambridge, Mass.: Society for the History of Technology and M.I.T. Press, 1966.

Ritter, Abraham. *Philadelphia and Her Merchants.* Philadelphia, 1869.

Roberts, Christopher. *The Middlesex Canal, 1793–1860.* Harvard Economic Studies, no. 61. Cambridge, Mass.: Harvard University Press, 1938.

Robinson, A. H. W. *Marine Cartography in Britain: A History of the Sea Chart to 1855.* Leicester: Leicester University Press, 1962.

Robinson, Eric H. "The Early Diffusion of Steam Power." *Journal of Economic History* 34 (1974): 91–107.

Robinson, Eric, and Musson, A. E. *James Watt and the Steam Revolution.* London: Adams & Dart, 1969.

Rolt, L. T. C. *Great Engineers.* New York: St. Martin's Press, 1963.

———. *Thomas Telford.* London: Longmans, Green and Co., 1958.

Rouse, Hunter, and Ince, Simon. *History of Hydraulics.* Ames, Iowa: Iowa Institute of Hydraulic Research, 1957.

Sanderlin, Walter S. *The Great National Project: A History of the Chesapeake and Ohio Canal.* The Johns Hopkins University Studies in Historical and Political Science, series 64, no. 1. Baltimore: Johns Hopkins Press, 1946.

Scharf, J. Thomas. *History of Baltimore City and County.* 2 vols. Philadelphia, 1881. Reprint. Baltimore: Regional Publishing Company, 1971.

Schubert, H. R. *History of the British Iron and Steel Industry from ca. 450 B.C. to A.D. 1775.* London: Routledge & Kegan Paul, 1957.

Singer, Charles; Holmyard, E. J.; Hall, A. R.; Williams, Trevor I., eds. *A History of Technology.* 5 vols. Oxford: Clarendon Press, 1954–58.

Smeaton, John. *A Narrative of the Building and a Description of the Construction of the Edystone [sic] Lighthouse with Stone.* London, 1791.

Snyder, Frank E., and Guss, Brian H. *The District: A History of the Philadelphia District U.S. Army Corps of Engineers, 1866–1971.* Philadelphia: U.S. Army Engineer District Philadelphia, 1974.

Stapleton, Darwin H., and Guider, Thomas C. "The Transfer and Diffusion of British Technology: Benjamin Henry Latrobe and the Chesapeake and Delaware Canal." *Delaware History* 17 (Fall-Winter 1976): 127–38.

Straub, Hans. *A History of Civil Engineering.* Translated by Erwin Rockwell. Cambridge, Mass.: M.I.T. Press, 1964.

Swank, James M. *Introduction to a History of Ironmaking and Coal Mining in Pennsylvania.* Philadelphia, 1878.

Thomson, David Whittet. "The Great Steamboat Monopolies, Part I: The Mississippi." *The American Neptune* 16 (1956): 28–40.

Thwaites, Reuben Gold, ed. *Early Western Travels.* 32 vols. Cleveland: A. H. Clark Co., 1904–7.

Vine, P. A. L. *London's Lost Route to Basingstoke: The Story of the Basingstoke Canal.* Newton Abbot, England: David & Charles, 1968.

Wilson, Samuel, Jr. Introduction to *Impressions Respecting New Orleans by Benjamin Henry Boneval Latrobe: Diary & Sketches, 1818–1820.* New York: Columbia University Press, 1951.

Wilson, Thomas. *Picture of Philadelphia, for 1824, Containing the "Picture of Philadelphia, for 1811, by James Mease, M.D.," With all its Improvements Since that Period.* Philadelphia, 1823.

II.

THE SUSQUEHANNA RIVER SURVEY MAP

Artz, Frederick B. *The Development of Technical Education in France, 1500–1850.* Cambridge, Mass.: Society for the History of Technology and M.I.T. Press, 1966.

Bagrow, Leo. *History of Cartography.* Revised and enlarged by R. A. Skelton. London: C. A. Watts & Co., 1964.

Bibliography

Ballard, Paul. "Cartographic Drawing Instruments: The Eighteenth and Nineteenth Centuries." *Cartography* 8 (1974): 140–46.

———. "Early Surveying Instruments in Relation to Australian Cartography." *Cartography* 8 (1974): 193–98.

"Barton, William Paul Crillon." In Allen Johnson and Dumas Malone, eds. *Dictionary of American Biography*. 20 vols. New York: Charles Scribner's Sons, 1927–36.

"Barton, William Paul Crillon." In *National Cyclopedia of American Biography*. 56 vols. New York: James T. White, 1891–1975.

Bedini, Silvio A. *Early American Scientific Instruments and Their Makers*. Washington: Museum of History and Technology, Smithsonian Institution, 1964.

Belidor, Bernard. *Architecture Hydraulique*. 4 vols. Paris, 1737–53.

Biswas, Asit K. *History of Hydrology*. New York: American Elsevier Publishing Co., 1970.

Brown, Lloyd Arnold. *Early Maps of the Ohio Valley: A Selection of Maps, Plans, and Views Made by Indians and Colonials from 1673 to 1783*. Pittsburgh: University of Pittsburgh Press, 1959.

Burnside, R. S., Jr. "The Evolution of Surveying Instruments." *Surveying and Mapping* 18 (1958): 59–63.

Chorley, Richard J.; Dunn, Anthony J.; and Beckinsale, Robert P. *The History of the Study of Landforms; Or The Development of Geomorphology*. London: Methuen, 1964.

Clark, William Bullock. *The Physical Features of Cecil County, Maryland*. Maryland Geological Survey. Baltimore: Johns Hopkins Press, 1902.

Daumas, Maurice. *Scientific Instruments of the Seventeenth and Eighteenth Centuries and Their Makers*. Translated by Mary Holbrook. London: B. T. Batsford, 1972.

Davies, Gordon L. *The Earth in Decay: A History of British Geomorphology, 1578–1878*. London: Macdonald & Co., 1969.

Davis, Raymond E., and Kelly, Joe W. *Elementary Plane Surveying*. 4th ed. New York: McGraw-Hill Book Company, 1967.

DeVorsey, Louis. "A Background to Surveying and Mapping at the Time of the American Revolution: An Essay on the State of the Art." In *The American Revolution 1775–1783: An Atlas of 18th Century Maps and Charts: Theatres of Operations*. Compiled by W. Bart Greenwood. Washington: Government Printing Office, 1972.

Dickinson, H. W. "A Brief History of Draughtsmen's Instruments." *The Bulletin of the Society of University Cartographers* 2 (1968): 37–52.

Ellicott, Andrew. *Several Methods by Which Meridian Lines may Be Found with Ease and Accuracy*. Philadelphia, 1796.

Fraser, Persifor. *The Geology of Lancaster County*. Pennsylvania Geological Survey, 2d series, 300. Harrisburg, Pa., 1880.

[Gallatin, Albert]. *Report of the Secretary of the Treasury on the Subject of Public Roads & Canals*. Washington, 1808. Reprint. New York: Augustus M. Kelley, 1968.

Gilpin, Joshua. *Pleasure and Business in Western Pennsylvania, The Journal of Joshua Gilpin, 1809*. Edited by Joseph E. Walker. Harrisburg, Pa.: The Pennsylvania Historical and Museum Commission, 1975.

Greenhood, David. *Mapping*. Rev. ed. Chicago: University of Chicago Press, 1964.

Hamlin, Talbot. *Benjamin Henry Latrobe*. New York: Oxford University Press, 1955.

Hammond, John. *The Practical Surveyor*. 2d ed. London, 1731.

Hayne, J. C. G. *Deutliche und ausfürliche Anweisung, wie man das militairische Aufnehmen nach dem Augenmaas ohne Lehrmeister erlernen könne*. Leipzig, 1794.

Heisey, M. Luther. "The Susquehanna Islands in Lancaster County." *Papers of the Lancaster County Historical Society* 59 (1955): 25–55.

Hindle, Brooke. *The Pursuit of Science in Revolutionary America*. Chapel Hill: University of North Carolina Press, 1956.

Hogrefe, Johann Ludwig. *Theoretische und praktische Anweisung sur militairischen Aufnahme oder vermessung im Felde*. Hanover, 1785.

Hooykaas, Reijer. "Catastrophism in Geology, Its Scientific Character in Relation to Actualism and Uniformitarianism." *Mededelingen der Koninklijke Nederlandse Akademie van Wetenschappen* 33 (1970): 271–316.

Humphreys, Arthur L. *Old Decorative Maps and Charts*. London: Halton & Truscott Smith, 1926.

Irwin, Daniel Richard. "The Development of Terrain Representation in American Cartography." Ph.D. dissertation, Syracuse University, 1972.

Jackson, Kern C. *Textbook of Lithology*. New York: McGraw-Hill Book Company, 1970.

Jones, Yolande. "Aspects of Relief Portrayal on 19th Century British Military Maps." *The Cartographic Journal* 11 (1974): 19–33.

Klinefelter, Walter. "Lewis Evans and His Maps." *Transactions of the American Philosophical Society*, new series 61(1971): 3–65.

Lemon, James T. *The Best Poor Man's Country: A Geographical Study of Early Southeastern Pennsylvania*. Baltimore, Md.: Johns Hopkins Press, 1972.

Leopold, Luna B.; Wolman, M. Gordon; and Miller, John P. *Fluvial Processes in Geomorphology*. San Francisco: W. H. Freeman and Company, 1964.

Lintner, Stephen F. "The Historical Physical Behavior of the Lower Susquehanna River, 1801 to 1976." Ph.D. diss. in progress, Johns Hopkins University, Baltimore, Md.

Livingood, James Weston. *The Philadelphia-Baltimore Trade Rivalry, 1780–1860*. Harrisburg, Pa.: The Pennsylvania Historical and Museum Commission, 1947.

Lorain, John. *Nature and Reason Harmonized in the Practice of Husbandry*. Philadelphia, 1825.

Love, John. *Geodaesia: or, the Art of Surveying and Measuring Land Made Easy*. London, 1744.

Love, John Barry. "The Colonial Surveyor in Pennsylvania." Ph.D. dissertation, University of Pennsylvania, 1970.

"Maclure, William." In *Dictionary of Scientific Biography*. Charles Gillespie, ed. 14 vols. New York: Charles Scribner's Sons, 1970–1976.

MacMinn, Edwin. *On the Frontier with Colonel Antes, or The Struggle for Supremacy of the Red and White Races in Pennsylvania*. Camden, N.J.: S. Chew & Sons, 1900.

"Maryland Historical Society." *Magazine of American History Illustrated* 14 (1885): 624–25.

Matthews, Edward B. "Submerged 'deeps' in the Susquehanna River." *Geological Society of America Bulletin* 28 (1917): 335–46.

Michel, Henri. *Scientific Instruments in Art and History*. Translated by R. E. W. Maddison and Francis R. Maddison. London: Barrie and Rockcliff, 1966.

Muller, John. *A Treatise Containing the Practical Art of Fortification in Four Parts. For the Use of the Royal Academy of Artillery at Woolwich*. London, 1755.

———. *A Treatise Containing the Elementary Part of Fortification, Regular and Irregular*. London, 1756.

Munsell Color Charts. Baltimore: Munsell Color Co., various dates.

Montesson, Dupain de. *La science de l'arpenteur*. Paris, 1766.

Prischer, J. D. C. *Coup d'oeil Militaire*. Berlin, 1755.

Reasonover, J. Roy. *Land Measures French, Spanish and English*. Houston, Texas: Privately printed, 1946.

Richeson, A. W. *English Land Measuring to 1800: Instruments and Practices*. Cambridge, Mass.: Society for the History of Technology and M.I.T. Press, 1966.

Bibliography

Robinson, A. H. W. *Marine Cartography in Britain.* Leicester: Leicester University Press, 1962.

Robinson, Arthur H., and Sale, Randall D. *Elements of Cartography.* 3d ed. New York: John Wiley & Sons, 1969.

Schneer, Cecil J., ed. *Toward a History of Geology.* Cambridge, Mass.: M.I.T. Press, 1969.

Sigafoos, R. S. *Botanical Evidence of Floods and Flood-Plain Deposition.* United States Geological Survey, Professional Paper 485-A. 1964.

"Smeaton, John." In Leslie Stephen and Sidney Lee, eds. *Dictionary of National Biography.* 21 vols. New York and London, 1885–1912.

Smith, John. *The Art of Painting in Oil to Which is Added the Whole Art and Mystery of Colouring Maps.* 9th ed. London, 1788.

Stevenson, Roger. *Military Instructions for Officers Detached in the Field.* Philadelphia, 1775.

Stose, G. W., and Jonas, Anna I. *Geology and Mineral Resources of York County, Pennsylvania.* Pennsylvania Geological Survey, 4th series, Bulletin C 67. Harrisburg, Pa.: Commonwealth of Pennsylvania, 1939.

Southwick, David L.; Owens, James P.; and Edwards, Jonathan, Jr. *The Geology of Harford County, Maryland.* State of Maryland Geological Survey. Baltimore: State of Maryland, 1969.

Thrower, Norman J. W. *Maps & Man: An Examination of Cartography in Relation to Culture and Civilization.* Englewood Cliffs, N.J.: Prentice-Hall, 1972.

Trcziyulny, Charles. *Report of Charles Trcziyulny, appointed to explore the River Susquehanna, in Pursuance of an Act of the General Assembly, with a View to its Improvement, from the New York to the Maryland Line.* Harrisburg, Pa., 1827.

Trimble, Stanley W. *Man-Induced Soil Erosion on the Southern Piedmont, 1700–1970.* Ames, Iowa: Soil Conservation Society of America, 1974.

Volney, Constantin François Chasseboef, Compte de. *Tableau du climat et du sol des États-Unis d'Amerique.* 2 vols. Paris, 1803.

INDEX

The index does not include illustrations, which may be found by using the list of illustrations, p. xvii.
In order to avoid a cumbersome index entry for Benjamin Henry Latrobe, the editor has restricted the entry to selected aspects of Latrobe's personal and professional life.

Akin, J., 247
Alexander, Robert, 36
American Philosophical Society, 11, 84; sponsors canal survey, 12, 62; rents portion of Centre Square Engine House, 33; BHL's election to, 70; BHL's report to on steam engines, 70, 157–58, 181, 182; secretary of, 137, 139
Antes, Colonel Frederick, 9, 76, 77
Antes, Henry, 77
Appomattox navigation, 8
Aquia quarries, 27
Archimedean screw pump, 20, 22, 152
Arks, on Susquehanna River, 75, 90–91

Baker, Nicholas D., 59, 61
Baldwin, Henry, 55
Baltimore City Council: invites proposals for improving Jones' Falls, 45; considers plans for Jones' Falls, 45–46, 48; authorizes port wardens to improve Jones' Falls, 48
Baltimore Waterworks, 30, 38
Bank of Pennsylvania: BHL appointed architect, 28; hoisting machine for, 240
Barber, John, 30
Barby (Moravian seminary), 3, 4
Barlow, Joel, 57
Barton, William, 78
Barton, William Paul Crillon, 78
Basingstoke Canal, 5, 6, 26
Bath, Peter, 14
Beelen, Anthony, 56, 59
Belidor, Bernard, 87, 205

Bindley, John, 14
Black powder, 67, 77
Bladensburg Road, 25
Blaney, Daniel, 12, 62
Blydensburg loom: demonstrated, 54; patent applications for, 54, 55; BHL designs mill for, 55, 66
Blydensburg, Samuel, 54–55, 66
Bollman, Eric, 50, 51, 70, 168
Bollman, Lewis, 50, 51
Boulton and Watt engines, 66, 157, 182
Brady, Peter, 30
Breillat, Thomas, 30
Brent, Daniel, 26
Brick, for construction, 64
Bridges: over canals, 25–26; over rivers, 26–27; designs by BHL, 26–27; BHL's patent application, 26–27; cast-iron, 65; suspension, 65; on Chesapeake and Delaware Canal feeder, 125; design for Harford Run, 245. *See also* names of specific bridges
Brindley, James, 8
Brooke, Robert, 14, 63, 69
Brown, John, 35n
Buffalo, 60, 61, 66
Bulfinch, Charles, 242

Caldwell, Elias B., 20
Canals, 7–24 passim, 31, 75, 76, 113, 137, 139. *See also* names of specific canals
Carey, Isaac, 56
Catawba Canal, 24
Caton, Richard, 84

Centre Square Engine House. *See* Philadelphia Waterworks, Centre Square Engine House
Chelmer and Blackwater Navigation, 65; BHL's role in planning alternate route, 6–7; maps of, 61, 113; BHL's promotion of, 69
Chelmer-Blackwater Navigation. *See* Chelmer and Blackwater Navigation
Chelsea Waterworks, 144
Chesapeake and Delaware Canal: feeder, 10, 12, 13, 14–17, 18; early plans for, 11–12; surveys for, 12, 62; route of main canal, 12–13; contractors, 14; construction of feeder, 14–17; dimensions of feeder and towpath, 17, 117; use of black powder for, 66; relation to Susquehanna improvement, 75, 76; Elk River aqueduct, 125; Cow Run aqueduct, 125; remains of feeder, 141; maps of surveys for, 243–44
Chesapeake and Delaware Canal Company: incorporation, 12, 76; financial crisis, 17; attempts to revive, 17–19
Chesapeake and Ohio Canal Company, 22
Chesterfield, Va., coal mines, proposed railroad at, 28, 139
Claiborne, W. C. C., 36
Clerks of the works, 13–14, 30, 63
Clermont, 57
Clinton, Governor DeWitt, 23, 69
Coal mines, near Richmond, Va., 28, 139
Cochran, James, 20, 24, 141
Cockerell, Samuel Pepys, 6
Colles, Christopher, 8

Columbia, Pa., 90–91
Columbia Pike, 25
Columbia Turnpikes Company, 25
Conewago Canal, 7
Contractors, BHL's use of, 63–64
Coombe, Griffith, 25
Cooper, Mahlon, 53
Cope, Thomas Pym, 34, 70
Copeland, Thomas, 56, 59
Coulter, Andrew, 40, 41
Craftsmen, BHL's reliance on, 63
Cramond, William, 51
Crane, for lifting stone columns, 242
Criddle, Jonathan, 38, 53, 56, 59
Cumberland Road. *See* National Road

Dagenham Gap, 208
Davis, John, 30, 38, 63, 69
Delaware and Schuylkill Canal, 7, 14
Delaware and Schuylkill Canal Company, 28
DeMun, Lewis, 13, 24, 64, 68
Dismal Swamp Canal, 8; BHL's plan for a better route, 23–24, 137
Dismal Swamp Canal Company, 137
Dismal Swamp Land Company, 23, 137
Drawings, making of, 62–63; Evans's praise of BHL's, 63
Drayton, John, 247
Dredging machine, 42
Drydocks, 46. *See also* Naval Drydock

Ellicott, Andrew, quoted, 79
Ellis, Samuel, 53, 54

253

Index

Engineering, state of in America, 7–8
Erie Canal, 23, 62, 69
Evans, George, 56
Evans, Oliver, 35, 66, 70, 71
Ewell, Thomas, 38, 54
Eytelwein, J. A., 208

Fabre, Jean Antoine, 86
Farrell, Mr., 26
Fascine works: proposed for New Orleans, 44, 208, 209; drawings of, 208–09
Fens, of Cambridgeshire and Lincolnshire, 5, 152
Ferguson, Eugene, 69
Forrest, Joseph, 25
Fort Nelson, in Virginia, 247
Fox, Samuel, 30
Foxall, Henry, 38
Frank's Island lighthouse, 245
Fulneck, Eng., 3
Fulton, Robert, 23, 59, 70, 71; gives BHL steamboat agencies, 57, 58; charters steamboat companies, 58; designs western steamboats, 58, 61; discharges BHL, 60; quarrels with stockholders, 61; corresponds with BHL on steam engines, 201

Gallatin, Albert, 24, 25, 69, 139, 141, 142
Germantown and Perkiomen Turnpike, 14
Gilpin, Joshua, 12; quoted, 90–91
Graff, Frederic, 68n, 196
Graff, Frederick, 30, 63, 175, 196; appointed Catawba Canal engineer, 24; career of, 68–69; drawings by, 158, 193; quoted, 181

Halle au Blé, 117
Harford Run, flood of 1817, 45
Harford Run Bridge, 26, 45, 246
Harriet, 60, 61
Hartshorne, William, 38, 64
Hauducoeur, Christian, 9, 76–84 passim, 99, 105
Havre de Grace, Md., 109

Hazlehurst, Robert, 78
Heth, Henry, 139
Hewitt, John, 34
Hoisting machine, 240
Hornblower, Josiah, 8
Howard, Cornelius, 12
Hurley, William, 59, 61, 63
Hydraulic cement, 14, 16, 64–65

Instruments, for surveying, 61–62
Iron, 65

James River Canal, 7, 23, 24, 139
Jeffers, Joseph, 45, 62, 247
Jefferson, Thomas, 9, 11, 12, 51, 70; asks BHL to estimate cost of drydock plan, 9; specifies design of drydock, 10, 11, 116, 117; instructs Tingey, 117
Jessop, William, 5
Joint Committee on Supplying the City with Water. See Watering Committee
Jones' Falls Improvement, 44–48; map of, 247

Kalorama, 57
Keelboats, on Susquehanna River, 75, 90
King, Nicholas, 117
Kneass, Samuel, 68n
Kneass, Strickland, 68n

Lancaster Turnpike, 7, 90, 141
Latrobe (LaTrobe), Anna Margaretta (Antes), 3
Latrobe (LaTrobe), Benjamin, 3
Latrobe, Benjamin Henry: early years and education, 3–4; engineering training in Germany, 4; returns to England, 4; employed at Stamp Office, 5; in John Smeaton's office, 5, 86, 152; works under William Jessop, 5; arrives in United States, 7; canal consultations, 6–7, 23; appointed engineer of the Chesapeake and Delaware Canal, 13; pupils of, 13, 24, 64, 68–69; appointed engineer of Washington Canal, 19–20; shareholder and director of Washington Canal Company, 23; applies for bridge patent, 26–27; knowledge of London Waterworks, 28–29; becomes consulting engineer of Philadelphia Waterworks, 29; establishes and operates foundry, 37–38; comments on operating a blacksmith shop, 38; dies at New Orleans, 41; promotes sheet iron roofing, 51; appointed Engineer to the Navy Department, 52; resigns position as Engineer to the Navy Department, 53–54; agent for Washington steamboat companies, 57; agent of Ohio Steamboat Company, 58; bankruptcy, 61; criticizes Fulton's western steamboats, 61; opinion on assistants, 62; drawing methods and skill, 62–63; difficulties with superiors, 64; theoretical knowledge, 67; permanence as a goal, 67–68, 70; engineering values, 68; promotes transfer of concepts and skills, 68–69; investments, 69; promotion of technical projects, 69–70; disputes with Evans, 70–71; role in the American technical community, 70–71; appointed to Susquehanna improvement, 76; geological concepts, 85–86; relies on European cartographic tradition, 100; use of European engineering writings, 208; installs Washington Navy Yard machinery, 234; interest in mechanization, 237
Latrobe, Benjamin Henry, Jr., 27, 48
Latrobe, Charles H., 84
Latrobe, Henry Sellon Boneval, 37, 39, 63, 66, 68, 201, 203; on Washington Canal, 20; on Columbia Turnpikes, 25; goes to New Orleans, 36–37; appointed engineer to repair levee, 44; drawings by, 196, 199, 245
Latrobe, John Henry Boneval, 41
Latrobe, Mary Elizabeth, 99
Law, Thomas, 19
Leitensdorfer, Eugene, 20, 25, 62
L'Enfant, Pierre, 8, 10, 19
Leslie, Robert, 11, 71
Lewis, John, 30

Library Company of Baltimore, 84
Lighthouse, at Frank's Island, 245
Little River Turnpike, 25
Livingston, John, 60
Livingston, Robert, 34, 57, 58
Locks, wooden, 130–31
London Bridge Waterworks, 196
Lorman, William, 54
Louisiana, territorial legislature, petitioned by BHL and Alexander, 36
Lower Appomattox Navigation, 8

McAdam, John Loudon, 141
Machinery, muscle-powered, 66
McKean, Governor Thomas, 76, 92
Maclure, William, 85
Madison, James, 20, 25
Maldon, Eng., 6, 113
Manchester, Va., 139
Maps, BHL's making of, 62
Mark, Jacob, 39
Mason, Jeremiah, 54
Mason and Dixon Line, 83, 102
Materials, used for construction, 64–65
Meade, Simeon, 20
Middlesex Canal, 7, 65
Mifflin, Samuel, 51
Mill, for hulling rice, 247
Miller, John, Jr., 28
Mills, Robert, 13, 68–69; plan for improving Jones' Falls, 45, 48; acting engineer of Chesapeake and Delaware Canal, 64
Mississippi Steamboat Company, 57, 58, 59, 60
Monticello, sheet iron for roofing, 51, 168
Moore, Thomas, 42, 43, 67
Moravian education, 3–4, 84
Morris, Ellwood, 68n
Morris, Lewis, 79
Mud Island bar: blocks part of Delaware River channel, 41; BHL's report on, 41–42; BHL's plan for controlling, 42, 205

National Road, 24–25; bridges on, 27; BHL's design for, 141; contractors for,

141; specifications for, 141; BHL's prediction verified, 142
Naval drydock: Jefferson's plans for, 9, 10, 116; BHL's plans for, 10, 11, 116–117
Navy yard. *See* Washington Navy Yard
Navy yards, Royal, 6
New Orleans, 57, 59, 61
New Orleans, La.: rapid growth, 36; yellow fever epidemics, 36, 40; flood at, 208
New Orleans City Council: petitioned by BHL, 36; grants franchise to BHL, 37; extends BHL's franchise, 38, 40; appoints Henry Latrobe to repair levee, 44
New Orleans levee, 44, 208
New Orleans Waterworks: insufficiency of water supply in New Orleans, 36; BHL's efforts to obtain a franchise, 36–37, 69; BHL's plan for, 37; franchise granted, 37; steam engines planned for, 38, 58; engines purchased, 39, 40; BHL takes charge of construction, 40–41; begins operations, 41; replaced by new system, 41; preliminary map of, 199; BHL design of engine for, 201; pier for suction pipe, 203
New Orleans Water Works Company, 41
Newsham, Richard, 35n
New York Western Navigation, 23, 62
Niagara canal, 23
Niesky (Moravian school), 3, 4
Noble, Mr., 39
Norfolk, Va., fort at, 247
North Carolina State Works, 23
North River, 57

Ogden, Francis, 56
Ohio River, canal at falls of, 23
Ohio Steamboat Company, 58–61, 69
Orr, Colonel Benjamin G., 40
Orth, Christopher, 57
Outram, Benjamin, 10, 117

Parkins, John, 69
Parliament, BHL's testimony in, 6, 7
Patapsco canal, 23
Patterson, Samuel, 55
Permanence, as a goal of engineering, 67, 70, 125
Perry, Captain John, 208
Personnel, BHL's selection of, 63
Philadelphia: health and sanitation problems, 28, 87; center of steam engine manufacture, 33–34, 35–36
Philadelphia Chamber of Commerce, 41, 42, 205
Philadelphia City Councils: contracts for two engines from Roosevelt, 29, 34. *See also* Watering Committee
Philadelphia Rolling Company, 50
Philadelphia Waterworks: problems with Philadelphia's water supply in 1790's, 28; BHL's proposal for, 29, 67; BHL becomes consulting engineer of, 29; BHL's assistants, 30; contractors, 30; Joint Committee approves BHL's plan, 30; basin, 30–31, 144, 146–47; canal, 31, 144, 146; tunnel, 31, 144, 146; well, 31, 165; brick conduit, 32, 154, 165; new works at Fairmount, 32, 33, 35, 158, 183; distribution of water, 33, 196; pipes, 33, 196; makes Philadelphia a center of steam engine manufacture, 33–34, 35–36; design and manufacture of engines for, 34–35; beginning of operations, 35; problems with engines, 35; use of black powder, 66, 144; Cope's opinion of BHL's performance, 70; BHL's fiscal responsibility questioned, 77; coffer dam, 146–47; Archimedean screw pump, 152
—Centre Square Engine House, 30, 33, 173, 175, 177, 179; reservoirs, 173, 179, 193; hoisting machine for, 240; plans for, 246–47; Centre Square engine, 179, 181–83, 189, 223
—Schuylkill Engine House, 30, 31–32, 154, 157; Schuylkill engine, 154, 157–59; purchased by city, 168; plans for, 245–46. *See also* Schuylkill rolling mill; Watering Committee
Pinkerton, John, 5
Pittsburgh Steam Engine Company, 56
Portsmouth, Eng., pulley block machines at, 53
Potomac Bridge Company, 26, 42
Potomac Canal: at Little Falls, 117; locks on, 242
Potomac Canal extension: BHL's plans for, 9–10, 117–18; dimensions of, 17; map of, 243
Potomac Company, 9–10, 24. *See also* Potomac Canal; Potomac Canal extension; Potomac River
Potomac River: attempt to control sedimentation, 42; sedimentation, 42, 43, 45; BHL's opinion on possibility of improving, 42–44, 45; locks on, 242; wharves on, 247
Potomac Steam Boat Company, 57–58, 59
Poussin, William Tell, 64
Power, sources of, 66–67
Pozzuolana, 64–65
Price, Mr., 56
Principio Furnace, 38
Pulley block machines, 53
Pulley block mill, at Alexandria, Va., 237
Pupils, of BHL, 13, 24, 64, 68–69
Pursell, Carroll, 71

Railroad, to coal mines near Richmond, Va., 139
Railroad car, 137
Railroads: BHL's use of, 17, 27, 66; BHL's knowledge of English, 27; BHL's report to Gallatin on, 27–28; proposed by BHL, 27, 28, 139
Ramsden, Jesse, 7, 62
Randall, Charles, 20, 24, 69, 141
Rennie, John, 5, 6, 7
Repton, Humphrey, 5
Repton, John, 5
Rhode, Frederick, 34
Roads, 24–25, 117. *See also* Turnpikes
Roanoke and James Canal, 23
Robinson, William, 37
Rolling mill, operated by Bishop and Malin, 38
Rolling mills: in Europe, 48–49; in the United States, 49, 167
Roman cement, 65
Roosevelt, Nicholas, 26, 39, 50, 70, 167, 173; and Soho Works, 29, 49, 50; patent with Smallman, 34; contracts for rolled copper, 49; and Schuylkill rolling mill, 51, 52, 168; agreement with Fulton, 57; and steamboats, 57, 58, 59
Rose, Louis, 53
Ross, James, 55
Rye Harbour, 5

Salem Creek Canal Company, 24
Santee Canal, 7, 23
Schneider, F. I., 35n
Schuylkill and Susquehanna Canal, 7
Schuylkill Engine House. *See* Philadelphia Waterworks, Schuylkill Engine House
Schuylkill Permanent Bridge, 7, 14, 26, 27, 240
Schuylkill rolling mill: English precedents, 48–49; plans by Roosevelt, Smallman, and Latrobe, 49–50; begins production, 50; prospectus for Philadelphia Rolling Company, 50; unprofitable, 50–51; part of the Schuylkill Engine House, 50, 154; products of, 51; makes sheet iron roofing, 51, 65; ceases operation, 51–52, 168; impact on transfer of steam rolling mill technology, 52; drawings of, 167–68
Sedimentation, of rivers, 42, 43, 45, 67, 86
Shade, Sebastian, 76, 77
Sheet iron roofing, 51, 65
Smallman, James, 37, 167, 214, 215, 223, 234; at Soho, 34; moves to Philadelphia, 35; knowledge of English rolling mills, 49, 50; builds and erects Washington Navy Yard engine, 52–53, 64, 234; designer of Philadelphia Waterworks engines, 157, 182; drawings of engines, 214–16, 223–24
Smeaton, John, 5, 6, 86; experiments with hydraulic cement, 14, 64–65; *Edystone Lighthouse*, 67
Soho works, in Birmingham, Eng., 34
Soho Works, in northern New Jersey, 29, 34, 157, 181
Steamboats: Stevens-Livingston, 29, 34; BHL's ideas for western, 61; proposed on Cumberland River, 69. *See also* Mississippi

255

Index

Steamboats (*continued*)
  Steamboat Company; Ohio Steamboat Company; Potomac Steam Boat Company
Steam engines: in England, 6; in United States in 1798, 29; Roosevelt-Smallman patent for, 34; use of iron, 65; as a power source, 66; BHL's fascination with, 66; BHL an intermediary for those interested in, 71; BHL's design for working gear, 201. *See also* Hartshorne, William; New Orleans Waterworks; Ohio Steamboat Company; Philadelphia Waterworks; Steubenville Woolen Mill; Washington Navy Yard
Stein, Albert, 41
Steubenville (O.) Woolen Mill, 55–57
Stevens, John, 34
Stone, for construction, 64
Stoner, George, 76, 77, 94
Stoudinger, Charles, 34, 157, 165, 167, 175, 181, 189
Strickland, John, 14
Strickland, William, 13, 18, 64, 68–69
Strutt family, 6, 113
Surveying: BHL's practice, 61–62; BHL's instruments for, 61–62, 79; BHL's method on Susquehanna River, 78; possibilities for error in, 79
Surveyors, American, 62
Susquehanna and Tidewater Canal, 78, 109
Susquehanna Canal Company, 75, 76, 77, 82
Susquehanna River: improvement for navigation, 8–9, 12, 76, 77–78; map of, 9, 82–84, 89–109; surveys by BHL and Hauducoeur, 62, 78–79, 102; use of black powder on, 67, 77; early use of, 75; keelboats on, 75, 90; arks on, 75, 90–91; projected cost of further improvement, 78; sounding the deeps of, 79–81; time spent on by BHL and Hauducoeur, 81–82; BHL's interest in geology of, 85–86, 104, 105; botany and plant ecology of, 86; and hydraulic engineering, 86–87; landscape change, 87; modern dams, 87; Turkey Hill Falls, 92–93; BHL's instructions to navigators on, 102
Susquehanna River map (by BHL with others): style of, 82, 100; making of, 83; place names, 83–84; history of, 84; geology on, 85–86, 104, 105; annotated, 89–109
Susquehanna River map (by Christian P. Hauducoeur), 83, 102, 105, 106, 108

Tarras, 16, 64, 65
Telford, Thomas, 6
Thomson, John, 12, 62
Thornton, William, 71
Tingey, Thomas, 9, 117
Tools: for canal excavation, 16, 20; for river clearance, 77
Transfer of concepts and skills, 68–69
Traquair, Adam, 30, 69, 84, 240
Traquair, James, 30
Trass, 16, 64
Trautwine, John C., 68*n*
Trautwine, John C., Jr., 68*n*
Tucker, William, 32
Turner, Joseph, 76
Turnpikes, 7, 14, 25, 67, 139; BHL's knowledge of, 141, 142. *See also* Lancaster Turnpike; National Road; Roads

U.S. Capitol: hoisting machines used for, 240; columns for House of Representatives, 242
U.S. Congress, investigates Potomac River improvement, 42
Upper Appomattox Navigation, 8

Vanreiper, John, 35*n*
Vaughn, John, 137, 139
Vesuvius, 58
Vickers, Thomas, 14, 15, 16, 30, 31, 32, 65, 69
Virginia Public Works, 23
Volney, Constantin F. C., comte de, 84

Washington, mayor of, 20
Washington Canal: early construction attempt, 19; survey by BHL, 19, 20, 62; Archimedean screw pump, 20, 22, 152; dimensions of, 21; route of, 21; locks, 21–22, 130–31; later years, 22–23; use of hydraulic cement, 65; BHL's goal of permanence for, 67–68; wharf near Capitol, 242; maps of, 244–45
Washington Canal Company (first), 10, 19
Washington Canal Company (second), 19, 22, 69
Washington Manufacturing Company, 55
Washington Navy Yard: plan for drydock at, 9–11, 116; BHL urges use of hydraulic cement at, 65; pulley block mill at, 237. *See also* Naval drydock; Potomac Canal extension; Washington Navy Yard steam engine
Washington Navy Yard steam engine: built and erected, 52; planned, 52; forge, 52, 53, 54, 234; destroyed, 53; operation begun, 53; pulley block mill, 53; sawmill, 53, 54; rebuilt, 54; preliminary drawings for, 214–16; final drawings for, 223–24
Watering Committee (of the Philadelphia City Councils): appointed, 28; BHL's relationship to, 30; report to by Cope, 34; 1812 report on Waterworks defects, 35; given protection of the Schuylkill Engine House, 52; hires Thomas Vickers, 65
Water power, BHL's attitude toward, 66–67
Water rice machine, in South Carolina, 247
Waterworks. *See* Baltimore Waterworks; New Orleans Waterworks; Philadelphia Waterworks; Waterworks, in London
Waterworks, in London, 28–29, 144, 158, 196
Wells, Bezaleel, 55
Werner, Abraham, 84
Western Inland Lock Navigation Company, 7
Weston, William, 7–8, 14
White, John, 20
Wilkes, Mr., of Derbyshire, 141
Wisler, Christian, 76, 77
Wood: BHL's attitude toward, 65; for construction, 65–66
Wright, Benjamin, 18–19

Yellow fever: in Philadelphia, 28, 29, 33; in New Orleans, 36, 40